£10.95

Logic 336
D 2 L

D1345129

A PROFILE OF
MATHEMATICAL LOGIC

HOWARD DELONG
Trinity College, Hartford, Connecticut

A PROFILE OF
MATHEMATICAL LOGIC

ADDISON-WESLEY PUBLISHING COMPANY
Reading, Massachusetts
Menlo Park, California · London · Amsterdam · Don Mills, Ontario · Sydney

This book is in the
ADDISON-WESLEY SERIES IN MATHEMATICS

Consulting Editor:
LYNN H. LOOMIS

Second printing, September 1971

Copyright © 1970 by Addison-Wesley Publishing Company, Inc. Philippines copyright 1970
by Addison-Wesley Publishing Company, Inc.

All rights reserved. No part of this publication may be reproduced, stored in a retrieval system,
or transmitted, in any form or by any means, electronic, mechanical, photocopying, recording,
or otherwise, without the prior written permission of the publisher. Printed in the United
States of America. Published simultaneously in Canada. Library of Congress Catalog Card
No. 75-109509.

ISBN 0-201-01499-8
CDEFGHIJKL-AL-79876

To Shirley

PREFACE

The general aim of this book is to describe mathematical logic: its historical background, its nature, and its philosophical implications. It is meant to appeal to anyone—students, mathematicians, linguists, computer specialists, philosophers—who would like a relatively brief and readable introduction to the subject. The book presupposes no knowledge of logic and only high school mathematics. The emphasis throughout is on understanding, not technique. In my judgment, not only is there an excess of good books on logic which emphasize elementary technique but, further, the learning of it is a relatively low-level operation, one which is easily adaptable to teaching machines and programmed texts. Hence the focus here is on topics not presently suitable for mechanical techniques of learning: Among these topics are the historical reasons why Aristotelian logic came into being, how it came about that after more than 2000 years traditional logic gave way to mathematical logic, the nature of the formal axiomatic method and the reasons for its use, the main results of metatheory, and the philosophic import of those results.

It is unusual for these subjects to appear in an introductory book. The content of most introductory texts consists entirely of material whose details can be thoroughly understood by a beginner. However, the presentation of such material is not the sole, or even the most important, function of a beginning text. The most important function of such a book is to stimulate long-lived interest by communicating some of the excitement and beauty of the subject. With such stimulation the learning of the technical and often complicated details of logic becomes a pleasure, instead of the chore that it too often is for many beginners. Thus my intention is to appeal to both the intellect and the imagination and, by so doing, to put the reader in a good position to learn, and to *want* to learn, more logic.

This *Profile*, however, is an outline or summary of many topics, and as such it is designed to be used by readers with a variety of different purposes, and by teachers in a variety of different classroom situations. To illustrate some of these possibilities, let us take a look at what the book encompasses.

vii

Chapter 1 presents a brief historical account of ancient logic. Emphasis is placed on the historical conditions which led to the development of ancient logic, including an account of the relation of ancient logic to both ancient mathematics and ancient philosophy. It is misleading to present traditional logic in any but a historical way, precisely because it does not have any great current theoretical interest. Furthermore, it is necessary to study ancient logic in its mathematical and philosophical context in order to correctly judge questions of motivation, presupposition, and adequacy.

Chapter 2 discusses in detail the historical reasons why a revolution in logic occurred in the nineteenth and early twentieth centuries. Emphasis is placed on the role in this process of non-Euclidean geometry, analytic geometry, set theory, and the paradoxes. The discussion of the motivation behind the development of mathematical logic is meant to put the reader in a better position to judge its import, as well as to prevent him from concluding that it was an arbitrary invention.

Chapter 3 attempts a careful description of the formal axiomatic method; the similarities and differences between it and Euclid's method are stressed. Axiomatic formulations of the propositional calculus, the predicate calculus, and set theory are then briefly presented. Every attempt is made to keep the symbolism as simple and standard as possible. Stress is put on the under-standing of concepts of logic, rather than on the development of the reader's *skill* in logical technique.

Chapter 4 presents a summary of the metatheory of logic. The major metatheorems are described and sketches (sometimes extensive ones) of their proofs are often given. These sketches are meant to serve a twofold function: On the one hand, they give the majority—who do not continue logic—an insight into the essential ideas of a proof. On the other hand, they facilitate—for the minority who continue the subject—the understanding of the com-plete, original proof. Examples of such sketches include Post's completeness theorem for the propositional calculus, Gödel's completeness theorem for the predicate calculus, and Gödel's incompleteness theorem for arithmetic. The concepts necessary to understand these theorems are defined and the motivation of proofs is stressed throughout. Every effort is made to keep the summaries as accurate as possible. This is especially true of Gödel's two famous metatheorems, which are often presented in a careless way in popular and philosophic accounts.

Chapter 5 presents some philosophical implications of mathematical logic. Here the philosophical problems chosen are those which arise because of the limitative theorems (Löwenheim-Skolem's, Gödel's, Church's, etc.). This choice was made to illustrate the double-edged impact of mathematical logic: To mathematically minded readers it emphasizes the sometimes philosophical motivation and evaluation of metatheorems; to philosophically minded readers it emphasizes the importance of technical results to philosophy. Furthermore, there are relatively few books which give the limitative theorems extended philosophical discussion; yet I know of no other set of problems more likely to produce that sense of wonder which excites both the intellect and the imagination.

Interspersed throughout the text are problems whose purpose is twofold: first, to get the reader to think carefully about the material being presented and, second, to instill a fascination with both the mathematical and philosophical aspects of logic. Answers are provided for all of them.

Finally, the Bibliography, which is annotated, gives many cross references. The purpose is to enable the reader to more easily enlarge his knowledge of mathematical logic in whatever direction he wishes. Special emphasis is put on references to popular and philosophical discussions of the limitative theorems.

Hence the book may be read *in toto* in an introductory or intermediate logic course given in either a philosophy or mathematics department. Depending on the teacher, it may be supplemented by either a programmed text or an orthodox text which emphasizes technique. Conversely, some teachers may wish to use it as a supplementary text in either a logic or a philosophy (foundations) of mathematics course. As a supplementary text, it may even be used in an advanced logic course, especially for the sketches of proofs in Chapter 4 and the philosophical comments of Chapter 5. Its use as a supplementary text is easily accomplished, as the chapters are relatively independent and the index is designed to facilitate cross reference.

In my judgment, most introductions to logic are deficient not so much in what they do, but in what they do not do: They omit historical considerations of both traditional and mathematical logic (and thus leave the reader in a poor position to judge the significance of either); they give no hint of what lies beyond elementary logic (and thus leave the reader in the dark about the enormously important results of metatheory); they say little or nothing about the new philosophical problems and perspectives created by mathematical logic (and thus leave many readers with the feeling that mathematical logic is no more relevant to philosophy than is long division); and finally, they give little or no indication of the openness of logic and the problematic nature of interpreting its results. The reader must judge for himself whether the present book succeeds at what most elementary texts do not attempt.

For their helpful comments and criticisms, I am indebted to Myron G. Anderson, W. Miller Brown, Peter Duran, William J. Frascella, D. Randolph Johnson, Paul Serafino, Brian Taylor, and numerous students in my logic, advanced logic, and philosophy of mathematics courses. The final contents of the book are, of course, solely my responsibility. Finally, I am indebted to the Trustees of Trinity College, Hartford, Connecticut, for their generous support of an earlier version of the book.

January 1970   H.D.
Hartford, Connecticut

ACKNOWLEDGMENTS

For permission to quote from the indicated sources, I wish to thank the following publishers:

The American Mathematical Society for "Recursively Enumerable Sets of Positive Integers and Their Decision Problems," by Emil Post, from *The Bulletin of the American Mathematical Society*, Volume 50, © 1944.

Cambridge University Press for *The Thirteen Books of Euclid's Elements*, 3 volumes, second edition. (Translated with introduction and commentary by T. L. Heath.)

Clarendon Press, Oxford, for *The Development of Logic*, by William and Martha Kneale; for *A History of Greek Mathematics*, by Sir Thomas Heath; for *Infinity: An Essay in Metaphysics, by* José A. Benardete; and for *The Oxford Translation of Aristotle*, translated under the editorship of W. D. Ross.

Doubleday and Co. for *The Birth of Tragedy and The Genealogy of Morals* by Friedrich Nietzsche (translated by Francis Golffing).

Holt, Rinehart, and Winston for *Introduction to Non-Euclidean Geometry*, by Harold E. Wolfe.

Harvard University Press for *From Frege to Gödel: A Source Book in Mathematical Logic*, edited by Jean van Heijenoort and for *A Source Book in Mathematics*, edited by David Eugene Smith.

John Wiley and Sons for *Mathematical Logic*, by Stephen Cole Kleene.

North-Holland Publishing Co. of Amsterdam for *Computer Programming and Formal Systems*, edited by P. Braffert and D. Hirschberg; for *Abstract Set Theory*, by Abraham A. Fraenkel; and for *Foundations of Set Theory*, by Abraham A. Fraenkel and Yehoshua Bar-Hillel.

Open Court Publishing Co., La Salle, Illinois, for *Contributions to the Founding of the Theory of Transfinite Numbers*, by Georg Cantor (translated with an introduction by Philip E. B. Jourdain), and for *Euclides ab Omni Naevo Vindicatus*, by Giralomo Saccheri (introduction and translation by G. B. Halsted).

Oxford University Press for *The Republic of Plato*, translated by F. M. Cornford.

Raven Press for *The Undecidable: Basic Papers on Undecidable Propositions, Unsolvable Problems, and Computable Functions*, edited by Martin Davis.

Viking Press for *The Ingenious Gentleman Don Quixote de la Mancha* by Miguel de Cervantes Saavedra (translated by Samuel Putnam) and for *The Portable Nietzsche*, translated by Walter Kaufmann.

CONTENTS

CHAPTER 1

HISTORICAL BACKGROUND OF
MATHEMATICAL LOGIC

§1 INTRODUCTION

Near the end of a work now called *Sophistical Refutations*, Aristotle apparently claims to have created the subject of logic [1928 183b 34ff].* The
nearest analog of such a claim in our century is no doubt Freud's statement
in 1914 that "... psychoanalysis is my creation; I was for ten years the
only person who concerned himself with it..." [1953 7]. It seems probable that Aristotle's claim is as true as Freud's. Although Freud's claim is
correct, it is nevertheless possible for the historian to find all kinds of hints
and anticipations of psychoanalysis in the works of earlier thinkers; so if the
works of Aristotle's predecessors were all intact, historians could no doubt
perform a similar feat.

For example, Plato makes the following statement in the *Republic*: "The
same thing cannot ever act or be acted upon in two opposite ways, or be two
opposite things, at the same time, in respect of the same part of itself, and
in relation to the same object" [1955 133 (436B)]. Aristotle claims that the
most certain of all principles is that "the same attribute cannot at the same
time belong and not belong to the same subject and in the same respect"
[1928a 1005b 18ff]. This latter principle is Aristotle's formulation of the
Law of Non-Contradiction, and it is tempting to say that Aristotle received
not only this law, but many of his ideas on logic, from his predecessors.
Nevertheless, one should resist this temptation because Plato makes this
remark only in passing and there is no evidence that he, or anyone else
before Aristotle, attempted to codify the rules of correct inference. Thus we
may accept Aristotle's claim and ask what led him to create the subject of
logic.

"All men by nature desire to know," Aristotle tells us in the famous
opening sentence of the *Metaphysics*. Both he and Plato believed that philosophy begins in wonder, and there can be little doubt that this motive was
strong in Aristotle's logical investigations. Yet it does not seem that this was
the only or even the most pressing motive. Rather two other related but more
practical aims were involved, one having to do with mathematics and the

---

* See the Bibliography for all references, as well as for explanation of reference system.

1

other with sophisms. If we wish to know what logic is all about we can do no better than to begin by asking about ancient Greek mathematics before Aristotle.

§2 MATHEMATICS BEFORE ARISTOTLE

The first Greek mathematician was Thales of Miletus (c.624–c.545 B.C.). Thales had visited Egypt and it is probable that he acquired some practical geometrical knowledge there. However, from what we now know of ancient Egyptian mathematics, it seems more likely that anything of value that the Greeks inherited in geometry they received ultimately from ancient Mesopotamia. The latter's geometric knowledge was vastly superior to Egypt's. If we can believe tradition, Thales must have been a very great mathematician indeed because he apparently was the first person both to conceive of general geometric propositions and to see the necessity of proving them. A number of geometric propositions are ascribed to him—for example, that the angles at the base of an isosceles triangle are equal—but unfortunately we do not have any idea how he proved them. The same must also be said for Pythagoras (c.566–c.497 B.C.), who according to tradition also visited Egypt and was a pupil of Thales. Pythagoras' most important contribution to Greek geometry is perhaps best summarized by Proclus (410–485 A.D.), who stated that after Thales, "Pythagoras transformed the study of geometry into a liberal education, examining the principles of the science from the beginning and proving the theorems in an immaterial and intellectual manner . . ." [quoted in Heath 1921 I 141]. If this be true, all the other mathematical achievements (real or alleged) of Thales and Pythagoras are insignificant by comparison, for it would mean that they were chiefly responsible for transforming geometry from an empirical and approximate science into a nonempirical and exact one. We do not, however, know enough to make this claim for them with any kind of assurance.

   In any case Pythagoras, who was probably born in Samos, moved to the Greek city of Croton in southern Italy. There he formed a religious brotherhood based on numerous ascetic practices and beliefs. The members apparently believed in the transmigration of souls and in numerology. The specifics of the doctrine are obscure in part because the followers were pledged to secrecy. Although the achievements of Pythagoras are uncertain, it is likely that his order had followers for over a century, and its doctrines certainly influenced many important thinkers, including Plato and Aristotle.

   One of the achievements of the school—perhaps even one of Pythagoras' achievements—was the first *proof* of the Pythagorean theorem, that is, the theorem which states that the sum of the squares of the sides of a right triangle is equal to the square of the hypotenuse. The *truth* of the theorem is

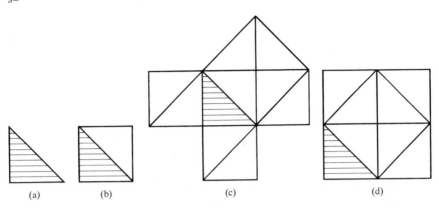

Figure 1

not very difficult to see—especially in some of its special cases—and either it or some special case of it was discovered independently in a number of cultures, for example in Babylonia, India, and China. We do not know how the Pythagoreans proved the theorem. However, it seems probable that the person who was offering the proof would draw a diagram while speaking and would ask the person listening to the proof if he agreed as he went along. This is the procedure of Socrates in Plato's *Meno* (*c.*390 B.C.), where a special case of the theorem is in fact proved. It also seems likely, as often happens in mathematics, that special cases were proved first and later generalized. Finally, it is probable that the assumptions of the proof were not first stated but were appealed to in the course of the proof, and—sometimes, at least— were not clearly understood by either party of the proof. Supposing all this, we might consider the following a likely story.

Pythagoras started with an isosceles right triangle (as shown in Fig. 1a) and made a construction on that triangle (Fig. 1b and c). The proof of the

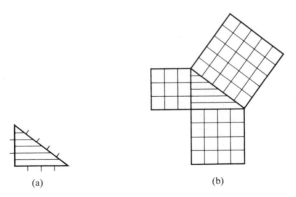

Figure 2

theorem can then be given by the process of counting the congruent triangles. Alternatively, he may have constructed one figure (Fig. 1d) and argued only after the construction. There is also evidence to suggest that he (like the ancient Babylonians who preceded him by 1200 years) knew that a 3, 4, 5 triangle is right. If so, he may have made constructions like those in Fig. 2, where it is possible to count unit squares.

Problem 1.*   Construct a square seven units by seven units, analogous to Fig. 1(d), such that the theorem's truth for a 3, 4, 5 triangle can be seen without any further construction.

If he proved the theorem in its full generality, he may have used a construction such as that given in Fig. 3(a). It is obvious that any right-angled triangle can be duplicated four times, as indicated in the figure. Now the area of the square of the hypotenuse is equal to the total area of the square minus the area of the four congruent triangles. Rearrange the triangles as indicated in Fig. 3(b). Clearly the sum of the squares of the two legs of the right triangle is equal to the total area of the square minus the area of the four congruent triangles. This is the most intuitively clear proof of the general Pythagorean theorem that has yet been discovered. But just for this reason it is unlikely that Pythagoras discovered it, since it often happens that the first proof of a theorem is far from the easiest.

Rather, he probably used his theory of *proportion*. In modern form this proof could be described by saying that we take triangle *ABC* (Fig. 4), which is right-angled at *B*, and drop a perpendicular to the hypotenuse from *B*.

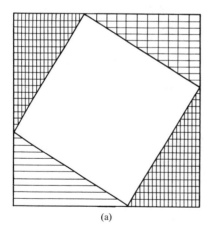
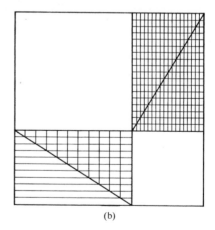

(a)   (b)

Figure 3

---

* The answers to all problems are found beginning on page 237.

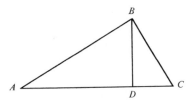

Figure 4

Now triangles *ABC, ADB, BDC* are all similar (that is, equiangular) and thus each have sides in the same ratio. Hence

$$AB : AD : : AC : AB,$$

$$BC : DC : : AC : BC.$$

Re-expressing these relations (by multiplying the means and extremes), we have

$$(AB)^2 = (AD) \cdot (AC),$$

$$(BC)^2 = (DC) \cdot (AC).$$

Adding, we get

$$(AB)^2 + (BC)^2 = (AD) \cdot (AC) + (DC) \cdot (AC)$$

$$= (AC) \cdot (AD + DC) = (AC)^2.$$

When we consider these possible proofs of Pythagoras, a number of features emerge. First, the assumptions are not clearly stated. The appeal at each point is merely to the intuitively obvious. Second, they are meant to be general, ideal, and exact. As Plato puts it,

[students of geometry] make use of visible figures and discourse about them, though what they really have in mind is the originals of which these figures are images : they are not reasoning, for instance, about this particular square or diagonal which they have drawn, but about *the* Square and *the* Diagonal; and so in all cases. The diagrams they draw and the models they make are actual things . . . while the student is seeking to behold those realities which only thought can apprehend [1955 225 (510 D, E)].

Third, no special notation was used. Except in Fig. 4, where the explanation would have become very long without it, only line shading was used, and Pythagoras may have used a similar device. However, he may not have known of the familiar device of labeling triangles with letters, as used in Fig. 4. This might seem utterly unimportant, but we know today that advances in science and mathematics have very often depended on advances in notation. Up to a point they seem merely a matter of convenience, but beyond this the notation itself serves the heuristic function of suggesting further developments of a substantive nature and of allowing a compactness of expression which makes understanding possible. We do not know who first thought of this simple device for naming points, lines, triangles, etc., but without it Euclid's *Elements* would not have been possible.

But the Pythagorean relation was not the most important mathematical theorem discovered by Pythagoras and his school; rather it was the discovery and proof of the existence of incommensurate lengths. *Commensurate* means having a common measure, and it is obvious that the unit is the common measure of all numbers; in fact, a number was understood as a multitude of units [cf. Euclid 1926 II 277]. The arithmetic unit was thought to be indivisible. Fractions such as $\frac{1}{2}$, $\frac{2}{3}$, $\frac{3}{4}$, etc., were not understood as representing a part of a unit, but always as 1 unit out of 2, 2 out of 3, 3 out of 4, etc. Given this arithmetic, it was probably obvious that there had to be an indivisible geometric entity so small that any length would be an even multiple of it. It would then follow that every two lengths would be in a definite fixed proportion to each other. What was meant by *definite fixed proportion* was that the relative size of any two lengths could be expressed as a ratio between numbers. That is,

$$\text{first length : second length : : } x : y,$$

where x is the number of indivisible entities in the first length and y the number in the second.

This theory of proportions makes understandable the creed of the Pythagoreans that the essence of things is numbers. For it was natural to identify the indivisible geometric entity with the numerical unit. But the Pythagoreans made a further identification: namely, that the unit (or indivisible entity) was also a physical atom. This belief prompted (or was prompted by) the Pythagoreans' discovery that all musical scales may be expressed as the ratio of the first four natural numbers: for example, octave 2 : 1, fifth 3 : 2, fourth 4 : 3. They saw special significance in the fact that $1 + 2 + 3 + 4 = 10$. As Aristotle tells us, Pythagoreans saw numbers everywhere:

In numbers they seemed to see many resemblances to the things that exist and come into being—more than in fire and earth and water (such and such a modification of numbers being justice, another being soul and reason, another being opportunity—and similarly almost all other things being numerically expressible); since, again, they saw that the modifications and the ratios of the musical scales were expressible in numbers;—since, then, all other things seemed in their whole nature to be modelled on numbers, and numbers seemed to be the first things in the whole of nature, they supposed the elements of numbers to be the elements of all things, and the whole heaven to be a musical scale and a number [1928a 985b 27ff].

Given this belief in number as the unifying principle of arithmetic, geometry, cosmology and philosophy—a belief which was instilled by all the artifice of religious practice—the discovery of incommensurate lengths must have been a real shock, the first of many clashes between science and religion in the West. It was said that the Pythagoreans were sworn never to reveal this discovery. Aristotle gives us a hint of how the existence of incommensurate lengths was first proved. The substance of the proof, in modern form, is as follows: Suppose we have unit square (see Fig. 1b) and consider the relation of the side (call it s) to the diagonal (call it d). According to the theory of proportions,

$$s : d : : x : y,$$

where x and y are natural numbers with no common divisor. Re-expressing this relation, we have

$$s/d = x/y,$$

from which, if we square both sides, we derive

$$s^2/d^2 = x^2/y^2.$$

By the Pythagorean theorem,

$$d^2 = s^2 + s^2 = 2s^2,$$

so we have

(1) $$s^2/d^2 = s^2/2s^2 = 1/2 = x^2/y^2.$$

That is, $y^2 = 2x^2$, from which it follows that y is even. Hence x must be odd, since x and y have no common divisor. If y is even, then $y = 2z$ and $y^2 = 4z^2 = 2x^2$, and so $x^2 = 2z^2$, from which it follows that x is even. Since it is impossible for a number to be both odd and even, s and d cannot be commensurate.

Problem 2. The proof was more general than required. In what way?

Since number was understood as a plurality of units, it followed that no number could correspond to the length of the diagonal of a unit square. Thus the harmony between numbers and lengths, or between arithmetic and geometry, was broken. The length of d ($= \sqrt{2}$) was neither a number, nor a ratio (cf. "rational") between two numbers. Rather, according to the Pythagoreans, d represented an *irrational* magnitude. Hence number cannot be the essence of geometry, much less of cosmology or philosophy.

The importance of this development for logic is that it represents the first scientific use of a *reductio ad impossibile* proof. In such a proof one derives a contradiction from a hypothesis and then concludes that the hypothesis is false. Its importance is that it enables one to refute a position held by either oneself or another. If what is derived is false, the argument is called a *reductio ad absurdum*. Thus the latter type of argument would include *reductio ad impossibile* arguments as well as arguments in which the derived conclusion is merely known to be false. This kind of distinction was no doubt not made until much, much later.

§3 ARGUMENTATION BEFORE ARISTOTLE

Mathematics developed in a number of significant ways between the time of the achievements of the early Pythagoreans and the time of Aristotle. However, from a logical point of view nothing really new was added to the proof and disproof procedure of the early Pythagoreans. But mathematics was not the only area which stimulated the development of logic; arguments in philosophy and the law courts did also.

The relevance of such arguments may be seen by considering the usefulness of developing a theory of logic in a situation in which there is both a lot of talk which is aimed at proving the truth of something or other and disagreement as to what the truth is. There would be no need to state logical principles if either there were no disagreements or only a small number of them (since in the latter case each could be considered individually). Conversely, the ability to elucidate logical principles presupposes agreement on some very simple arguments. Without this it is unlikely that communication would be possible at all.

Ancient Athens of Aristotle's time provided a great variety of viewpoints and thinkers. Some of the thinkers were from Athens, but many were from Greek colonies; either they would visit Athens or reports of their views would be brought by their disciples. Further, there were extant writings or oral traditions of a philosophical heritage that even then was well over 200 years old. Thales argued that the basic stuff of the world is water, Anaximander that it is not one thing but an indeterminate something or other; Heraclitus that all things are in motion, Parmenides that no things are; Protagoras that our ethical judgments are relative, Socrates that they are not, and so on. Thus, in order to refute the arguments of the sundry Sophists and philosophers whose conclusions Aristotle found either false or paradoxical, he tried to devise a set of principles by which one could determine whether any given argument is a good one.

Zeno of Elea was a typical example of a Pre-Socratic whose arguments Aristotle tried to refute. According to Plato, Zeno "has an art of speaking by which he makes the same things appear to his hearers like and unlike, one and many, at rest and in motion" [1937 I 265 (261D)]. We have no extant writings of Zeno, and it is even possible (although unlikely) that he wrote nothing at all. Nevertheless it is clear that Zeno devised a good number of puzzles which have philosophical interest. Commentary and criticisms of these riddles appeared early and the literature is still growing at a good pace. A typical example of one of his paradoxes is the so-called Achilles argument against motion. Aristotle tells us that

... it amounts to this, that in a race the quickest runner can never overtake the slowest, since the pursuer must first reach the point whence the pursued started, so that the slower must always hold a lead [1941 335 (239b 14–17)].

One of the reasons that there has been so much commentary on the puzzles is the (for the most part) cryptic descriptions we have of them. The above reference is typical in this respect, but a probable reconstruction might be as follows:

Achilles, born of a goddess and the fastest of human runners, cannot catch even a tortoise, the slowest of moving creatures. For suppose we have a race in which the tortoise is given a lead. However fast Achilles runs to reach the point at which the tortoise was, he must take some time to do it. In that time the tortoise will move forward some (smaller) distance. But now we may repeat our argument again and again and again. It is clear that Achilles may get closer and closer to the tortoise but he cannot catch the tortoise.

There is no way to be certain, but it is possible that this *reductio ad absurdum* argument against the existence of motion was inspired by the *reductio ad impossibile* arguments of Pythagorean mathematics. At any rate, we know of no earlier use of this form of argument in philosophy. Its importance is that once the *reductio* form is learned, it tends to breed discussion and dispute rather than disciples who faithfully accept and promulgate the master's teaching. The creation of a heritage of discussion rather than one of truths laid down by authority is perhaps the most important contribution of the Pre-Socratic philosophers to our civilization, and is surely one of the greatest cultural achievements of all time.

Unfortunately, we know very little about the origins of this contribution. It is probable, however, that it originated in the playful element in human nature. Perhaps one of its first forms was the riddle such as that posed by the Sphinx in the Oedipus myth. "What creature," the Sphinx asked, "goes on four feet in the morning, on two at noonday, on three in the evening?" Oedipus' correct answer depended on an ambiguity: "Man, because in childhood he creeps on hands and feet; in manhood he walks erect; in old age he helps himself with a staff." The flash of insight saved Thebes, and it is likely that much of the teaching of the Pre-Socratic philosophers was similar: aphoristic wisdom which comes "out of the blue."

However, another play form developed in which there was a game whose object was to defeat one's opponent by words. Perhaps Zeno was an important influence in the development of it. We do know that there existed a class of teachers who came to be known as sophists. These sophists would travel, much like wandering minstrels, and for a fee would teach their students how to speak persuasively on many different kinds of topics. Sophists were also prepared to defeat any opponent in a public argument. The competitiveness of such a spectacle must have been very keen and the arguments often dramatic, so that we can understand why the arrival of an important sophist in town was the occasion of much excitement and why sophists were often able to command large fees.

Protagoras, often considered to be the greatest of the sophists, would no doubt be thought a great thinker if his works had survived. He is best known for his saying that "man is the measure of all things" and his humanism probably exhibited itself in ways we consider uniquely modern. The following ancient story about him, although probably apocryphal, indicates the kind of verbal pyrotechnics of which the sophists were capable. Protagoras had contracted to teach Euathlus rhetoric so that he could become a lawyer. Euathlus initially paid only half of the large fee, and they agreed that the second installment should be paid after Euathlus had won his first case in court. Euathlus, however, delayed going into practice for quite some time. Protagoras, worrying about his reputation as well as wanting the money, decided to sue. In court Protagoras argued to the jury:

Euathlus maintains he should not pay me but this is absurd. For suppose he wins this case. Since this is his maiden appearance in court he then ought to pay me because he won his first case. On the other hand, suppose he loses the case. Then he ought to pay me by the judgment of the court. Since he must either win or lose the case he must pay me.

Euathlus had been a good student and was able to answer Protagoras' argument with a similar one of his own:

Protagoras maintains that I should pay him but it is this which is absurd. For suppose he wins this case. Since I will not have won my first case I do not need to pay him according to our agreement. On the other hand, suppose he loses the case. Then I do not have to pay him by judgment of the court. Since he must either win or lose the case I do not have to pay him.*

Problem 3. Construct arguments for the defense and prosecution similar to those of Euathlus and Protagoras in the circumstances of the following story. It is taken from Cervantes' *Don Quixote.* Sancho Panza, the governor of the island of Baratavia, has the following case brought before him by a foreigner:

My Lord ... there was a large river that separated two districts of one and the same seignorial domain—and let your Grace pay attention, for the matter is an important one and somewhat difficult of solution. To continue then: Over this river there was a

---

* I have altered this story in inessential ways in order to bring out its logical form. For those who are interested in looking up the original story, see Gellius 1927 404 ff.

bridge, and at one end of it stood a gallows with what resembled a court of justice, where four judges commonly sat to see to the enforcement of a law decreed by the lord of the river, of the bridge, and of the seignory. That law was the following: "Anyone who crosses this river shall first take oath as to whither he is bound and why. If he swears to the truth, he shall be permitted to pass, but if he tells a falsehood, he shall die without hope of pardon on the gallows that has been set up there." Once this law and the rigorous conditions it laid down had been promulgated, there were many who told the truth and whom the judges permitted to pass freely enough. And then it happened that one day, when they came to administer the oath to a certain man, he swore and affirmed that his destination was to die upon the gallows which they had erected and that he had no other purpose in view ... [1949 842].

It is clear that to straighten out such puzzles one has to inquire into *general* procedures of argument. The motive for such an inquiry might not be just to find out the truth but also to defeat one's opponent. Surely this latter motive influenced Plato (427–347 B.C.), who wrote a series of dialogs which leave him unsurpassed both as a thinker and as a writer. Plato was aware of the sportive element in his enterprise and at times—especially in later life—felt it, and artistic endeavor in general, to be unworthy of a true philosopher, who should seek truth without art or playfulness. The young Nietzsche captured Plato's élan:

What ... is of special artistic significance in Plato's dialogues is for the most part the result of a contest with the art of orators, the sophists, and the dramatists of his time, invented for the purpose of enabling him to say in the end: "Look, I too can do what my great rivals can do; indeed, I can do it better than they. No Protagoras has invented myths as beautiful as mine; no dramatist such a vivid and captivating whole as my *Symposion*; no orator has written orations like those in my *Gorgias*—and now I repudiate all this entirely and condemn all imitative art. Only the contest made me a poet, a sophist, an orator" [1954 37–8].

In perhaps no dialog is the playfulness and competitive spirit of Plato better revealed than in the *Protagoras*. The heart of the dialog consists in a long conversation between Protagoras and Socrates before an audience. It concerns the question of whether or not virtue can be taught. Protagoras asserts that it can be; Socrates questions this. This leads to the question of what virtue essentially is. This dialog reveals more than just playfulness, however, when Plato has Socrates state at the end that

... the result of our discussion appears to me to be singular. For if the argument had a human voice, that voice would be heard laughing at us and saying: 'Protagoras and Socrates, you are strange beings; there are you, Socrates, who were saying that virtue cannot be taught, contradicting yourself now by your attempt to prove that all things are knowledge, including justice, and temperance, and courage, which tends to show that virtue can certainly be taught; for if virtue were other than knowledge, as Protagoras attempted to prove, then clearly virtue cannot be taught; but if virtue is entirely knowledge, as you are seeking to show, then I cannot but suppose that virtue is capable of being taught. Protagoras, on the other hand, who started by saying that it might be taught, is now eager to prove it to be anything rather than knowledge; and if this is true, it must be quite incapable of being taught.' Now I, Protagoras, perceiving this terrible confusion of our ideas, have a great desire that it should be cleared up [1937 I 129–30 (361A ff)].

This idea of an argument having a life of its own, the conclusions of which may be unexpected or unwanted by the formulator of the argument, was probably not new with Socrates. But it leads to the open quality of the Socratic dialogs, which in turn heightens interest because the outcome is unknown. It seems clear that Socrates was the first to give his supreme loyalty to inquiry itself rather than to some specific proposition which might be the result of inquiry. "... for you will come to no harm," he says on one occasion to his partner in philosophical conversation, "if you nobly resign yourself into the healing hand of the argument as to a physician without shrinking..." [1937 I 535 (475D)]. It is from Socrates that even today we get the ideal of honest inquiry, in which the inquirer follows the argument wheresoever it leads—even if this brings him embarrassment or disadvantage. The development of this attitude, although not a contribution to logic proper, is nevertheless of utmost importance for it. For without it a large part of the desire to argue correctly is dissipated.

This attitude leads to a reverence for philosophical conversation which keeps its integrity and does not degenerate into mere quibbling. Socrates believes that people

... often seem to fall unconsciously into mere disputes which they mistake for reasonable argument, through being unable to draw the distinctions proper to their subject; and so, instead of a philosophical exchange of ideas, they go off in chase of contradictions which are purely verbal [1955 151 (454A)].

Many examples of purely verbal contradictions and quibbling are given in Plato's dialogs, especially in the *Euthydemus*. In that dialog extravagant arguments are put in the mouths of speakers; arguments whose conclusions are, for example, that no one can tell a lie, that there is no such thing as a contradiction, that Socrates knows everything. An illustration is the famous exchange between Dionysodorus and Ctesippus:

... You say that you have a dog.

Yes, a villain of a one, said Ctesippus.

And he has puppies?

Yes, and they are very like himself.

And the dog is the father of them?

Yes, he said, I certainly saw him and the mother of the puppies come together.

And is he not yours?

To be sure he is.

Then he is a father, and he is yours; ergo, he is your father, and the puppies are your brothers [1937 I 161 (298D–E)].

Problem 4.   Give two different definitions of one of the terms in the argument which would remove the ambiguity on which the argument's plausibility depends.

It would be foolish to think that Plato was not aware of the invalidity of arguments like those mentioned in the *Euthydemus*. Yet we are unaware of just how much difficulty he had in deciding about the validity of many of the arguments put in the mouths of his characters. Nevertheless Plato did preserve a large body of arguments in writing, and this provided a significant part of the corpus of arguments from which Aristotle developed his logic.

It is clear that this writing-down of arguments was one of Plato's main contributions to the prehistory of formal logic. In the course of the dialogs a few logical principles are enunciated, but the real impetus to logic of his writings is the portrayal of that peculiarly Socratic turn of mind which states that we should be

careful of allowing or of admitting into our souls the notion that there is no health or soundness in any arguments at all. Rather say that we have not yet attained to soundness in ourselves, and that we must struggle manfully and do our best to gain health of mind . . . [1937 I 475 (90E)].

Such an attitude requires that one not allow a philosophical argument to degenerate into *mere* quibbling, but it certainly does not require that kind of seriousness which excludes playfulness.

We may now summarize Aristotle's motivation in inventing logic. First, there is the desire to know the truth about the nature of argument, an intellectual curiosity which needs no further account or justification. Second, there is the desire to know the conditions under which something is proved. This question was perhaps most clearly focused in the case of geometry: How are we going to decide when a mathematical relationship really holds? But the problem was wider and was also, in metaphysical questions, acute. Zeno's arguments provide a good example of the latter. Third, there is the desire to refute opponents. Here there is perhaps an analogy with the invention of probability theory. The theory was initiated when Chevalier asked Pascal to solve certain problems having to do with odds in gambling. But probability theory is vastly more comprehensive, applying to many, many different areas, including all physical and social sciences. Similarly, logic is vastly more comprehensive and useful than merely a device which may be used to show that an opponent is wrong. Yet we should not overlook the egoism and spirit of competitiveness which marked its origin.

§4 ARISTOTLE'S LOGIC

A. Preliminaries

Aristotle (384–322 B.C.) must have been a man of almost boundless energy; the range of his intellectuality is astounding. He contributed in important ways to biology, physics, astronomy, political theory, and ethics. His achievement in logic was just one of many others, and even if it were all wrong we would nevertheless consider him an intellectual giant. But just because of the great diversity of his interests, as well as his conviction that logic is a *techné*, an art or a tool, it is unlikely that he would approve of our considering just this aspect of his thought. Indeed his logical theory is embedded in a number of other related metaphysical doctrines which, if that theory is to be considered comprehensively, would have to be discussed in detail. To avoid being diverted from our purpose of providing a background for mathematical logic, we shall forgo such a discussion.

Although Aristotle did not give a definition of *argument*, it is clear that he meant by it what we mean; namely, a set of propositions of which one is claimed to follow from the others. That is, the one is claimed to be true *if* the other proposition or propositions are also true. The proposition which follows from the other proposition or propositions is called a *conclusion*; that from which it follows is the *premiss* or *premisses*. Now a *valid* argument is one in which *if* the premisses are true the conclusion must necessarily also be true. An *invalid* argument is any argument which is not valid. The validity of an argument is in general independent of the truth or falsehood of the premisses. It is perfectly possible for a valid argument to have a false conclusion and for an invalid argument to have a true conclusion.

Problem 5. Construct an example of each kind of argument.

Since the requirements of validity and truth are in general independent, we shall follow standard logical practice in applying 'valid-invalid' to arguments only and 'true-false' to either the premisses or conclusion. This departs somewhat from ordinary speech, in which it is permissible to say 'valid premiss' or 'true argument'. But by avoiding such language we shall be avoiding some of the confusions of ordinary talk about arguments.

The distinction between valid and true and invalid and false was understood by Aristotle, but its first clear and accurate formulation was made by the Stoics. A similar statement must be made about the distinction between a sentence and a proposition, although on this distinction Aristotle at times seems confused. In English we normally make a distinction between 'sentence' and 'proposition' but it is not a hard and fast one. A good example would be the use of the word 'proposition' in Lincoln's Gettysburg address:

Fourscore and seven years ago, our fathers brought forth upon this continent a new nation, conceived in liberty, and dedicated to the proposition that all men are created equal [cf. Church 1956 26].

One does not dedicate oneself to a sentence. Similarly, if one refers to the fifth postulate of Euclid, one does not generally mean something in the Greek language. Consider an English sentence. Translate it into a dozen languages. We would then have thirteen sentences, but only one proposition. A proposition is what is expressed by a sentence.

Aristotle does draw a distinction between a sentence and a proposition, but different from the one just described:

Every sentence has meaning ... by convention. Yet every sentence is not a proposition; only such are propositions as have in them either truth or falsity. Thus a prayer is a sentence, but is neither true nor false [1928 17a 1 ff].

In order to keep closer to contemporary usage, we shall not make the distinction in this way, but rather speak of declarative, interrogative, imperative, and exclamatory sentences or propositions.

Consider Socrates' prayer at the end of the *Phaedrus*:

Beloved Pan, and all ye other gods who haunt this place, give me beauty in the inward soul; and may the outward and inward man be as one. May I reckon the wise to be the wealthy, and may I have such a quantity of gold as a temperate man and he only can bear and carry [1937 I 282 (297E)].

We shall say not that there are two sentences and no propositions, but rather that there are two Greek imperative sentences to which correspond two English (or German or French, etc.) sentences and which all express two imperative propositions. On the other hand, one sentence can be used to express two different propositions. Thus the sentence 'I am pleased with the results of the election' might be true if one person says it and false if another does. For this reason we shall say that the two utterances do not express the same proposition. Declarative propositions will be considered true or false in the primary sense, declarative sentences in the secondary sense (that is, they are true if they express true propositions); interrogative, imperative, and exclamatory sentences or propositions will not be understood as true or false. Logic, as we shall understand it, is primarily about propositions, and only secondarily about sentences. Further, we shall consider only declarative propositions. Relatively little work has been done on the logic of nondeclaratives.

B. Immediate Inferences

By a *simple proposition*, Aristotle means "a statement, with meaning, as to the presence of something in a subject or its absence" [1928 17a 23 f]; by a *composite proposition*, he means one that is made out of simple propositions. A proposition is *universal* if it affirms or denies the predicate to every instance of the subject; *particular* if it affirms or denies the predicate to some (that is, at least one) instance of the subject; *singular* if it affirms or denies the predicate of a subject which denotes exactly one thing. Aristotle considers four general types of propositions: the universal affirmative, the universal negative, the particular affirmative, and the particular negative. Medieval logicians later gave the names **A, E, I, O**, respectively, to these general types of propositions, apparently from the Latin _affirmo_ (I affirm) and _nego_ (I deny). We shall use these names for convenience. Examples, respectively, are 'Every man is white', 'No man is white', 'Some man is white', and 'Some man is not white'. The **A** proposition is often expressed by a variant using 'all' instead of 'every'. For example, 'All men are white'.

Singular propositions may be reduced to universal ones. For example, the singular affirmative proposition 'Socrates is white' may be rewritten as 'Every individual identical with Socrates is white'. The singular negative proposition 'Socrates is not white' may be rewritten as 'No individual identical with Socrates is white'. Hence singular propositions need not be considered separately.

Two propositions are _contradictories_ when if either is true the other must be false, and if either is false the other must be true. **A** and **O** as well as **E** and **I** are thus contradictories when they have the same subject and predicate. _Contrary_ propositions are such that, while both may be false, they cannot both be true. Thus **A** and **E** are contraries when they have the same subject and predicate. A diagram may be used to make Aristotle's doctrine clearer, although no diagram is found in his text.

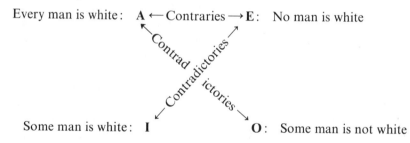

Every man is white: **A** ← Contraries → **E**:   No man is white

Some man is white: **I**   **O**:   Some man is not white

Aristotle assumes that **A** implies **I**, and that **E** implies **O**. Thus, for example, 'every man is white' implies 'some man is white'. In later times the name _subalternation_ was given to this relationship, the universal proposition being called the _superaltern_, its corresponding particular proposition being called the _subaltern_. Aristotle also indicates that he is aware of the relationship between the **I** and **O** propositions, although he does not explicitly define it nor give it a name. Later logicians have called the **I** and **O** propositions _subcontraries_.

Problem 6. Give a definition of the term _subcontraries_ analogous to the one given for _contraries_. If you think about it you will see that the relationships already given allow only one possible definition for this term.

If you add the names *subalternation* and *subcontraries* to the above diagram in the appropriate places, you have the classical *Square of Opposition*; the word *opposition* being used as a generic name for all the possible logical relationships that the **A, E, I, O** propositions might have with one another. Alternatively, one could use a grid to make all these relationships explicit (Table 1).

Table 1

	A	E	I	O
1 **A** given as T				
2 **E** given as T				
3 **I** given as T				
4 **O** given as T				
5 **A** given as F				
6 **E** given as F				
7 **I** given as F				
8 **O** given as F				

In Table 1, 'T' is an abbreviation for 'true', 'F' for 'false'. Consider row (1). If **A** is T, then **A** will be T (given), **E** will be F (since **A** and **E** are contraries), **I** will be T (since **A** implies **I**), **O** will be F (since **A** and **O** are contradictories). Not all spaces can be filled in with a 'T' or 'F'. In such cases we shall use 'U' for 'undetermined'. Thus in row (5), where **A** is F, **E** is U. This can be seen by considering an example. If 'every man is white' is false, then 'no man is white' may be either true or false. It would be false under the condition that some men are white and some men are not white.

Problem 7. Fill in each space in Table 1 with a 'T' or 'F' or 'U'. Use an example if you are uncertain.

Aristotle was the first to use symbols in logic. It can be easily seen from what has been said that it is not the content of a proposition that is important but the form. In geometry, also, it was the triangle *qua* triangle that was important and the geometer had to abstract from the triangle he drew by ignoring any of its particular conditions. Apparently early in his career

Aristotle followed a similar procedure in logic. Thus he would use indiscriminately such statements as 'every man is white', 'every man is an animal', 'every pleasure is good', etc. At some time, however, he came to see that he could use a letter to stand for any term. This practice was perhaps suggested by the use of letters for points, lines, triangles, etc. in geometry. In any case this procedure not only reveals the structure of the proposition more clearly but it is also more convenient to use. Thus Aristotle began using what are now called *propositional forms*: 'Every S is P', 'No S is P', etc. The term *formal logic* arose to designate the study of the various kinds of relationships which these forms might have.

One of these relationships which Aristotle studied is called *conversion*, that is, the interchanging of the subject and predicate term. If 'no S is P', then it follows that 'no P is S'. Similarly, if 'some S is P', then 'some P is S'. With the universal affirmative proposition the situation is more complicated. If 'every S is P', it does not follow that 'every P is S'. But Aristotle asserted that 'some S is P' follows. Later logicians gave the name *limitation* to the procedure of changing an **A** (or **E**) proposition to an **I** (or **O**) proposition (with the subject and predicate remaining constant). Thus, by Aristotle's scheme, the converse of an **A** proposition does not follow, but the converse by limitation does. The inference from an **O** proposition to its converse is not a valid one. For if 'some S is not P', then it doesn't follow that 'some P is not S'. This can be seen by considering Aristotle's example of letting 'S' stand for 'animal' and 'P' for 'man'.

Aristotle also considered changing a term to what is now called its *complement*. If 'animal' is the term, its complement is 'that which is not an animal'. The latter would include the number seven, the Parthenon, and the sun—in fact, everything which is not an animal. By way of notation, if 'S' is a term, we shall use 'non-S' to stand for its complement. Aristotle noted that if 'every S is P' it does not follow that 'every non-S is non-P', but that 'every non-P is non-S' does follow. This process is an example of what was later to be called *contraposition*; that is, the procedure of changing both terms to their complements and then converting.

Problem 8. Of the remaining categorical propositions—that is, **E, I, O**—it happens that for one contraposition is valid, for another only contraposition by limitation is valid, and the third has no valid contrapositive. Verify this last statement by determining which is which.

C. Syllogistic Theory

The kind of inference of Aristotle that we have been considering came to be called *immediate*, presumably because the inference depended on only one premiss, and thus the conclusion followed immediately without the introduction of any other premiss. Aristotle's main contribution to logic was not this theory of the immediate inference, but rather the theory of the *syllogism*, which involves *mediate inference*, that is, an inference depending on more than one premiss.

At first Aristotle understood a syllogism to be any argument, but later he took it to mean an argument with two premises and a conclusion. He concentrated most of his attention on what later came to be known as *categorical syllogisms*, that is, those in which each of the three propositions involved is one of the four we have considered.

A syllogism must contain exactly three terms, each of which occurs twice (but not twice in the same proposition), and each term must be used in the same sense in each occurrence in the syllogism. An example would be the following: If 'all mammals are animals' and 'all men are mammals', then 'all men are animals'. Aristotle always called the predicate term of the conclusion the *major term*, the subject term of the conclusion the *minor term*, and the other term the *middle term*. Thus, in our example, 'animals' is the major term, 'man' the minor, and 'mammals' the middle. The *major premiss* is the one containing the major term and is always stated first; the *minor premiss* is analogously defined.

When we consider the possible arrangements of the middle term in the premisses, there are four. This can best be seen in a diagram such as Table 2, in which we use '*P*' for the major term, '*S*' for the minor, and '*M*' for the middle.

Table 2

	Figure 1	Figure 2	Figure 3	Figure 4
Major premiss	*M-P*	*P-M*	*M-P*	*P-M*
Minor premiss	*S-M*	*S-M*	*M-S*	*M-S*
Conclusion	*S-P*	*S-P*	*S-P*	*S-P*

Aristotle gave the name *figure* to these various arrangements, although (for reasons which are now of interest only to the historian of logic) he did not recognize the fourth figure. For ease of exposition we state each premiss or conclusion on a separate line, although Aristotle himself generally understood the syllogism as one 'if...then' proposition. Our example above about animals, mammals, and men is in the first figure. An example in the second figure is: If 'all animals are mammals' and 'all men are mammals', then 'all men are animals'. This syllogism is invalid, whereas the former was valid.

Now for each figure with a given conclusion there are 16 possible combinations of premisses:

AA, AE, AI, AO

EA, EE, EI, EO

IA, IE, II, IO

OA, OE, OI, OO

Since there are four possible conclusions there are 64 (4 × 16) possible syllogisms for each figure, and therefore 256 (64 × 4) possible syllogisms in all. Only 24 of these represent valid arguments according to the Aristotelian framework. (See Table 3, which indicates the 24 valid forms (V = valid).)

Armed with the theory of the syllogism, Aristotle felt that he could answer the question as to the conditions under which something is proved and thus refute the philosophical doctrine that scientific knowledge (that is, knowledge based on demonstration) is either impossible or circular. It was impossible if, in order to demonstrate a conclusion, one also had to demonstrate the premisses, and so on *ad infinitum*. It was circular if the conclusion was allowed 'in the premisses, which really comes to 'if *A*, then *A*'. "A simple way of proving anything," Aristotle remarks sarcastically.

To understand why anyone would hold either of these two views, we have to recall the situation in geometry when Aristotle was writing. It is likely that there were many treatises on the subject, and that what was assumed without proof in one was proved in another. As we have seen, the initial assumptions or axioms were probably not all made explicit. In such circumstances it is not unlikely that some thinkers would come to believe in the futility of demonstration because the demonstrator must always assume something he hasn't proved; whereas other thinkers would believe circular demonstration to be legitimate, since it was obvious that the geometers had knowledge.

"Our own doctrine," Aristotle tells us, "is that not all knowledge is demonstrative" [1928 72b 18]. Accordingly he provided a philosophical basis for demonstration. Demonstration must start with self-evident truths which are themselves not demonstrable. They must be clearly true and better known than anything that is subsequently proved from them. Aristotle's belief in such truths explains what seems to be an inversion of the relationship between demonstration and syllogism:

Syllogism should be discussed before demonstration because syllogism is the more general: the demonstration is a sort of syllogism, but not every syllogism is a demonstration [1928 25b 28f].

Table 3

Figure	Conclusion								Premisses								
		AA	AE	AI	AO	EA	EE	EI	EO	IA	IE	II	IO	OA	OE	OI	OO
1st	A	✓															
	E					✓											
	I	✓		✓													
	O					✓		✓									
2nd	A																
	E		✓			✓											
	I																
	O		✓		✓	✓		✓									
3rd	A																
	E																
	I	✓		✓						✓							
	O					✓		✓						✓			
4th	A																
	E		✓														
	I	✓								✓							
	O		✓			✓		✓									

A syllogism—even a valid one—might have false premisses, whereas a demonstration is a valid argument based on true premisses. Today we would say that the syllogism can represent but one type of demonstration from premisses, whereas for Aristotle, who apparently believed that all correct reasoning could be made syllogistic (he wasn't always consistent in this belief), demonstration became a species of syllogistic reasoning. A demonstration, for Aristotle, was "a syllogism productive of scientific knowledge." A demonstration is productive of scientific knowledge because the premisses are true and the conclusion necessarily follows. Aristotle went further than this and asserted that the premisses in a demonstration must not only be true, but they must necessarily be true. This, as well as other considerations, makes it likely that he had in mind demonstration in geometry, although clearly he didn't mean to so limit it. In any case, he gave an answer to the question as to when something is proved in geometry, and his doctrine of the nature of demonstration influenced Euclid when he was organizing the *Elements*.

But this ideal of demonstrative knowledge did not just influence others; it brought about the theory of *reduction.* One syllogism can be reduced to another if it can be shown that the first will be valid if the second is. For example, consider an **EAE** syllogism in the second figure:

No *P* is *M*

Every *S* is *M*

Therefore no *S* is *P*

If we convert the major premiss we have:

No *M* is *P*

Every *S* is *M*

Therefore no *S* is *P*

This syllogism is in the first figure. Making use of more involved techniques, Aristotle is able to show that all valid syllogisms can be reduced to either **AAA** or **EAE**, both in the first figure. In other words, if we consider the latter two syllogisms as axioms, the other valid syllogisms can be derived from them. Aristotle's system of syllogisms represents, then, the first body of knowledge presented in an axiomatic way.

There are two other aspects of Aristotle's logic which should be mentioned, if only in passing. The first is that Aristotle developed a theory of modal logic in connection with his theory of the syllogism. *Modal logic* concerns inferences involving such notions as necessity, contingency, and possibility. An example would be the following:

It is possible that some white things are men.

It is necessary that all men are animals.

Therefore it is possible that some white things are animals.

Aristotle's theory of modal logic is quite complex and much of it is wrong. It represents, however, an attempt to deal with notions that any logic which aims at completeness must encompass.

The other aspect of Aristotle's theory that should be mentioned is his theory of definition. "A 'definition'," Aristotle tells us, "is a phrase signifying a thing's essence" [1928 101b 38]. Objects have properties that are both essential and accidental, but only the former enter into a definition of an object. Because Aristotle considered the difference between essential and accidental properties to be an objective fact, a definition, to him, was also objective and if correct, was necessarily true. In this he was thoroughly Socratic; indeed Aristotle gave Socrates the credit for being "the first to raise the problem of universal definition" and said that "it was natural that Socrates should be seeking the essence, for he was seeking to syllogize, and 'what a thing is' is the starting point of syllogisms..." [1928a 1078b 19 ff]. However, Aristotle did not consider a definition true until it had been shown that the word defined referred to something that existed and until the definition had indicated the thing's essential properties in terms which were better known than the word being defined. A definition, as Aristotle pointed out in a number of places, does not assert the existence of the thing defined. This must be either assumed or proved.

All in all, Aristotle's logic is a magnificent achievement; he started with virtually no predecessors and invented a theory which today is considered in many respects right and even complete. If from today's vantage point it also seems limited, it must be remembered that the discovery of its limitations is a rather recent achievement, and that it was 2000 years before anyone besides the Stoics made substantial progress in formal logic. Aristotle by no means claimed that his syllogistic theory covered all kinds of arguments. He was aware of others. For example, early in his career he formulated his famous argument on the necessity of philosophizing:

Either we ought to philosophize or we ought not. If we ought, then we ought. If we ought not, then also we ought [i.e., in order to justify this view]. Hence in any case we ought to philosophize [cf. Kneale 1962 97].

Later in his career—indeed here and there throughout his logical writings— he considered or mentioned in passing a number of principles or examples or techniques which do not fit in with his main theory. There is even a hint that in some of his lost writings he dealt with asyllogistic inference in some detail. Thus when Aristotle's logical system was denounced in Renaissance times and later, the attack applied not so much to Aristotle as to those less-imaginative thinkers who followed him and who were not imbued with the Socratic spirit. This spirit was not absent even in his logical inquiries:

...all syllogism, and therefore *a fortiori* demonstration, is addressed not to the spoken word, but to the discourse within the soul... [1928 76b 23 ff].

When Aristotle made this statement he was merely betraying his Socratic allegiance, for Socrates says:

... the soul when thinking appears to me to be just talking—asking questions of herself and answering them, affirming and denying ... [To] form an opinion is to speak, and opinion is a word spoken—I mean, to oneself and in silence, not aloud or to another ... [1937 II 193 (190A)].

Dante, in a famous phrase, called Aristotle "the master of those who know." It would perhaps be more accurate to call him "a master of those who inquire."

§5 GREEK MATHEMATICS AND LOGIC AFTER ARISTOTLE

A. Mathematics

There are really only two further developments in Greek mathematics which turned out to be important for logic: The first is the systemization of geometrical knowledge, the second is the formulation of geometrical problems which the Greeks themselves could not solve. We shall consider each in turn.

The great system builder of ancient geometry is, of course, Euclid. Unfortunately nothing is known about him. Like the name *Bourbaki* in contemporary mathematics, it is possible that the name *Euclid* refers not to one man, but several. In any case, he (henceforth we shall ignore the possibility that the proper pronoun might be *they*) probably lived about 300 B.C. Tradition says he was a Platonist. If this is true, the following ancient story about him gains credibility, since Plato's theories did not emphasize practical applications:

Some one who had begun to read geometry with Euclid, when he had learnt the first theorem, asked Euclid, "But what shall I get by learning these things?" Euclid called his slave and said, "Give him threepence since he must make gain out of what he learns" [cf. 1926 I 3].

Euclid wrote other works beside his great *Elements*, but only the latter is important for the development of logic. The nature of Euclid's aim in the *Elements* is not altogether clear. The book contains no explanation of its purpose. Of course, Euclid wished to systematize geometrical knowledge, but he may have wished to do more than this. To see what this further aim may have been, let's return to Plato.

Plato established an Academy whose purpose was the advancement of knowledge. Above the entrance Plato is reported to have inscribed: "Let no one ignorant of geometry enter my door." Plutarch reports that Plato said that God continually geometrizes. These hints, together with several passages in the dialogues, confirm that Plato valued geometry highly. Indeed he emphasized the importance of geometry in education, and at one point [cf. 1937 II 573–574 (819E–820C)] extravagantly emphasized the importance of knowing about the incommensurables. There is evidence to suggest that, since the Pythagoreans had failed in their attempt to base their cosmology on arithmetic, Plato intended to base his cosmology on geometry. Apparently he thought his attempt was incomplete and only partly successful. In this light, Euclid's *Elements* may be looked on as an attempt to continue the work of Plato, that is, to base arithmetic and cosmology on geometry. Euclid dealt with arithmetic from a geometrical point of view, and the last section of the *Elements* concerns the five regular solids which played a part in Plato's cosmology. As we shall see [cf. §10A], Euclid showed that geometry could succeed, where arithmetic had failed, in dealing with the incommensurate. Whatever the lack of clarity of Euclid's motives, his achievement is clear : He produced the greatest textbook of all times. (On the topics of this paragraph, see Popper 1952 and Szabó 1967.)

The book starts out with a series of 23 definitions, 5 postulates and 5 common notions. From these Euclid deduced a large number of propositions or theorems. It is important to understand the difference between these four categories (definitions, postulates, common notions, theorems). They indicate the structure of the axiomatic method, which was destined to be closely tied to the development of mathematical logic. Only some of the definitions are omitted in the following extensive quotation from the opening passages of the *Elements* [1926 I 153–155, 241–242]:

Definitions

1. A *point* is that which has no part.
2. A *line* is breadthless length.
3. The extremities of a line are points.
4. A *straight line* is a line which lies evenly with the points on itself.
5. A *surface* is that which has length and breadth only.
7. A *plane surface* is a surface which lies evenly with the straight lines on itself.
10. When a straight line set up on a straight line makes the adjacent angles equal to one another, each of the equal angles is *right*, and the straight line standing on the other is called a *perpendicular* to that on which it stands.
15. A *circle* is a plane figure contained by one line such that all the straight lines falling upon it from one point among those lying within the figure are equal to one another.
16. And the point is called the *centre* of the circle.
19. *Rectilineal figures* are those which are contained by straight lines, *trilateral* figures being those contained by three, *quadrilateral* those contained by four, and *multilateral* those contained by more than four straight lines.
20. Of trilateral figures, an *equilateral triangle* is that which has its three sides equal, an *isosceles triangle* that which has two of its sides alone equal, and a *scalene triangle* that which has its three sides unequal.

23. *Parallel* straight lines are straight lines which, being in the same plane and being produced indefinitely in both directions, do not meet one another in either direction.

Postulates

Let the following be postulated:

1. To draw a straight line from any point to any point.
2. To produce a finite straight line continuously in a straight line.
3. To describe a circle with any centre and distance.
4. That all right angles are equal to one another.
5. That, if a straight line falling on two straight lines make the interior angles on the same side less than two right angles, the two straight lines, if produced indefinitely, meet on that side on which are the angles less than the two right angles.

Common Notions

1. Things which are equal to the same thing are also equal to one another.
2. If equals be added to equals, the wholes are equal.
3. If equals be subtracted from equals, the remainders are equal.
4. Things which coincide with one another are equal to one another.
5. The whole is greater than the part.

Propositions

Proposition 1. On a given finite straight line to construct an equilateral triangle.

Let *AB* be the given finite straight line (Fig. 5).

Thus it is required to construct an equilateral triangle on the straight line *AB*.

With centre *A* and distance *AB* let the circle *BCD* be described;   [   ]
again, with centre *B* and distance *BA* let the circle *ACE* be described;   [   ]

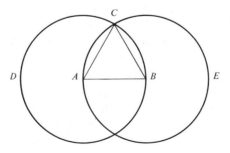

Figure 5

and from the point C, in which the circles cut one another, to the points A, B let the straight lines CA, CB be joined.   [ ]

Now, since the point A is the centre of the circle CDB, AC is equal to AB.   [ ]

Again, since the point B is the centre of the circle CAE, BC is equal to BA.   [ ]

But CA was also proved equal to AB; therefore each of the straight lines CA, CB is equal to AB.

And things which are equal to the same thing are also equal to one another;   [ ]
therefore CA is also equal to CB.

Therefore the three straight lines CA, AB, BC are equal to one another.

Therefore the triangle ABC is equilateral; and it has been constructed on the given finite straight line AB.

(Being) what it was required to do.

Problem 9. In each of the empty brackets at the right in the proof of Proposition I, place the number of the Definition, Postulate, or Common Notion which "justifies" the preceding statement.

Aristotle's view of the nature of demonstrative science apparently decisively influenced Euclid when he was organizing the *Elements*. In any case, what Aristotle has to say about definitions, postulates and common notions clarifies what Euclid was trying to do. We have seen that Aristotle considered definitions to be objective and true. They are objective in the sense that they may not violate established usage. Thus it would be wrong to define *circle* as that which is both equilateral and right-angled. They are true in a derivative sense. Strictly speaking, a definition refers only to a thing's essence, and says nothing about its existence. However, a definition becomes true if it is proved that the thing corresponding to the definition exists and that it has the essential properties which the definition requires. Thus, for example, Definition 20 gives a definition of *equilateral triangle*. It is objective in the sense that it fits in with established usage. It becomes true in conjunction with Proposition 1, which proves that equilateral triangles exist with the properties specified. Of course, one cannot *prove* that everything corresponding to definitions exists; according to Aristotle, in geometry one must *assume* the existence of some things: namely, points and lines.

One of the functions of the postulates is to make these assumptions explicit. Thus Postulate 1 asserts the existence of straight lines, Postulate 3 of circles. It is true that Euclid does not explicitly assume the existence of points, but we must either assume that he does this indirectly through the third definition (in which case it must be demonstrated that the first and third definitions coincide) or assume that this lack represents a failure of the *Elements* to meet Aristotle's requirements.

A second function of the postulates is to assert the possibility of certain constructions. For example, we are guaranteed by Postulate 1 the construction of not only a line but a straight line. Finally, the postulates also assert basic relationships which are needed to prove the theorems of the subject matter under investigation. Postulates 4 and 5 are examples of this sort.

Common notions—Aristotle uses the term *common opinions*—are basic assumptions that are common to a number of sciences, whereas postulates make use of terms that are specifically geometric. Thus, for example, Common Notion 1, "Things which are equal to the same thing are also equal to one another," is as applicable to arithmetic as it is to geometry. None of the postulates are applicable to arithmetic, at least not without changing the meaning of the terms used.

Finally, the propositions—or theorems—are meant to be derived from the definitions, postulates, and common notions without any additional assumptions. It is extremely difficult to do this in a systematic way. For example, even Euclid's Proposition 1 is deficient in that it makes an assumption that is not derivable from the definitions, postulates, and common notions: namely, that the constructed circles will meet at point C. However, this is surely a slip, and Euclid would no doubt have added a postulate to take care of this case, had he noticed it.

There is a final Aristotelian requirement for an axiomatic system: that the axioms be better-known than, simpler than, more certain than—in short, epistemologically prior to—the theorems. Otherwise the system commits the fallacy of *begging the question*. A person commits this fallacy either when he tries to prove what is already epistemologically prior (in which case the best that can be done is to covertly assume what one is trying to prove) or when he assumes as an axiom what is not prior (in which case simpler and more certain things than the axiom would follow from it). In the latter case the proof is fraudulent, since what is proved is less in doubt than the original premises. Euclid's system seems to satisfy this last requirement in almost all respects: The postulates and common notions are not proved; indeed, there is no discussion whatever as to why we should accept them as true. The common notions are clearly self-evident. The first three postulates seemingly propose the simplest possible constructions; the fourth postulate, the determinateness of right angles. It is only with respect to the fifth postulate that he apparently failed, for two reasons: First, if it is true, it does not seem to be epistemologically prior; that is, it seems in need of proof from other statements that would be essentially simpler. Second, although it is plausible, it might not even be true, since other lines are known (for example, a hyperbola and its asymptote) which converge but do not meet.

These objections were raised but not answered in antiquity, and one of the legacies of Greek mathematics was to solve the problem raised by Euclid's fifth postulate; that is, either to prove it true or to get around its difficulties by, for example, giving another definition of *parallel* (for example, two straight lines are *parallel* if they are everywhere equidistant).

Problem 10. See if you can find where the mistake occurs in the following ancient argument, which apparently shows not only that it is possible for the lines in Postulate 5 not to meet but that they can't meet!

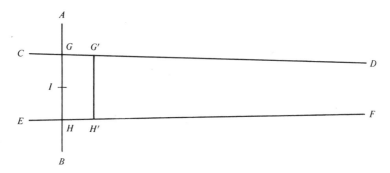

Figure 6

Let *AB* in Fig. 6 be the straight line which falls on two straight lines *CD* and *EF* such that the interior angles *AHF* and *BGD* are together less than two right angles. Bisect *GH* at *I* and make *GG'* and *HH'* each equal to *GI*. Form line *G'H'*. Lines *CD* and *EF* will not meet at a point both nearer to *AB* than *G'* on *CD*, and nearer to *AB* than *H'* on *EF*, for if they did meet at some nearer point *L*, then *GL* plus *LH* would be shorter than *GH*, which is impossible, since two sides of a triangle cannot be shorter than the third. Similarly, lines *CD* and *EF* cannot meet on *G'H'*, since two sides of a triangle cannot be equal to the third. Now repeat the same argument for *G'H'* falling on *CD* and *EF* by forming (in an analogous way) *G"* and *H"*. It is obvious that the argument can be repeated indefinitely, and thus lines *CD* and *EF* cannot meet.

Three other famous geometric problems were also proposed by the Greeks. First, the duplication of the cube, that is, constructing another cube with twice the volume of a given cube. Second, squaring the circle, that is, constructing a square with an area equal to that of a given circle. Third, trisecting an angle, that is, dividing an arbitrary given angle into three equal parts. In each case the problem was to make an exact construction with the aid of a straightedge and compass alone. Greek geometers failed to solve any of these problems under the condition that only a straightedge and compass be used; but by using more complex methods, they solved all of them. Some of the ancient Greeks seem to have been convinced that the problems were impossible to solve under the condition stated. However, no one, so far as we know, conceived of the idea that it might be open to proof that such a construction was impossible. So the Greeks left to posterity the task of deciding whether their failure to solve these problems under the condition stated was due to lack of ingenuity or to the intrinsic nature of the problems themselves. As we shall see, it wasn't until the nineteenth century that geometers made any real advance beyond the fundamentals set forth by the Greeks.

B. Logic

Ancient Greece developed another tradition in logic beside the Aristotelian. This was the Stoic logic, which evolved from the logic of the Megarians, which itself was developed from the philosophy of Parmenides and Zeno. Euclides (*c.* 430–*c.* 360 B.C.), who founded the Megarian school, had a pupil named Eubulides. This Eubulides formulated numerous paradoxes, only one of which is of interest today. However, this one is of great interest, and was destined to play an important part in the development of mathematical logic. It is called the *Liar paradox*, and is mentioned by St. Paul:

One of themselves, a prophet of their own, said, "Cretans are always liars, wily beasts, lazy gluttons." This testimony is true. [*Titus* I : 12–13]

This prophet was probably Epimenides, who lived in the sixth century B.C. Sometimes the Liar is even called the Epimenides, but because he, like St. Paul, apparently did not see its logical interest, the credit for its formulation as a paradox should go to Eubulides. We do not have any ancient books on the subject of the Liar (though we know that many were written), and the references we do have are somewhat indirect and cryptic. Nevertheless it is probable that the ancient Greeks were familiar with such versions of the Liar paradox as the following:

1)   Cretans always lie [uttered by a Cretan].

2)   Whosoever says "I lie" lies and speaks the truth at the same time.

3)   This proposition is not true.

The difficulty with the Liar paradox is that it leads to a contradiction, whether the proposition involved is considered true or false. Consider (3). Suppose that it is true. Then what (3) says is correct, and it says that it is not true. Hence, by *reductio ad impossibile*, (3) cannot be true. On the other hand, suppose that (3) is not true. Then it says of itself that it is not true, and this is of course true under the assumption. Hence, by *reductio ad impossibile*, (3) cannot be not true. In sum, if (3) is true, it is not true and if not true, it is true.

Problem 11. Show that (2) would require an additional (but plausible) hypothesis in order to generate a contradiction; that (1) would also require an additional but less plausible hypothesis in order to generate a contradiction.

In the formulation of this paradox the Megarians showed themselves to be the true heirs of Zeno. Of ancient resolutions of it we know little, but one of Chrysippus has come down to us. Chrysippus (280–207 B.C.) was the leading figure of the Stoic school of logic. If we possessed his works, it is probable that we would consider him, in logic, the equal of Aristotle. Perhaps we would even consider him Aristotle's superior, as some of the ancients did. In any case, he suggested that the Liar proposition—for example, (3)—has no meaning at all, and this anticipates one of the modern resolutions.

Problem 12. It must not be thought that the suggestion that (3) is meaningless easily resolves the antinomy. Determine at least an initial objection to this resolution by assuming the plausible principle that something meaningless is neither true nor false.

Stoic logic was surely very rich. For example, as we have seen, Aristotle was aware of the distinction between true and valid, but the first explicit and clear statement of it was made in Stoic writings. Nevertheless, since none of the modern innovators of mathematical logic knew anything of Stoic logic (Peirce is somewhat of an exception), it does not form part of the historical development of mathematical logic. Indeed it was not until 1927 that the Polish logician Łukasiewicz showed that in many ways Stoic logic was the unknown forerunner of contemporary logic. Thus much of what had been thought to be recent discoveries were really rediscoveries. We can get some idea of the extent of the Stoics' anticipation of modern logic by considering the fact that Stoic logicians defined inclusive and exclusive disjunction, discussed different kinds of implication and defined material and strict implication, devised an equivalent of our truth tables, discovered (in substance) the deduction theorem, developed a propositional calculus that may have been complete, made a distinction which is the virtual equivalent of the sense-denotation distinction of Frege, and indicated an awareness of the language-metalanguage distinction. We shall discuss most of these ideas in this book, when we consider mathematical logic.

Stoic logic, then, was a very great achievement. It is fundamentally simpler than Aristotelian logic, which cannot even be *systematically* presented without the equivalent of the Stoic logic of propositions. One is tempted to apply Santayana's scornful dictum, "Those who cannot remember the past are condemned to repeat it," to the innovators of mathematical logic. Before doing so, however, we should remember that although references to Stoic logic were scattered throughout many ancient writings, they are for the most part fragmentary; and that mathematical logicians not only rediscovered ancient ideas, but (because of a better mathematical heritage) went as far beyond the ancients in logic as contemporary physicists have outstripped their predecessors in physics.

§6 LOGIC FROM THE STOICS TO THE NINETEENTH CENTURY

Our ignorance of the past applies to medieval and Renaissance logic even more than to ancient logic. With respect to ancient logic, historians have at least checked all available material, and although our knowledge of it is inadequate, this is due to the destruction of our sources rather than to lack of interest or effort by historians. With respect to the history of post-ancient logic, however, many manuscripts are known to exist which even today have not been read, let alone translated and produced in critical editions. This is true even if we leave aside Arabian and Jewish logicians, about whom also relatively little is known, due to lack of competent investigators. In such a situation it is not surprising that post-ancient logic had practically no influence on the first formulators of mathematical logic, and it would perhaps be justifiable to omit here any reference to this logic at all. Most of the work that was done (of which we have knowledge) would belong to what is now known as the philosophy of logic. In purely formal logic, we know of little that we would today consider important. There were many reasons for this, the principal ones being the tradition-bound mentality of medieval writers, and the fact that many post-medieval thinkers scorned logic as a mere tedious game.

Nevertheless it might be worth while to briefly note a few of the developments that did take place in the medieval period. In a somewhat arbitrary manner we shall choose just one topic: the methods of deciding whether a syllogism is valid. This is mainly of historical interest, but it will serve as a useful contrast to what mathematical logicians later achieved when they tried similar things.

One of the ways was simply to introduce a mnemonic poem, which goes as follows:

> Barbara celarent darii ferio baralipton
> Celantes dabitis fapesmo frisesomorum;
> Cesare camestres festino baroco; darapti
> Felapton disamis datisi bocardo ferison.

This poem was apparently first published in the thirteenth century by William of Sherwood (c. 1205–c. 1268). To get some idea of the complications of this verse, consider part of its explanation: Each of the words represents a valid syllogism; the mood is represented by the first three vowels; the figure is indicated by the position of the punctuation mark, the first before the first semicolon, etc.; the first letter of each word (after the first four words) indicates which of the first four it should be reduced to; 's' indicates that one uses conversion (on the proposition that is indicated by the letter preceding 's') in the reduction, 'p' indicates that one similarly uses conversion by limitation, 'm' between the first two vowels of a word means that the order of the premisses is to be reversed, 'c' occurring after one of the first two vowels means that the respective premiss should be exchanged for the contradictory of the conclusion for a *reductio ad impossibile* type of reduction. Exactly why this poem seems to recognize 9 valid moods in the first figure, or why it recognizes only 19 valid syllogisms (when under the presuppositions made there should be 24) is unimportant from a logical point of view. However, the very complexity of this little poem gives some insight into the enormous labyrinth of medieval logic and indicates why, when the mood of the times became less tradition-bound, logic was treated with so much scorn by many thinkers.

Another less-artificial method was used to determine the validity of syllogisms. A series of rules was devised such that, if a syllogism satisfied all the rules, it was considered valid, and vice versa. To state the rules the technical term *distribution* is needed: A proposition *distributes* a term if it gives information (either positive or negative) about everything denoted by the term. Thus the **A** proposition (for example, 'all men are mortal') distributes the subject term but not the predicate term (since it says something about all men but not about all mortals); the **E** proposition (for example, 'no men are mortal') distributes both the subject and predicate terms (since it says that each and every man is not mortal and every mortal is not a man); the **I** proposition (for example, 'some men are mortal') distributes neither term (since it neither gives us information about all men or about all mortals), the **O** proposition (for example, 'some men are not mortal') distributes the predicate term (since it tells us that some men are excluded from the entire class of mortals). We can now state the rules for validity of a categorical syllogism. A categorical syllogism is valid when it satisfies the following conditions:

a) It does not contain two negative premisses.

b) It contains one negative premiss if and only if the conclusion is negative.

c) If the conclusion distributes a term it is also distributed in its premiss.

d) The middle term is distributed in at least one premiss.

Justification of these rules (which we shall here omit) was also given. What is interesting from the viewpoint of mathematical logic is that these rules represent an attempt to decide validity from a purely formal point of view. Moreover, it could be done in a few steps in a relatively short time.

Problem 13. Using Table 3 of §4C, verify that every invalid syllogism breaks at least one rule, and that the 24 valid ones break no rule. (Time: approximately 20–30 minutes.)

Problem 14. Suppose that condition (b) were changed to 'If it contains one negative premiss the conclusion is negative'. It is a curious fact that if the other conditions were kept the same, only one invalid syllogism would satisfy all the conditions. Find it. [*Hint*: Begin by ruling out an **E** conclusion by appealing to conditions (c) and (d).]

Nevertheless it might seem that these rules are somewhat arbitrary. This was the opinion of an Italian Jesuit priest named Gerolamo Saccheri (1667–1733), who published a book in 1697 called *Demonstrative Logic*. He wished to put logic on a basis as sound as geometry's:

When I speak of demonstrative logic I wish you to think of geometry—that rigorous method of demonstration which grudgingly admits first principles and allows nothing that is not clear, not evident, not indubitable [cf. Emch 1935 58].

Following Euclid's scheme, he founded his system of Aristotelian logic on a series of definitions, three common notions, and one postulate. However, after proving a large part of Aristotelian theory, he began to feel that the postulate was not really self-evident. He thus tried another method:

It is now my intention to follow another and, as I think, a very beautiful way of proving these same truths without the help of any assumption. I shall proceed as follows: I shall take the contradictory of the proposition to be proved and elicit the required result from this by a straight-forward demonstration [cf. Kneale 1962 346].

The essence of Saccheri's method is as follows: Suppose that a proposition A is true. Then, of course, A is true. Suppose that A is false. Now a demonstration is given which shows that the truth of A follows from that assumption. Since A must be either true or false, it follows in either case that it is true. Therefore it must be necessarily true. This argument was given the name *consequentia mirabilis*. An example which almost fits this form we have already seen from the young Aristotle:

Either we ought to philosophize or we ought not. If we ought, then we ought. If we ought not, then also we ought [i.e., in order to justify this view]. Hence in any case we ought to philosophize [cf. Kneale 1962 97].

This is not an exact example of the *consequentia mirabilis*, since the conclusion does not follow from its contradictory but only from the process of establishing the contradictory. The Stoics also used similar arguments to refute sceptics. Exact instances of the *consequentia mirabilis* can be found in a number of writers before Saccheri, including Euclid [cf. Problem 16 (§9)].

It appears, however, that Saccheri was the first to apply this form of argument to syllogistic theory. Here is an example:

Every syllogism with a universal major premiss and an affirmative minor premiss is an argument with a valid conclusion in the first figure.

No **AEE** syllogism is a syllogism with a universal major premiss and an affirmative minor premiss.

Therefore no **AEE** syllogism is an argument with a valid conclusion in the first figure.

This argument has true premisses and is itself an **AEE** syllogism in the first figure. Now if an **AEE** syllogism in the first figure is invalid, it is of course invalid. On the other hand, if an **AEE** syllogism in the first figure is valid, then the conclusion of the above argument is true (since the premisses are true) and thus an **AEE** syllogism in the first figure must be invalid. In sum, the assumption of the validity of such a syllogism implies its invalidity, and thus it is necessarily invalid.

Arguments of this form have a kind of self-reflective quality which at first makes one doubt that they are valid. However, not only are they valid but they provide an elegant method of proof in logical theory. And, as we shall see, another important use was made of this method which was significant for the development of mathematical logic [cf. §8].

Problem 15. What is *wrong* with the following argument?

All syllogisms which satisfy the four rules for the syllogism are valid.

All **AAA** syllogisms in the first figure are syllogisms which satisfy the four rules for the syllogism.

Therefore all **AAA** syllogisms in the first figure are valid.

§7 SUMMARY

Aristotle, with little help from his predecessors, formulated a rather extensive theory of logic, which consisted of the theory of syllogism, together with connected doctrines such as the theory of definition and modal logic. It was this theory which was to have an enormous influence on philosophy and theology for more than 2000 years. When a valid argument was found which could not be accounted for by that theory, it was either stated without developing any formal theory to account for it (as, for example, in the writings of Aristotle himself), or if some theory was provided it had little influence (as in the case of the works of the Stoics). Indeed, we even have the example of the physician Galen (*c.* 129–*c.* 199), who gave examples of arguments which could be accounted for by neither Aristotelian nor Stoic logic (for example: If 'Sophroniscus is father to Socrates', then 'Socrates is son to Sophroniscus'). Nevertheless the degree to which non-Aristotelian logic was known and was influential may perhaps be judged by the opinion of Immanuel Kant, often thought to be the greatest of modern philosophers. In 1787 he stated:

That *Logic*, from the earliest times, has followed this secure method [that is, of science], may be seen from the fact that since *Aristotle* it has not had to retrace a single step, unless we choose to consider as improvements the removal of some unnecessary subtleties, or the clearer definition of its matter, both of which refer to the elegance rather than to the solidity of the science. It is remarkable also, that to the present day, it has not been able to make one step in advance, so that, to all appearance, it may be considered as completed and perfect [1961 501].

In 1800, he added:

Aristotle has omitted no essential point of the understanding; we have only to become more accurate, methodical, and orderly [1885 11].

We can contrast this respectful and reverent attitude of Kant's with that of an important contemporary logician, Willard Van Orman Quine. The first sentence of one of Quine's books is: "Logic is an old subject, and since 1879 it has been a great one" [1949 vii].*

Now Kant's statement is incorrect and reveals an ignorance of the history of logic; and Quine's is surely an exaggeration. Logic was a "great" subject when Aristotle and Chrysippus were devising their respective theories. Nevertheless, as an indication of what many important thinkers believe both then and now, the statements are perhaps not misleading.

This radical change in attitude took place because of certain developments in the nineteenth and early twentieth centuries. This is the period of transition to which we now turn.

---

* 1879 was the date of publication of Frege's *Begriffsschrift*.

CHAPTER 2

PERIOD OF TRANSITION

§8 INTRODUCTION

The nineteenth and early twentieth centuries are called the *period of transition* because they were considered as such by the innovators of mathematical logic. Even today most logicians would probably consider this designation correct. Yet the transition was probably not as smooth and neat as this label might suggest. We know that a number of supposedly original discoveries of this period were only rediscoveries. For example, logicians of this period, without having any knowledge of Stoic logic, rediscovered much of the content of Stoic logic. If the gaps in our knowledge of logic from the Stoics to the nineteenth century were to be filled, no doubt we would find still more foreshadowings of modern logic. Thus the transition from that fusion of Aristotelian and some Stoic logic—often called *traditional logic*—to mathematical logic was rather complicated, and the various parts of the development did not take place simultaneously or at the same rate. Nevertheless, for our purposes, it is perhaps easier to organize this account around developments that took place in this period of transition, noting the known anticipations of these developments as they are described. And rather than sticking to a strict chronological account of them, we shall divide them into topics in order to discuss them more systematically.

If we wished to summarize the situation most simply, we could say with Quine that logic again became a "great" subject in the nineteenth century because of certain developments in mathematics. The progress of mathematics confronted mathematicians with new and profound logical problems, for which the traditional logic was of no help. These problems centered mainly around the nature of axiomatic systems and the elimination of contradiction from a given system. The attempts to solve these problems led to the application of mathematical methods to the problems. Thus it appears that logic became rejuvenated because of one of the same considerations that led to its initial development: the need to solve certain logical problems in mathematics. A new and more sophisticated logic was needed to equal the more sophisticated mathematics of the nineteenth century. Yet it must not be thought that the traditional logic was adequate for all the logical problems that arose before then. We have already cited an example of Galen's which fits no ancient framework. Here is a mathematical example of Galen's:

Theon has twice as much as Dio, and Philo twice as much as Theon; therefore Philo has four times as much as Dio [cf. Kneale 1962 185].

Nevertheless such arguments could be (and were) safely ignored, since the progress of mathematics did not depend on having a theory that would account for such an argument.

The case was otherwise in the nineteenth century. For then there arose in mathematics certain puzzling and surprising questions to which neither intuition nor the then current logical theory had any answer. Everyone can "see" that Galen's argument follows even if there is no logical theory to account for it. But the rise of a geometry different from Euclid's was a real surprise, and neither mathematicians nor philosophers were quite sure what to make of it. The same holds true both for the discovery of different sizes of the infinite and the logical inconsistency of some axiomatic systems with seemingly innocuous axioms. The latter occurred in a mathematical discipline called *set theory*. To understand what led to mathematical logic, we must first consider in turn each of these disciplines: non-Euclidean geometry and set theory.

§9 NON-EUCLIDEAN GEOMETRY

The reasons why a man continues to hold to a belief sometimes has little to do with what made him adopt the belief in the first place. Very often these reasons have to do with pride or self-esteem. This is especially true with regard to a belief with which a person identifies himself, as in the case of philosophical or religious beliefs. In such a case the person may develop a prejudice, that is, a refusal to consider impartially evidence on both sides of the question at hand. If the belief is widely held, the society may develop customs to make it more unlikely that the belief be questioned. We may then speak of a social prejudice.

In this sense there was in Europe at the end of the eighteenth century a social prejudice consisting of the belief that Euclidean geometry was true. There was good evidence for the truth of this system of mathematics. Was not Euclidean geometry more than 2000 years old? Yet the intellectual climate was such that a questioning of this tradition met not so much with refutation as with derision.

The security of those who laughed at the innovators of non-Euclidean geometry was greatly enhanced by the philosophy of Kant (1724–1804), which then prevailed. Further, it was the views that Kant presented in his *Critique of Pure Reason* (1781) that were influential, rather than some of his earlier views which, with the advantage of hindsight, we can see were more compatible with the development of geometrical knowledge than those that were laid down in the *Critique*. According to the *Critique*, all judgments can be divided into two kinds: The first kind is *analytic*, that is, the predicate belongs to the subject because it is overtly or covertly contained in it. The second kind is *synthetic*, that is, the predicate is connected with the subject but not contained in it. For example, the judgment that all bodies are extended is analytic since, according to Kant, only logical analysis of the concept body is needed in order to see its truth. However, the judgment that all bodies are heavy cannot be found to be true through analysis of the concept body, and thus is synthetic. Synthetic judgments extend our knowledge, whereas analytic judgments do not; they only clarify what we already know.

Kant tells us that all judgments based on experience (*a posteriori* judgments, as he calls them) are synthetic. However, just because they are based on experience they are not necessarily true, for at any time we might have a different sort of experience which would prove them to be false. Necessary judgments are all *a priori*, that is, independent of experience. In fact, since it is also true that all *a priori* judgments are necessary, we may use the criterion of necessity to decide whether or not a given judgment is *a priori*.

We have then seemingly four possible kinds of judgments: analytic *a priori*, analytic *a posteriori*, synthetic *a priori*, synthetic *a posteriori*. The second category, however, does not exist, since if a judgment is analytic no experience is needed to see that it is true. The problem as it appeared to Kant can be put thus: Analytic *a priori* judgments are necessary but they do not add to our knowledge. Synthetic *a posteriori* judgments add to our knowledge but are not necessary. How is it possible to have a judgment which adds to our knowledge and is necessary, that is, a synthetic *a priori* judgment? Kant never doubted that we had synthetic *a priori* knowledge; in particular, he believed most mathematical judgments to be of this sort. We may summarize the point reached so far with a diagram (Table 4) in which more examples are given:

Table 4

Judgments	Analytic	Synthetic
A priori	All bodies are extended. Every effect has a cause. *A* is *A*.	Seven plus five equals twelve. A straight line is the shortest distance between two points.
A posteriori		All bodies are heavy. Aristotle was a student of Plato.

Kant's view of geometrical knowledge was that he believed all judgments of geometry to be synthetic *a priori*, with the exception of "some few propositions." These latter are analytic. Although he did not explicitly say so, the

apparent distinction he had in mind was the distinction between the common
notions and postulates in Euclidean geometry. That is, the common notions
are analytic, whereas the postulates—and those propositions which can be
deduced from them—are all synthetic *a priori*. He said that "the whole is
greater than its part" is analytic, whereas "that the straight line between two
points is the shortest, is a synthetical proposition." His reasoning here was
as follows:

> ...my concept of *straight* contains nothing of magnitude (quantity), but a quality only.
> The concept of the *shortest* is, therefore, purely adventitious, and cannot be deduced
> from the concept of the straight line by any analysis whatsoever. The aid of intuition,
> therefore, must be called in, by which alone the synthesis is possible [1961 529–30].

What this intuition is is difficult to characterize. The word is used in
contrast to conception, which is discursive, whereas intuition is immediate.
As Kant puts it:

> ...I construct a triangle by representing the object corresponding to that concept
> either by mere imagination, in the pure intuition, or, afterwards on paper also in the
> empirical intuition, and in both cases entirely *a priori* without having borrowed the
> original from any experience. The particular figure drawn on the paper is empirical, but
> serves nevertheless to express the concept without any detriment to its generality,
> because, in that empirical intuition, we consider always the act of the construction of
> the concept only, to which many determinations, as, for instance, the magnitude of the
> sides and the angles, are quite indifferent, these differences, which do not change the
> concept of a triangle, being entirely ignored [1961 421–2].

Kant argues that the conditions which determine this intuition are the
conditions of having any experience whatsoever. An analogy might be
useful in explaining this: One of the necessary conditions of sight in general
is the existence of an eye. There is no sight without eyesight. So Kant is
saying that one of the conditions of experience is that we experience things
in space, in particular, in Euclidean space. Thus when we say 'a straight line
is the shortest distance between two points', it is *a priori* because we know
this to be one of the conditions of experience, and therefore it is not part of
experience. It is the structure of our mind which makes space Euclidean.
The world around us we must understand in terms of that structure and thus
it too is Euclidean.

This then is Kant's answer to the problem of how synthetic *a priori*
judgments are possible. To the question 'of what are Euclidean propositions
true?' Kant now had an answer. They are true of our *a priori* intuitions.

It is easy to see how such a view might hinder the development of a
geometry different from Euclid's. For if such a view were influential it would
be unlikely that anyone would look for a non-Euclidean geometry. If
someone did, it would take intellectual daring to assert the existence of a new
geometry (remember that Euclidean geometry was thought necessarily true

of the world), and finally it might take courage to publicly proclaim its existence. One can easily exaggerate the importance of Kantian philosophy in hindering the discovery of non-Euclidean geometry. The ancient authority of Euclidean geometry, as well as its seemingly unimpeachable presentation, were surely stronger factors. Nevertheless Kant's influence was not insignificant in this regard.

But what is non-Euclidean geometry? The story of its development goes back to the dissatisfaction with Euclid's fifth postulate. This postulate states 'that, if a straight line falling on two straight lines makes the interior angles on the same side less than two right angles, the two straight lines, if produced indefinitely, meet on that side on which are the angles less than the two right angles'. As we have seen, the doubts about this postulate were really two: First, if it is true, we should surely be able (so it was thought) either to prove it from the other common notions and postulates or to find an essentially simpler postulate from which it can be proved. Second, how can we be sure it is true, since other lines are known (for example, a hyperbola and its asymptote) which converge but do not meet?

So far as we know, no one who came between the time of Euclid and the nineteenth century doubted the truth of the postulate.* Thus efforts focused on proving the postulate or getting an essentially simpler substitute. However, all efforts at proving the postulate failed. We know today that they had to fail, since it cannot in fact be proved from the other assumptions. As for getting around the postulate by using simpler substitutes, two methods were tried: The first was to make a change in definition, the second to propose a simpler postulate. An example of the first method is the following: It was proposed that the definition of parallel straight lines be changed from 'straight lines which, being in the same plane and being produced indefinitely in both directions, do not meet in either direction' to 'straight lines which are in the same plane and are everywhere equidistant'. The difficulty with this approach lay in proving that such lines exist and in showing that lines which are not parallel do meet somewhere. It was found that to prove these things one had to make an assumption logically equivalent to the fifth postulate. *Logically equivalent* here means that if we assumed the fifth postulate we could derive the new assumption with the help of the other common notions and postulates, whereas if we replaced the fifth postulate with the new assumption we could, using the latter (and, of course, the other common notions and postulates), deduce the fifth postulate. Thus from the Aristotelian point of view no proof would be given. All that would be shown would be a logical connection between the two statements; to claim a proof would be to commit the fallacy of begging the question.

---

* For some evidence that the postulate may have been doubted before Euclid, see Tóth 1969.

Of the many substitutes that have been made for the fifth postulate, the most famous and widely used is *Playfair's postulate*, so named because John Playfair (1748–1819), a Scottish mathematician, called attention to it, although it was known in ancient times. It states that 'through a given point, only one parallel can be drawn to a given straight line'. This may be illustrated by the following diagram:

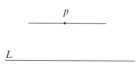

That is, if we are given a line L and a point p not on L, there is only one line through that point parallel to L. (That at least one parallel line can be drawn can be proved from the other common notions and postulates; at least, it can be proved if the second postulate is understood as asserting the infinitude of straight lines.) Playfair's postulate is logically equivalent to Euclid's.*

Thus, again, if we are to avoid begging the question, we cannot say that we have advanced epistemologically from Euclid's geometry. If we had doubts about the fifth postulate we should have equivalent doubts about Playfair's postulate.

All attempts at either proving or finding a substitute for the fifth postulate failed until, in 1733, Saccheri finally tried a new method, in a book called *Euclid Freed of Every Flaw*. After studying what numerous commentators on Euclid had said before him (including even Arabic writers such as Omar Khayyam and Nasiraddin) and remembering the success of the *consequentia mirabilis* in his *Demonstrative Logic*, Saccheri tried to solve the problem by this indirect form of argument. Euclid himself had used this kind of argument, which, as we have seen, consisted essentially in proving a proposition by assuming the falsehood of that proposition. Since Saccheri believed that every "primal verity" could be thus established, it was natural for him to try it on the fifth postulate.

Problem 16. Saccheri said that Euclid used this form of argument to prove Proposition 12 in Book IX of the *Elements*. In modern form that proposition states the following: If in a finite geometrical progression†

---

* Euclid's fifth postulate, although it does not even contain the word 'parallel', is often called the *parallel postulate*; presumably because, on one hand, the whole Euclidean theory of parallels rests on it, and on the other, it has this equivalent which does involve the word *parallel*.

† A geometrical progression $(g_1, g_2, g_3, \ldots, g_m)$ is a sequence of numbers in continued proportion, that is, $g_i : g_{i+1} :: g_{i+1} : g_{i+2}$. Since the above sequence begins with 1, we have $1 : a_1 :: a_1 : a_2$, that is, $a_2 = a_1 \cdot a_1$. In general, $a_i = a_1 \cdot a_{i-1}$.

beginning with $1\,(1, a_1, a_2, \ldots, a_n)$, a prime number ($b$) evenly divides a_n, it (b) also evenly divides a_1. Euclid assumed that b evenly divided a_n but did not evenly divide a_1. He went on to show that nevertheless b did divide a_1. In his proof he also assumed (something proved earlier in the text) that if $c : d : : e : f$ and c and d had no common divisor, then c evenly divided e, and d evenly divided f. From these assumptions, see if you can prove Proposition 12. [*Hint:* Start with the assumption that b evenly divides a_n and show that this in conjunction with the other assumptions implies that b evenly divides a_{n-1}.]

In order to bring out the logical structure of the problem as he saw it, Saccheri considered a quadrilateral constructed as follows (see Fig. 7): On a line segment AB construct two equal perpendicular lines at the extremities of AB, that is, construct AC and BD. Join points C and D with a straight line.

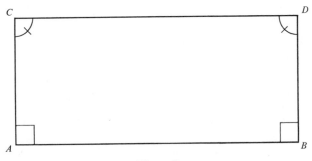

Figure 7

Without assuming Euclid's fifth postulate, Saccheri was able to prove that angles ACD and BDC were equal. Then Saccheri considered three hypotheses: First, that the angles were obtuse; second, that they were right; third, that they were acute. It can be shown that the hypothesis of the right angle is logically equivalent to the fifth postulate. Suppose that we deny this hypothesis and accept the hypothesis of the obtuse angle. Saccheri showed that the fifth postulate follows from this assumption. To do this he had to understand Euclid's second postulate—'to produce a finite straight line continuously in a straight line'—as asserting the infinitude of a line. Euclid also understood the postulate in this way, so that he was justified on a historical basis. Note, however, that the postulate does not explicitly state the infinitude of a line; it only suggests it in a more or less ambiguous way. In any case Saccheri was able to prove the fifth postulate from the hypothesis of the obtuse angle, which is contrary to the fifth postulate. That is, as Saccheri put it in his Proposition XIV:

The hypothesis of the obtuse angle is absolutely false, because it destroys itself [1733 59].

He then had only to dispose of the hypothesis of the acute angle in order to prove the fifth postulate. In this he failed, as today we know he had to, since there was no contradiction to be found. When he came to Proposition XXXIII he said:

The hypothesis of the acute angle is absolutely false; because repugnant to the nature of the straight line [1733 173].

What is "repugnant to the nature of the straight line" is the existence of two straight lines

AX, BX, existing in the same plane, which produced *in infinitum* toward the parts of the points X must run together at length into one and the same straight line, truly receiving, at one and the same infinitely distant point a common perpendicular in the same plane with them [1733 173].

He then went on to state and prove Proposition XXXVIII: "The hypothesis of the acute angle is absolutely false, because it destroys itself" [1733 225]. The proof, however, is erroneous. Saccheri himself apparently felt some uneasiness, for he states that there is

...a notable difference between the foregoing refutations of the two hypotheses. For in regard to the hypothesis of the obtuse angle the thing is clearer than midday light; since from it assumed as true is demonstrated ... the absolute falsity of this hypothesis .... But on the contrary I do not attain to proving the falsity of the other hypothesis, that of the acute angle, without previously proving that the line, all of whose points are equidistant from an assumed straight line lying in the same plane with it, is equal to this straight [line], which itself finally I do not appear to demonstrate from the viscera of the very hypothesis, as must be done for a perfect refutation [1733 233, 235].

To sum up, Saccheri thought he could prove the fifth postulate by means of the *consequentia mirabilis*. Some evidence that it is capable of being proved is given by the fact that Euclid himself proves the converse of the fifth postulate (Euclid's Proposition 17). Had Saccheri found the contradiction he was looking for, he would have proved the fifth postulate. But this would not have shown Euclid to be "freed of every flaw." On the contrary, it would have shown Euclid to have made an unnecessary assumption. When an initial assumption (or its negation) is derivable from other initial assumptions, it is called *dependent*; otherwise it is called *independent*. Saccheri would thus have showed the fifth postulate to be dependent on the other assumptions.

Saccheri (in his *Demonstrative Logic*) was perhaps the first person in history to study the idea of independence and to provide a criterion for it: Show that it is possible for all the other assumptions to be true and the one in question to be false. If it follows that the assumption in question must necessarily be false, the assumptions taken as a whole are inconsistent; that is, it is possible to derive a contradiction from them. The test for consistency of a set of assumptions then is: Is it possible for all the assumptions to be true? If so, the set is consistent; if not, not. Saccheri tried to prove that a contradiction followed from the hypothesis of either the acute or the obtuse angle. Thus, Saccheri tells us,

> ... not without cause was that famous axiom assumed by Euclid as known *per se*. For chiefly this seems to be as it were the character of every primal verity, that precisely by a certain recondite argumentation based upon its very contradictory, assumed as true, it can be at length brought back to its own self [1733 237].

Ironically, it is only because Saccheri failed in the latter attempt that Euclid might be freed from the charge of an unnecessary or false assumption.

Problem 17. Consider two different axiom systems: the first Euclid's, in which the fifth postulate is replaced by its converse; the second Euclid's, in which the fifth postulate is replaced by the hypothesis of the obtuse angle. Determine, on the basis of information given in this section, whether the new postulate is independent and whether each system is or is not consistent.

In the course of trying to find the two contradictions he needed, Saccheri stated and proved a number of theorems which later became part of the body of non-Euclidean geometry. By taking the pose of being initially neutral about the fifth postulate, he was able to prove statements which no one else would be able to prove for nearly a century. Here are three examples:

1.   According as the hypothesis of the obtuse angle, or right angle, or acute angle is proved true in a single case, it is respectively true in every other case.

2.   According as the hypothesis of the obtuse angle, or right angle, or acute angle is true, the sum of the angles of a triangle is respectively greater than, equal to, less than, two right angles.

3.   According as the hypothesis of the obtuse angle, or right angle, or acute angle is true, two straight lines always intersect; two straight lines intersect except in the single case in which they share a common perpendicular; two straight lines may or may not intersect, but (except in the case in which they share a common perpendicular) they either intersect or they mutually approach each other.

If we consider the last result and a few other things Saccheri either proved or easily knew, we can show that the hypotheses of the obtuse, right, and acute angles are respectively equivalent to the following statements:

Given a line *L* and a point *p* off that line, there are no lines through the point parallel to the given line; there is exactly one line through the point parallel to the given line; there are two lines through the point parallel to the given line (see Fig. 8 at the top of page 46).

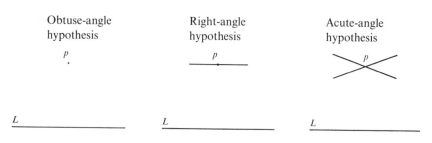

Figure 8

Saccheri showed that if there were two lines parallel, as in the case of the hypothesis of the acute angle, there were then an infinite number of such lines, namely, all those "between" the two given lines. He further showed there were always two determinate straight lines which separated those which are parallel from those which are not.

Nevertheless, because Saccheri didn't understand the nature of his discoveries, he—unlike Columbus— is not given credit for the discovery of a new-found land. The credit usually goes to three men: Gauss (1777–1855), who was German, Bolyai (1802–1860), who was Hungarian, and Lobachevski (1793–1856), who was Russian. Although this credit is not undeserved, the story of the development of non-Euclidean geometry from Saccheri to these men is complex. Saccheri's work was known in those days, but his influence is uncertain. The full story would involve more than a dozen names. To give just one example: Karl Schweikart (1780–1859), a lawyer, sent Gauss in 1818 a memorandum in which he asserted (among other things) that there were two kinds of geometry, Euclidean and another in which the sum of the angles of a triangle is less than two right angles. Schweikart did not publish any of his results. If he had, he might be given credit for the discovery of non-Euclidean geometry.

Gauss also did not publish his writings on the subject. However, his genius and versatility in mathematics were such that perhaps only Archimedes and Newton were in his class. This fact no doubt has led historians to credit mathematician Gauss with the discovery of non-Euclidean geometry, often without mentioning lawyer Schweikart. Gauss delayed publishing his results, probably due to the influence of both conservative mathematicians and Kantians. In a letter of 1824 Gauss states that

... the assumption that the sum of the three angles [of a plane triangle] is less than [two right angles] leads to a curious geometry, quite different from ours (the Euclidean), but thoroughly consistent, which I have developed to my entire satisfaction, so that I can solve every problem in it with the exception of the determination of a constant, which cannot be designated *a priori*. The greater one takes this constant, the nearer one comes to Euclidean Geometry, and when it is chosen infinitely large the two coincide. [This fact had already been contained in Schweikart's memorandum which Gauss read in 1818.] The theorems of this geometry appear to be paradoxical and, to the uninitiated, absurd; but calm, steady reflection reveals that they contain nothing at all impossible. ...All my efforts to discover a contradiction, an inconsistency, in this non-Euclidean Geometry have been without success, and the one thing in it which is opposed to our conceptions is that, if it were true, there must exist in space a linear magnitude, *determined for itself* (but unknown to us). But it seems to me that we know, despite the say-nothing word-wisdom of the metaphysicians, too little, or too nearly nothing at all, about the true nature of space, to consider as *absolutely impossible* that which appears to us unnatural. If this non-Euclidean Geometry were true, and it were possible to compare that constant with such magnitudes as we encounter in our measurements on the earth and in the heavens, it could then be determined *a posteriori*. Consequently in jest I have sometimes expressed the wish that the Euclidean Geometry were not true, since then we would have *a priori* an absolute standard of measure [cf. Wolfe 1945 46–47].

Other evidence exists which shows that Gauss was delayed in announcing his results because of the anticipated reaction of the Kantians. But in February 1832 he received a letter from a friend of his, Wolfgang Bolyai (1775–1856). This contained an appendix which his son, Johann Bolyai (1802–1860), had written to a book of Wolfgang's. This appendix contained a description and many theorems on non-Euclidean geometry. Johann had been working on this approach on and off since 1823. He was a very excitable young man and correctly thought he had achieved something very important; "*out of nothing I have created a strange new universe,*" he wrote his father in 1823. Because of Gauss' fame, his answer was eagerly awaited. It stated:

If I begin with the statement that I dare not praise such a work, you will of course be startled for a moment: but I cannot do otherwise; to praise it would amount to praising myself; for the entire content of the work, the path which your son has taken, the results to which he is led, coincide almost exactly with my own meditations which have occupied my mind for from thirty to thirty-five years. On this account I find myself surprised to the extreme.

My intention was, in regard to my own work, of which very little up to the present has been published, not to allow it to become known during my lifetime. Most people have not the insight to understand our conclusions and I have encountered only a few who received with any particular interest what I have communicated to them. In order to understand these things, one must first have a keen perception of what is needed, and upon this point the majority are quite confused. On the other hand it was my plan to put all down on paper eventually, so that it would not finally perish with me.

So I am greatly surprised to be spared this effort, and am overjoyed that it happens to be the son of my old friend who outstrips me in such a remarkable way [cf. Wolfe 1945 52].

Johann was far from pleased with this reply, for it showed that either he had been preceded by Gauss or Gauss was lying. Sixteen years later he received another blow when he learned that Nikolai Lobachevski had published essentially the same results a few years before he wrote his own appendix. Johann wrote:

If Gauss was, as he says, surprised to the extreme, first by the *Appendix* and later by the striking agreement of the Hungarian and Russian mathematicians: truly, none the less so am I [cf. Wolfe 1945 56].

Again, it seems quite likely that Lobachevski discovered non-Euclidean geometry quite independently of Gauss and Bolyai. Since Lobachevski was the first to publish his work, this form of geometry is called *Lobachevskian*.

Lobachevskian geometry results when the fifth postulate of Euclid is replaced by the hypothesis of the acute angle or by the statement 'given a line L and a point p off that line, there are two lines through the point parallel to the given line'. We can examine the situation by dropping a perpendicular from p to L (see Fig. 9).

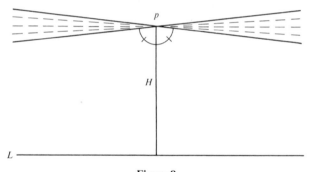

Figure 9

The lines which pass through p are divided into two classes: those which intersect L and those which do not. For convenience only those two lines which divide these two classes are called *parallel* (represented in the figure by solid lines); the others (represented by dashed lines) are called *non-intersecting*. That such parallel dividing lines exist was proved by Lobachevski (as it had been previously by Saccheri). Now the angle which one of these parallels makes with the perpendicular (Lobachevski shows them to be equal) is called the *angle of parallelism*. It would be interesting to know what this angle is. Lobachevski showed that it is acute and that it depends on the height H of the perpendicular. As the height of the perpendicular approaches infinity, the angle approaches 0; as the height of the perpendicular approaches 0, the angle approaches a right angle. Lobachevski shows that to every length *H* there corresponds exactly one angle of parallelism, and vice versa. Thus in Lobachevskian geometry there is an absolute measure of length. By *absolute* here is meant a standard implicit in the axioms themselves. For example, in Euclidean geometry angles have an absolute measure, whereas lengths do not. This follows from Euclid's definition of *right angle*, "when a straight line set up on another straight line makes the adjacent angles equal to one another, each of the equal angles is right," and his fourth postulate, "that all right angles are equal to one another." Thus in Euclidean geometry it does not matter how large or small a triangle is, the sum of the angles of a triangle is equal to two right angles. It is otherwise in Lobachevskian geometry, for there, although the sum of the angles of a triangle is less than two right angles, the sum approaches two right angles as the area of the triangle gets smaller, and approaches 0 as the triangle gets larger.

We can now understand what Gauss meant by saying that if this non-Euclidean geometry were true,

there must exist in space a linear magnitude, *determined for itself* (but unknown to us). ... I have sometimes expressed the wish that the Euclidean Geometry were not true, since then we would have *a priori* an absolute standard of measure.

This standard would be the length *H* of the perpendicular which corresponds to the angle of parallelism for this world.

Lobachevskian geometry has as a limiting case Euclidean geometry. That is, the difference between Lobachevskian geometry and Euclidean becomes arbitrarily small depending on the choice of certain constants which occur in it. There is no *a priori* way of determining the constants. Thus the interesting question arises: Which of the two geometries is true? But the important question to notice here is: true of what? Since Euclidean and Lobachevskian geometry cannot both simultaneously be true, it follows that at most one of them can be true. Yet once Lobachevskian geometry is known it is hard to argue, as Kant did before it was known, that Euclidean propositions are true of our *a priori* intuitions. For what reasons can be *a priori*, given that Lobachevskian geometry is false? There don't seem to be any.

Thus it is very tempting to say that what geometrical statements are true of is the physical world. It becomes an empirical problem to find out whether Euclidean or Lobachevskian geometry is true. But one must be careful here. In a certain sense we are sure that neither Euclidean nor Lobachevskian geometry is true of the physical world. For consider the first two definitions of Euclid: "a *point* is that which has no part," and "a *line* is breadthless length." There seems to be nothing in our experience or the physical world which corresponds to these definitions. Of course, there are a lot of things which for one purpose or another might serve as a point or a line. But none will serve exactly, and in mathematics one is interested in absolute exactitude. This fact does not disturb the logical relationship between the axioms and the theorems, but it still leaves up in the air the question of truth.

To get at this question, let us say that a geometry is true of the physical world if it is presupposed by theoretical physics, that is, if the geometry serves the purpose of the theoretical physicist. The first hint of an attempt to consider geometry an empirical (that is, *a posteriori*) science appears in the work of Saccheri, who speaks of physico-geometric demonstrations; but Gauss was apparently the first person to attempt experiments. Gauss had been commissioned to make a survey of the earth around Hanover, and in the course of this work measured the triangle formed by three distant mountain peaks. He found that the difference between his measurement and two right angles was well within experimental error. Thus it seemed that Euclidean geometry was true of the world, but not true *a priori*. As he put it in one of his letters:

I keep coming closer to the conviction that the necessary truth of our geometry cannot be proved, at least *by* the human intellect *for* the human intellect. Perhaps in another life we shall arrive at other insights into the nature of space which at present we cannot reach. Until then we must place geometry on an equal basis, not with arithmetic, which has a purely *a priori* foundation, but with mechanics [cf. Wolfe 1945 57].

Quite independently Lobachevski came to a similar view: In 1825 he said:

The fruitlessness of the attempts made, since Euclid's time, for the space of 2000 years, aroused in me the suspicion that the truth, which it was desired to prove, was not contained in the data themselves; that to establish it the aid of experiment would be needed, for example, of astronomical observations, as in the case of other laws of nature [Bonola 1955 92].

This viewpoint led Lobachevski to work the calculations for a triangle one of whose sides was the radius of the earth's orbit and the opposite vertex a star. However, again the calculations showed that even for so large a triangle the deviation from the Euclidean would be within observational error.

To sum up the position reached by Gauss, Lobachevski and a few others by 1830: There are two mathematical theories of space, each consistent with itself but incompatible with the other. From a purely mathematical point of view, neither is superior to the other. On the other hand, there is only one theory of physical space; that is, that mathematical theory which must be presupposed by physics. Which of these two geometries is true of the world cannot be determined *a priori*. However, measurements seem to indicate that for all purposes—including theoretical physics—Euclidean geometry is sufficiently accurate. This did not preclude the possibility that further work in physics might, for some purposes, lead us to use Lobachevskian geometry, although this seemed at the time unlikely.

There the matter stood until Gauss was blessed with another pupil of superlative genius: Bernhard Riemann (1826–1866). Riemann had originally intended to become a theological student, but changed his mind and decided to go into mathematics and physics. Gauss (who was not easily impressed by the mathematical efforts of others) was very enthusiastic about Riemann's doctoral thesis, which he submitted when he was 25. Shortly thereafter Riemann sought a lectureship at the University of Göttingen, where it was the custom to have applicants present a trial lecture which the faculty attended. Riemann submitted three topics, for the first two of which he was well prepared. The third was on the foundations of geometry, and he did not expect it to be picked. But Gauss, who was on the faculty and at the very height of his fame and influence, saw to it that this was the topic selected. So on June 10, 1854, when he was 27, Riemann gave his lecture, "On the Hypotheses which Lie at the Foundations of Geometry." The lecture was very general and suggestive (the audience was not composed exclusively of mathematicians), but was incredibly profound. Gauss—who was to die within a year—was in the audience, and was quite excited by the lecture. Riemann's achievement of exciting this "prince of mathematicians" by lecturing on a topic on which he had spent 60 years was no mean feat.

What Riemann did was to propose a very general mathematical point of view from which physical space is merely a special case. As he put it:

... I have proposed to myself ... the problem of constructing the concept of a multiply extended magnitude out of general notions of quantity. From this it will result that a multiply extended magnitude is susceptible of various metric relations and that space accordingly constitutes only a particular case of a triply extended magnitude. A necessary sequel of this is that the propositions of geometry are not derivable from general concepts of quantity, but that those properties by which space is distinguished from other conceivable triply extended magnitudes can be gathered only from experience. There arises from this the problem of searching out the simplest facts by which the metric relations of space can be determined, a problem which in the nature of things is not quite definite; for several systems of simple facts can be stated which would suffice for determining the metric relations of space; the most important for present purposes is that laid down for foundations by Euclid. These facts are, like all facts, not necessary but of a merely empirical certainty; they are hypotheses; one may therefore inquire into their probability, which is truly very great within the bounds of observation, and thereafter decide concerning the admissibility of protracting them outside the limits of observation, not only toward the immeasurably large, but also toward the immeasurably small [Smith 1929 411–412].

On this last topic Riemann commented later in the lecture that

... the empirical notions on which spatial measurements are based appear to lose their validity when applied to the indefinitely small, namely the concept of a fixed body and that of a light ray; accordingly it is entirely conceivable that in the indefinitely small the spatial relations of size are not in accord with the postulates of geometry, and one would indeed be forced to this assumption as soon as it would permit a simpler explanation of the phenomena [Smith 1929 424].

Problem 18.   Can you see any connection between this latter quotation and the Achilles argument of Zeno?

It was from the general point of view suggested by the above quotations that Riemann was able to see the difference between the unbounded and the infinite. As Riemann expressed it:

> When constructions in space are extended into the immeasurably great, unlimitedness must be distinguished from infiniteness; the one belongs to relations of extension, the other to those of measure. That space is an unlimited, triply extended manifold is an assumption applied in every conception of the external world; by it at every moment the domain of real perceptions is supplemented and the possible locations of an object that is sought for are constructed, and in these applications the assumption is continually being verified. The unlimitedness of space has therefore a greater certainty, empirically, than any experience of the external. From this, however, follows in no wise its infiniteness, ... [Smith 1929 423].

The geometry suggested by this lecture is one in which straight lines are unbounded, but not infinite. Precisely such a geometry results if we replace Euclid's fifth postulate by the hypothesis of the obtuse right angle and if we understand Euclid's second postulate—"To produce a finite straight line continuously in a straight line"—as asserting only the unboundedness of a straight line and not its infinity. However, in this geometry—now called *Riemannian geometry*—it is not universally true that given two points exactly one straight line may connect them. Thus, if a consistent geometry is to result, we must understand Euclid's first postulate—"To draw a straight line from any point to any point"—as meaning 'at least one straight line', not 'exactly one straight line'.

If these changes are made, Euclid's geometry is transformed into a consistent Riemannian geometry. The similarities and differences of the three plane geometries which so far have been distinguished may be summed up in a table (see Table 5). When we look at the table, one of the most interesting questions which arises is: How long are all the finite straight lines? In Riemannian geometry, as in Lobachevskian, the measurement of lengths as well as angles is absolute. It would be interesting to know the length of straight lines, as we would then have an absolute measure of size, that is, in terms of the length of a straight line. This length, however, cannot be stated *a priori*; it depends on the assignment of a constant. Well then, how about empirical observation? Is Riemannian geometry true of the world? Unfortunately very few people in the nineteenth century took this question seriously. Riemann did, William K. Clifford (1845–1879) in England did, and perhaps one or two others. But Riemann died at 39 and Clifford at 34. Had they

Table 5 Geometries

Euclidean	Lobachevskian	Riemannian
	Comparison of some postulates	
1. (Postulate 1) To draw a [that is, exactly one] straight line from any point to any point.	1. To draw a [that is, exactly one] straight line from any point to any point.	1. To draw at least one straight line from any point to any point.
2. (Postulate 2) To produce a finite straight line continuously in an [infinite] straight line.	2. To produce a finite straight line continuously in an [infinite] straight line.	2. To produce a finite straight line continuously in an unbounded straight line.
3. (Logical equivalent of Postulate 5) Given a line and a point off that line, there is exactly one line through that point parallel to the given line.	3. Given a line and a point off that line, there are at least two lines through that point parallel to the given line.	3. Given a line and a point off that line, there are no lines through that point parallel to the given line.
	Comparison of some theorems	
4. The sum of the angles of a triangle is equal to two right angles.	4. The sum of the angles of a triangle is less than two right angles, the sum approaches two right angles as the area of the triangle approaches 0, and the sum approaches 0 as the area of the triangle approaches the maximum area for a triangle.	4. The sum of the angles of a triangle is more than two right angles, the sum approaches two right angles as the area approaches 0, and the sum approaches six right angles as the area of the triangle approaches the maximum area for a triangle.
5. The ratio of the circumference of a circle to its diameter is π.	5. The ratio of the circumference of a circle to its diameter is greater than π; the ratio approaches π as the area of the circle approaches 0.	5. The ratio of the circumference of a circle to its diameter is less than π; the ratio approaches π as the area of the circle approaches 0.
6. Similar figures of different areas exist.	6. Similar figures of different areas do not exist.	6. Similar figures of different areas do not exist.
7. Straight lines are infinite.	7. Straight lines are infinite.	7. Straight lines are finite and they all have the same length.

both had long lives, perhaps they would have preceded Einstein in the Theory of Relativity, for that theory proposes that a generalized version of Riemannian geometry is true of physical space.

In any case, after Riemann's lecture it seemed clear that there were at least three different geometries, and the question became more insistent: Are they each consistent? The method which Saccheri used to test the consistency of a system—and few, if any, before him thought much about the problem—was to ask: Are the axioms true? Any collection of statements all of which are true is necessarily consistent. However, the converse is not correct, namely, that any consistent collection of statements is true. In particular it was believed that Lobachevskian and Riemannian geometries— although false of the physical world—were nevertheless consistent. But how could this be proved? There was no answer to the question on the basis of traditional logic. The fact that no one had found a contradiction— even though a number of very gifted men had worked on finding one— was not sufficient. Perhaps tomorrow someone would find an inconsistency in one or both of these geometries. Was there any way to settle the question once and for all?

In the nineteenth century great progress was made in doing just this. The method chosen was to reduce one system to another, that is, to show that the system in question had the same logical structure as a system believed to be consistent. This was really a *reductio ad absurdum* proof. If the system in question is inconsistent, then this known system is inconsistent. But this consequence is false. Therefore the system is consistent. Efforts focused, therefore, on reducing Lobachevskian and Riemannian geometry to known mathematical systems, in particular, to Euclidean geometry and to algebra. Unfortunately, the methods by which these reductions were made are quite complex. The following account of one possible reduction is a simplification, but perhaps it will not be misleading.

Consider the geometry of the surface of a sphere. We may define a *great circle* to be a circle with a radius equal to the radius of the sphere, and whose center is the same as the center of the sphere. Now one of the properties of a straight line in plane Euclidean geometry is that it is the shortest distance between two points. On the surface of a sphere, great circles have this property. This is why airlanes—especially those relatively close to the poles—are not straight. It is easy to see that one could extend Euclidean geometry to such a surface. We might even write down the postulates for such a geometry:

Postulates

Let the following be postulated:

1.   To draw a great circle from any point to any point.

2. To produce an arc of a great circle continuously in an unbounded great circle.

3. To describe a circle with any center and distance.*

4. All right angles are equal to one another.

5. Given a great circle and a point off that great circle, there are no great circles through that point parallel to the given great circle.

All these postulates appear to be true of the surface of a Euclidean sphere. They are, however, written in such a way that the analogy to Riemannian geometry leaps to the eye. In fact, if the phrase 'great circle' is replaced by 'straight line' and 'arc of a great circle' by 'finite straight line', the postulates become those of Riemannian geometry. Provided the definitions can be made to correspond (and they can be) it is easy to see that all the theorems will also correspond. Under this scheme, if someone had worked out a large body of theorems for Riemannian geometry, the theorems and their proofs could be made into theorems and proofs of this system by a systematic change of names.

Now suppose Riemannian geometry were inconsistent. This would mean one could prove a statement and another statement incompatible with the first. But the same result would then also be true of the above postulates applied to a Euclidean sphere. Hence if Riemannian geometry is inconsistent, then the other also is. In a way analogous to this† it was shown that if any of the three geometries is consistent, they all are; and if any inconsistent, they all are.

But a more general insight came from studies such as these; namely, that what a mathematical system is "about" is irrelevant insofar as one's interest is in deducing consequences of the given postulates. In fact, one could just as well put dummy letters (that is, letters which in context don't stand for anything) in place of 'great circle', etc. Then, after the system was developed, one could consider those things which it could be about, that is, those interpretations of the dummy letters which make the axioms true. From the time of Aristotle there have always been some thinkers who knew that in an argument form (for example, if 'all M is P' and 'all S is M', then 'all S is P'), it does not matter if the premises are true, or what is substituted for the dummy letters (that is, what the argument is about). So long as the substitution is systematic and meaningful, the conclusion will follow from

---

* Just as in plane Euclidean geometry 'distance' means 'distance along a straight line,' so here it means 'distance along a great circle'.

† The above must not be taken for an exact mathematical proof. The situation is more complicated than is indicated. In fact, it can be proved that in Euclidean space there are no complete surfaces which satisfy all the properties of non-Euclidean planes [cf. Bonola 1955 146].

the premisses, that is, *if* the premisses are true the conclusion is also true. The problem of the consistency of the non-Euclidean geometries led some mathematicians to suspect that a good deal (if not all) of mathematics was concerned with such activity. Thus inquiries were launched into the relation of mathematics and logic and the details as to how one can go about developing systems without any commitment as to what they are about.

One fact soon became apparent: that if one adopted this very abstract point of view, one was less likely to make certain kinds of errors in reasoning. For example, in Euclidean geometry errors of reasoning often occur due to an incorrectly drawn diagram or a misleading one. We have seen how even in Euclid's first proposition he (unconsciously) assumed something not in the postulates or common notions. About the turn of the twentieth century, the pursuit of the abstract axiomatic point of view described above finally led to Euclidean geometry being put on a completely sound logical basis, that is, an abstract system was developed such that the dummy letters could be interpreted so that the system became equivalent to Euclidean geometry, but the system itself was developed without such reference.

Problem 19. One of the ways in which the *Elements* of Euclid is logically defective is with respect to the notions of inside, outside, and between. The following "proof" depends on carelessness with respect to these concepts. See if you can find the error.

Theorem: Every triangle is isosceles.

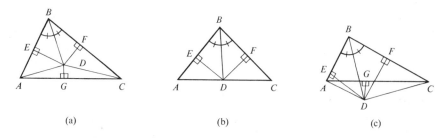

(a)   (b)   (c)

Figure 10

Suppose that triangle *ABC* is not isosceles (see Fig. 10). Bisect angle *B*. This bisector is not perpendicular to *AC*, since if it were, the triangle would be isosceles. The perpendicular bisector of *AC* cannot be parallel to the bisector of *B*, and must intersect it, say, at a point *D*. This point must be either inside (part a of Fig. 10), on (part b), or outside (part c) the triangle. The argument which follows applies to any of the three possibilities. From *D* draw perpendiculars *DE* and *DF*, respectively, to *AB* and *BC*. Then connect points

A and *D* and *D* and *C* with straight lines (if *D* is not on *AC*). In the right triangles *BED* and *BFD*, two angles (the right angles and the ones formed by the bisector) are equal and one side (*BD*), so the triangles are congruent. In right triangles *AED* and *CFD*, side *ED* equals *FD* (since *D* is on the bisector of angle *B*) and side *AD* equals *CD* (since *D* is on the perpendicular bisector of *AC*). Thus the triangles are congruent. From these two pairs of congruent triangles, we get *BE* equals *BF* and *EA* equals *FC*. Adding equals, we get *BA* equals *BC*.

It wasn't until the latter half of the nineteenth century that Euclidean geometry was axiomatized in a modern way. This was done by a number of men, but the version given by the German mathematician David Hilbert (1862–1943) was the most famous. Hilbert's axiomatization contained eight dummy words, or as they are usually called, *primitive terms*: 'point', 'straight line', 'plane', 'incidence of a point on a line', 'incidence of a point on a plane', 'betweenness of a point between two other points on a straight line', 'congruent line segments', 'congruent angles'. Hilbert purposely left these terms undefined (since it is possible to give them more than one interpretation which would make his axioms true). However, he defined all other terms (for example, 'triangle') in terms of them. Hilbert's axiomatization contained five groups of axioms: seven axioms of connection, five axioms of order, one axiom of parallels (that is, an equivalent of the fifth postulate), six axioms of congruence, and one axiom of continuity. He didn't attempt to prove these axioms, not because they are too fundamental to prove but because they are incapable of proof, since they contain dummy words. They are not "about" anything, and thus not the sort of thing which could be true or false. To be less misleading, instead of using loaded words like 'point', 'straight line', etc., one should use meaningless symbols (such as '⊠' or '◎') or neutral signs (such as 'thing of sort 1,' 'relation of sort 4,' etc.). This is precisely what those who developed mathematical logic did.

From the time of Riemann on, geometry developed at a rapid pace. Many different non-Euclidean* geometries were developed. In fact, Riemann's paper was the inspiration for the development of differential geometry. To anyone familiar with this development, the characterization of non-Euclidean geometry given so far in this book must seem rather quaint. Geometry developed in such far-reaching ways that the very question of what we mean by *geometry* became a problem. Applications of geometry also became more diverse. To mention one intriguing one: It has been suggested that the mathematical theory that best represents what we see

---

* Gauss was the first to use this term for what is now known as Lobachevskian geometry. Logically the term should be applied to any geometry which differs from the Euclidean. However, it is sometimes used to indicate only geometries that differ from Euclid's on the question of parallels. In this book we shall be "logical."

(that is, the way in which persons of normal vision understand their visual space) is Lobachevskian geometry. For example, railroad tracks (when one is between them and looks off into the distance) appear to converge in one direction and diverge in the other [cf. Luneburg 1947 and, for further discussion and references, Grünbaum 1963 152–157]. As soon as we try to mathematically represent our perceptions as a whole, the problems get even more complicated. As Bertrand Russell has noted, the space of sight and the space of touch are quite different [cf. 1929 120]. Even if we limit ourselves to physical space, complications can easily ensue. For example, Hans Reichenbach suggests that "it might happen that the geometry of light rays differs from that of solid bodies" [1962 137]. All this need not concern us, for we are now in a position to understand the influence of the growth of non-Euclidean geometries on the development of mathematical logic. There were at least four different ways in which the rise of non-Euclidean geometries was influential in that development.

First, the possibility of non-Euclidean geometries was a blow to the belief in intuition, in a twofold way. In the first place, the conviction of the Kantians that Euclidean geometry is *a priori* true of our intuitions was shown to be incorrect. This does not mean that Kantian philosophy as a whole, or even the general tenor of Kant's theory of space, is incorrect. It means only that this particular doctrine is. Even so, this was a powerful blow to the authority of metaphysicians who were claiming a necessary synthetic *a priori* truth, which didn't turn out to be *a priori* and might not even be true. In the second place, the reliance on intuition as a criterion for truth received a decisive blow. For centuries, any contrary to the fifth postulate seemed false to all who investigated the matter. Now it was found that some of these are individually consistent with the other axioms and thus possibly true. Henceforth mathematicians became sceptical of the claim of obviousness to prove truth.

Second, due in part to the rise of non-Euclidean geometries a new view of the nature of mathematical statements presented itself. To the question *Of what are geometrical propositions true?* a number of answers had been given. For example, they are true of patterns fixed in the nature of things (a Platonistic answer), they are approximately true of the physical world (an empiricist answer), they are true of our *a priori* intuitions (a Kantian answer). Now the possibility arose that they might not be true at all, not because they are false, but because they are not kinds of things which can be true. From this point of view the work of the pure mathematician is mainly or completely that of the logician; that is, the mathematician, like the logician, deduces conclusions from assumed premises. Thus pure mathematics is not true of anything at all because it is not about anything. This point of view was (and is) far from universal among mathematical logicians; some of the very best logicians would deny it. Nevertheless it was (and is) influential.

It led to inquiry into the exact relationship of logic and mathematics, and further, to an effort to create a logic that would be adequate for nineteenth and twentieth century mathematics (as traditional logic was demonstrably not).

Third, non-Euclidean geometries focused attention on the problem of the consistency of mathematical systems. The method of showing that the axioms of a given system are true is certainly sufficient to prove consistency. But unfortunately it doesn't work for the interesting systems. For example, how can one show that Euclidean geometry is true? The implicit reference to infinity—both the infinitely large and the infinitely small—would seem to rule out any non-probabilistic assertion of its truth. The new method developed, which reduces one system to another, is quite useful but only relative. That is, it says one system is consistent *if* another is. But how are we to decide about the latter system? Could one develop an absolute proof of consistency? Such a problem stimulated interest in logic.

Fourth and finally, the appearance of non-Euclidean geometries focused attention on impossibility proofs. If it is proved that the fifth postulate is independent of the others, then we have a proof that it is impossible to prove the fifth postulate from the others. Henceforth mathematicians, who saw or experienced repeated failure at proving something, thought more readily of the possibility that it might be impossible to prove it. A proof that no proof is possible is then called for. A number of the most important results of mathematical logic are of this kind; that is, they prove that it is not possible to prove certain kinds of things. It is unlikely that these results would have been found had not other areas of mathematics provided successful examples of impossibility proofs.

§10 MATHEMATICS AND ARGUMENTATION

From the point of view of logic, besides non-Euclidean geometries the other major development in mathematics in the nineteenth century was Cantor's theory of sets. However, before turning to this topic, we shall consider two other developments which are important for logic: the enlargement of the concept of number and the increased use of analytic techniques.

A. Numbers

In ordinary life we use numbers for two purposes: counting and measurement. The numbers we first learn about are called *natural numbers*, that is, 1, 2, 3, and so on. The first mathematical theory most people learn is that of the natural numbers; it is called *arithmetic*. People learn to add and multiply, often making use of simple addition or multiplication tables. These concepts appear at first to apply merely to discrete objects such as stones or money.

Thus we could represent the addition of a collection of three stones to four stones as in Fig. 11.

Multiplication of three by four could be represented by Fig. 12.

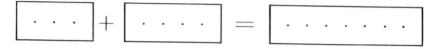

Figure 11

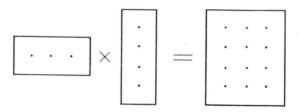

Figure 12

Subtraction (of a smaller number from a larger) and division (where there is no remainder) could be represented in an analogous way. However, the introduction of these two operations also introduces some difficulties. For example, three minus three doesn't equal any natural number. Neither does three minus four. The first way of dealing with this problem was apparently to deny the meaningfulness of subtraction in a case in which the subtrahend is equal to or larger than the minuend. After all, it does not seem to make any sense. What could it possibly mean to say that you have three stones and someone takes away four? It seemed that no number at all corresponds to this situation. The same seemed to be true for the case in which, say, three stones were taken away from three stones. That is, no number at all corresponded. Today we would express this by saying that the operation of subtraction is not defined in the domain of natural numbers when the subtrahend is equal to or larger than the minuend.

The case of division is different. Here there are many repeatable situations in which new numbers seemed called for. If we divide five apples between two people by having each take two and cutting the remaining apple in two equal parts, we need a method of expressing this situation. Furthermore it should be systematic so that it can be extended to a variety of situations. This problem was solved by the invention of fractions.

Once this was done it appeared that things had "gone about as fur as they could go." In particular the combination of the natural numbers and the fractions seemed to be sufficient not only for discrete but also for continuous objects. Thus when numbers were used for measuring it appeared that the natural numbers and fractions were sufficient even for a theoretical absolute accuracy of measurement. This apparent close harmony between numbers and magnitudes was thought secure. We even have numbers being defined in terms of the kind of magnitudes they can represent. Thus four is a *squared* number, eight a *cubed* number. As we have seen, this harmony was upset by the discovery, in the Pythagorean school, of irrational magnitudes.

Yet it seemed clear that magnitudes as well as numbers stood in fixed determinate ratios. We can say that the ratio of the length of the side of a square to its diagonal is equal to the ratio of the side and diagonal of some other square. But what can a ratio be if not a relationship between two natural numbers? It is one of the very greatest achievements of Greek mathematics to have solved this problem. Eudoxus' theory of proportions contains the solution. This theory is believed to be reproduced in Book V of Euclid's *Elements*. There the following very important definition is given:

Magnitudes are said to *be in the same ratio*, the first to the second and the third to the fourth, when, if any equimultiples whatever be taken of the first and third, and any equimultiples whatever of the second and fourth, the former equimultiples alike exceed, are alike equal to, or alike fall short of the latter equimultiples respectively taken in corresponding order [1926 II 114].

The meaning of this definition is easier to understand if we consider an example: the corresponding ratios of the side and diagonal of two squares (Fig. 13).

 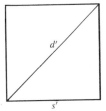

Figure 13

Now to assert that the ratio of the side to the diagonal is the same in both squares is to say that

$$s : d : : s' : d'.$$

Now consider two lines: the first is marked in units of *s* below the line and units of *d* above; the second is similarly marked for *s'* and *d'* (Fig. 14).

Consider the upper line. If *s* and *d* were commensurable, somewhere a line on the lower part would exactly match a line on the upper part. Suppose

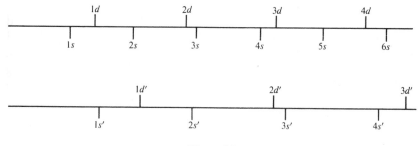

Figure 14

this happened at $5d$ and $7s$. Then, of course, $5d = 7s$ and $s : d : : 5 : 7$. It would then follow that $s' : d' : : 5 : 7$, if our ratios of sides to diagonals were the same. However, we know that the sides are not commensurable, so that this ratio does not exist. Note, however, that we can improve our accuracy by going farther out on the line. From the line as drawn we can easily obtain the following series of inequalities:

$$1s < 1d < 2s$$

$$2s < 2d < 3s$$

$$4s < 3d < 5s$$

$$5s < 4d < 6s$$

$$\begin{matrix} \cdot & \cdot & \cdot \\ \cdot & \cdot & \cdot \end{matrix}$$

By dividing each inequality throughout by s and the corresponding multiple of d we get

$$1 < \frac{d}{s} < 2$$

$$1 < \frac{d}{s} < \frac{3}{2}$$

$$\frac{4}{3} < \frac{d}{s} < \frac{5}{3}$$

$$\frac{5}{4} < \frac{d}{s} < \frac{3}{2}$$

$$\begin{matrix} \cdot & \cdot & \cdot \\ \cdot & \cdot & \cdot \end{matrix}$$

It can be proved that the difference between the large and smaller fractions approaches zero as the number of inequalities approaches infinity. What this means is that we can get any desired degree of accuracy we want.

We are now in a position to understand Euclid's (Eudoxus'?) definition. What it means in our particular case is as follows: Suppose that we multiply d by any natural number we want, say m. Suppose that m is such that $ns < md < (n + 1)s$. Then if $s : d : : s' : d'$, it will follow that $ns' < md' < (n + 1)s'$. What we have then is the equality of ratios between two magnitudes being defined in terms of the possibility of making the error of a ratio expressed in natural numbers as small as we please. If the magnitudes are incommensurable, any stated ratio between two natural numbers will always be either too small or too large.

Problem 20. In decimal notation $\sqrt{2} = 1.41421\ldots$. If $s = 1$, then $d = \sqrt{2}$. For this particular case, write out two series of inequalities analogous to those above, such that the first line in your series corresponds to the first above, the second to the tenth, the third to the hundredth, etc.

In this way the Pythagorean conception of ratio was enlarged to include irrational magnitudes. It was now possible to do with geometric methods what the Pythagoreans had failed to do with arithmetic methods; namely, compare any two magnitudes. The Greeks, however, never quite succeeded in enlarging the concept of number (although they came very close). The enlargement was to come much later, and was the work of many men in a number of different countries. For our purposes we need note only the result of their labors, not how it came about.

To the natural numbers 0 was added. This was useful in describing certain situations (for example, how much money do you have?) as well as providing a way of expressing the value of x in such equations as $3 + x = 3$. But there were other simple equations which caused trouble; for example, $3 + x = 2$. To solve such equations, as well as to express conditions such as being in debt, the negative integers were proposed. Negative fractions then followed easily, in order to solve such equations as $3x + 2 = 1$. The name *integers* was given to the class consisting of negative whole numbers, zero, and natural numbers. The name *rational numbers* was given to all numbers expressed by fractions, both positive and nonpositive.

It wasn't until the nineteenth century that adequate definitions were proposed for the concept of an irrational number. All the proposed definitions turned out to be equivalent. One of these—that of Richard Dedekind (1831–1916)—was quite similar to the definition of equal ratios given by Euclid. According to Dedekind, an irrational number can be defined whenever there exists a rule which will put any given rational number into exactly one of two classes A and B subject to the following conditions: first, every rational number in A is smaller than any rational number in B; second, there is no last number in A and no first in B. The irrational number in question then is that one number which is larger than all numbers in A and smaller than all in B.

This is a very abstract definition, but a concrete example of it may be had by considering the decimal system. We get a decimal representation of a fraction by dividing the numerator by the denominator. Thus

$$\tfrac{1}{2} = 0.5, \ \tfrac{1}{3} = 0.\overline{3}, \ \tfrac{2}{5} = 0.4, \ \tfrac{1}{6} = 0.1\overline{6}, \ \tfrac{4}{7} = 0.\overline{571428}.$$

A line over some of the decimals indicates that they are infinite periodic decimals. For example,

$$0.\overline{3} = 0.333333\ldots, \ 0.\overline{571428} = 0.571428571428571428\ldots.$$

It can be shown that every rational number has a terminating or infinite periodic decimal expansion and every terminating or infinite periodic decimal expansion represents a rational number. Now consider $\sqrt{2}$. Since it isn't a rational number it cannot be represented by a terminating or infinite periodic decimal expansion. Suppose that we start working out the decimal expansion of $\sqrt{2}$ by the method of finding square roots. This rule would enable us to put any given rational number into one of two classes A or B according as it is smaller or larger than $\sqrt{2}$. All we need do is to find the decimal expansion of the given rational number and see where it differs from the decimal expansion of $\sqrt{2}$. For example, suppose the given rational number is $\tfrac{707}{500}$. Its decimal expansion is 1.414. Thus, since $\sqrt{2} = 1.41421\ldots$, it is in class A. It is clear that this can be done for any repeating or terminating decimal.

Problem 21.   The above proof depends on the fact that the decimal expansions for irrational numbers are unique; that is, two decimal expansions of irrational numbers are not equal unless they are identical, digit by digit. Can you prove this statement?

The concept of number was extended still further with the introduction of what was called an *imaginary number*: $\sqrt{-1}$. Complex numbers consisting in part of real numbers and in part of imaginary numbers could then be formed. They have the general form of $a + bi$, where a and b are real numbers and i is $\sqrt{-1}$. Still further extensions of the number concept were made, but we shall not consider them here because they are irrelevant for our purposes.

In any case the type of numbers we have considered forms a hierarchical scheme: Natural numbers correspond to a special case of integers, integers to a special case of rational numbers, rational numbers to a special case of real numbers, and real numbers to a special case of complex numbers. Now with the exception of complex numbers, the geometrical analogs of each of the kinds of numbers seem obvious. On the other hand, the representation of geometrical relations in terms of relations between numbers did not come until the seventeenth century, with the invention of analytic geometry by Fermat (1601–1665) and Descartes (1596–1650). Let us now turn to this subject.

Problem 22. The first sentence of the preceding paragraph mentions a number of "special cases." Describe each of these.

B. Analytic Geometry

In order to say something about analytic geometry, we must first explain the idea of a function. The concept of a function in mathematics is quite old and it has been successively enlarged in the course of the history of mathematics in a manner analogous to the way the concept of numbers has been enlarged. We shall here not use the most general concept of function, but one which nevertheless is sufficient for the purposes of analytic geometry. It is basically due to Lejeune Dirichlet (1805–1859). We shall understand a numerical variable to be a symbol which can stand for any one of a collection of numbers. We shall say then that y *is a function of* x if x and y are variables and so related by a rule that the assignment of a number to x determines a unique assignment of a number to y. Here x is called the *independent variable*, y the *dependent variable*. The collection of numbers which x may separately represent is called the *domain of definition*; the analogous collection for y is called the *domain of values*. Let's consider an example: $y = x^2$. If the domain of definition for this function is the integers, the domain of values is the nonnegative integers that are perfect squares. If the domain of definition is the real numbers, the domain of values is the nonnegative reals.

Problem 23. What is the domain of values for the function $y = x^2$ if the domain of definition is the imaginary numbers, that is, all numbers of the form ai, where a is a real number and i is $\sqrt{-1}$?

It is convenient to have a further notation for a function, namely, '$f(x)$' (read 'f of x'). '$f(x)$' stands for the value of a given function of x. Thus, for example, we have: $y = f(x) = x^2$. In some circumstances it is important to distinguish between $f(x)$ and f. 'f' stands for the function, '$f(x)$' for the value of the function of x. To take our example: 'f' stands for the squaring function. However, if $f^1(u) = u^2$ (the domain of definition being the same as for f), then it would follow that $f = f^1$; but it would not follow that $f^1(u) = f(x)$. To see this, let x and u be unequal numbers. Another notation which is sometimes used for the function is: '$\lambda x f(x)$' (read 'lambda $x f$ of x'). Thus $f = \lambda x f(x) = \lambda u f^1(u)$.

Our example of a function was a special case of a general type of function called a *polynomial*, that is, a function of the form

$$f(x) = a_n x^n + a_{n-1} x^{n-1} + \cdots + a_1 x + a_0,$$

where n is a nonnegative integer and $a_0, a_1, \ldots, a_n$ are real numbers. In our

case, $a_0 = 0$, $a_1 = 0$, $a_2 = 1$, $a_i = 0$, where $i \geq 3$. Other examples of polynomials are $f(x) = \sqrt{17x^{43}} + 13x + 3$, $f(x) = x^2 - 2x$, $f(x) = 4$. There
are many other types of functions: rational functions, trigonometric functions, functions with complex variables, etc.

We are now in a position to describe analytic geometry. The main idea
is the transformation of every geometrical figure or operation into a relationship of numbers. That is, it was shown that there is a logical analogy between
relationships in geometry and relationships between numbers. The crucial
idea—discovered independently by both Fermat and Descartes early in
the seventeenth century—is the exploitation of the harmony between real
numbers and a line. This was extended to the plane by the introduction of
what is now known as a *Cartesian coordinate system*. It consists of two fixed
and calibrated perpendicular lines. These are generally called the *x-axis*
and the *y-axis*, and are marked with the same unit (see Fig. 15).

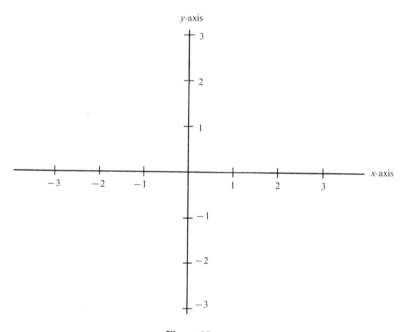

Figure 15

The point of intersection is called the *origin*. Using this idea, one can
assign a unique pair of numbers (called *coordinates*) to each point, and vice
versa. Consider a point p in Fig. 16.

The x-coordinate is always stated first. Thus the coordinates for
p, q, r, s, respectively, are $(2, 1)$, $(3, -2)$, $(-2, -2)$, $(-1, 2)$. The distance

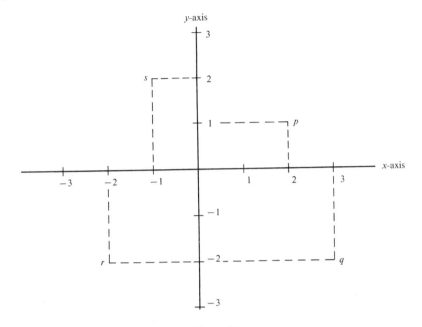

Figure 16

between any arbitrary points p_1 and p_2 [with coordinates, respectively, of (x_1, y_1), (x_2, y_2)] can be found by making use of the Pythagorean theorem. The correct equation representing this distance can be plausibly derived by considering Fig. 17. Thus $d = \sqrt{(x_2 - x_1)^2 + (y_2 - y_1)^2}$.

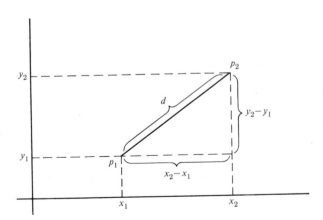

Figure 17

Making use of this equation, we may find the equation of a circle. Given that p is an arbitrary fixed point with coordinates a and b, we can find the equation for the circle with fixed radius r. For what we want is an equation containing x and y such that exactly those values of x and y whose corresponding point is at distance r from p satisfy the equation. Again this can be made plausible by considering a diagram (Fig. 18). Since

$$r = \sqrt{(x - a)^2 + (y - b)^2},$$

we know that any values of x and y which satisfy that equation will represent a point which lies on the circle.

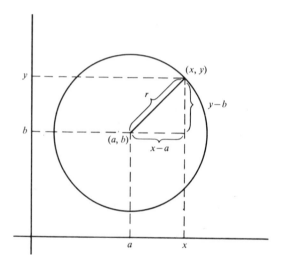

Figure 18

If we now return to Euclid, we can consider his third postulate—"to describe a circle with any center and distance"—from an analytic point of view. It means simply that p may be an arbitrary point (that is, a and b an arbitrary pair of real numbers) and that r may also be an arbitrary positive real number.

Elaboration of these techniques enables us to do the entire Euclidean geometry by analytic techniques. Table 6 gives a few of the more familiar geometrical figures and their corresponding equations.

What developed by the elaboration of these techniques was something much more powerful than plane Euclidean geometry. For example, it is easy to extend the above kind of analysis from two-dimensional Euclidean space to three. In the latter, corresponding to a point there is a triplet of numbers, corresponding to a line there is a pair of equations of the form

Table 6

Geometrical object	Corresponding analytic form
Point	Ordered pair of real numbers
Line	$ax + by + c = 0$
Triangle	A set of three equations for lines which are not parallel
Circle	$r = \sqrt{(x - a)^2 + (y - b)^2}$
Ellipse	$\dfrac{x^2}{a^2} + \dfrac{y^2}{b^2} = 1$
Hyperbola	$\dfrac{x^2}{a^2} - \dfrac{y^2}{b^2} = 1$
Parabola	$y = ax^2 + bx + c$

$ax + by + cz + d = 0$. Once this point of view is understood, there seems to be nothing sacred about a *triplet* of numbers. Why not have a quadruplet, quintuplet, or in general an *n*-tuplet ordered collection of numbers which could serve as a "point"? The equations of a line, circle, etc., could be similarly generalized. The corresponding geometries would then be 4-dimensional, 5-dimensional, ..., *n*-dimensional. Since they are all extended from Euclidean geometry, they are called *n-dimensional Euclidean geometries*. There is even an infinite-dimensional Euclidean geometry. Further, analogous techniques can render Lobachevskian or Riemannian geometry analytic. In fact, this is the way those geometries are very often presented. For example, it is perfectly possible to have a 4-dimensional Riemannian geometry.

Using this powerful method of analysis, nineteenth century scientists were finally able to solve the three famous outstanding geometric problems which the Greeks passed on to posterity: the duplication of the cube, the squaring of the circle, and the trisection of an arbitrary angle. The problem was to make an exact construction with the aid of a compass and straightedge alone. It was shown in each case that it is impossible to solve the problem under the condition stated. *Impossible* here means *logically impossible.* That is, given certain assumptions, it was shown that it is as impossible to solve those problems as it is to draw a square circle. The proofs are complex—especially the one for the squaring of the circle—but they nevertheless settle the problems once and for all.

We may now summarize the importance of analytic geometry for the development of logic. First, the idea of mapping logical relations of one area (geometry) into those of another (algebra) proved extremely fruitful. Not only was this idea exploited in many areas of mathematics, it became the basis for one of the most powerful techniques of mathematical logic [cf. §23].

Second, the exploitation of the algebraic techniques of symbolizing problems with various kinds of signs and then manipulating the signs proved extremely powerful. Geometric problems whose solutions had gone begging for thousands of years were solved; new ideas were formulated whose very existence depended on the pursuit of these techniques. For example, in the mid-nineteenth century George Boole (1815–1864) conceived the idea of an *algebra of logic*. This was one of the crucial ideas—perhaps even *the* crucial idea—which allowed logic to become a "great" subject again.

Third, there was a renewed focus on impossibility proofs. A number of important outstanding problems were now proved impossible to solve. New motivation was therefore provided to consider seriously that a problem which has repeatedly resisted solution might not be solvable. As mentioned above, a number of the most important results of mathematical logic are of this sort.

Fourth, the techniques of analytic geometry reintroduced the possibility of harmony between geometry and other areas of mathematics. This harmony, as we have seen, was broken by the Pythagorean discovery of incommensurables. With the enlargement of the concept of number to include the real numbers, and with the mapping technique of analytic geometry, the harmony was reintroduced. Not only were algebraic methods applied to geometrical problems but, in time, the axiomatic technique of geometry was applied to algebra. With these developments the idea of the unity of all mathematics arose—a unity based on logic (as suggested by the developments in non-Euclidean geometry [cf. §9]). It was this background which led to attempts to systematize all of mathematics. The possibility of such systematization is the topic of several of the most important theorems of mathematical logic, and the impact of these theorems has a significance that is more than logical or mathematical.

We may note that, beginning in the seventeenth century, the use of the mapping technique for the illumination of logical problems was explored with respect to geometrical analogs by a number of individuals. This exploration of geometrical analogs is still going on. Nevertheless, it is fair to say that although the use of diagrams is often useful as a pedagogical technique it has not been helpful in the discovery of new and important ideas about logic. The reason for this is that the geometrical representation of the difficult logical problems is either impossible by a particular technique or so complicated as to be of little value. The use of algebraic techniques has proved much more useful [cf. Gardner 1958 and Edwards 1967 ("Logic Diagrams" by Gardner)].

§11 SET THEORY

Gauss once said in an often-quoted letter,

I protest against using infinite magnitude as something consummated; such a use is never admissible in mathematics. The infinite is only a *façon de parler*: one has in mind limits which certain ratios approach as closely as is desirable, while other ratios may increase indefinitely [cf. Fraenkel 1953 1].

It was Georg Ferdinand Ludwig Philipp Cantor (1845–1918) who brought about a change in this orthodox attitude, by his creation of set theory, but the change was slow in coming and didn't take place until many years after Cantor first published his work beginning in 1872.

Cantor suggested that the word *infinite* in mathematics has two meanings. The first is a magnitude (indicated above in the quotation of Gauss) which increases beyond any indicated limit. This is called by Cantor the *improper infinite*, presumably because the magnitude is always finite although variable. It was represented by the symbol '∞'. The second meaning of the word is that of the *proper* or *completed infinite*. This use of the word comes up in an essential way with respect to the idea of a real number. Previous to the nineteenth century a real number was associated—following the Pythagoreans—with a point on a line in geometry. Thus the reference to a completed infinity—for example, a line is an infinite collection of points—was covert and apparently, for the most part, unconscious. With the development of the analytic temper in mathematics it became increasingly obvious that real numbers could not be defined without reference to a completed infinite aggregate. It was this problem as well as other technical problems in mathematics that led Cantor to investigate the general theory of classes. Also it should be added that Cantor was a Christian (although of Jewish lineage on both sides) and a serious student of medieval theology. This factor appears to be not uninfluential in his motivation to investigate the infinite.

Cantor stated that by a *set** "we are to understand any collection into a whole M of definite and separate objects m of our intuition or our thought" [Cantor 1895–1897 85]. The separate objects m are called *elements* of M. This conception of a set is very abstract and includes many different sorts of things. Included are such things as a set of dishes, a set of rational numbers between 0 and 1, and a set consisting of the even prime number, the sun, and Socrates. Two sets M and N are *equal* if and only if every element of M is an element of N, and vice versa. A set M is a *subset* of a set N if and only if every element of M is an element of N. It follows from these two definitions that if a set M is equal to set N it is also a subset. Where there is at least one element of N which is not a member of M, and M is nevertheless a subset,

---

* The terms *set, class, aggregate, collection* will be understood as being synonymous.

M is called a *proper subset*. For example, if M is the set of numbers between 0 and 1 which can be represented by terminating or infinite periodic decimals, and N is the set of rational numbers between 0 and 1, M and N are equal. If R is the set of rationals between $\frac{1}{2}$ and $\frac{1}{3}$, it is a subset of M (or N).

Sets may contain one or even no elements. Thus if M is the set of even prime numbers it contains only one element, if N is the set of even prime numbers greater than 2, it contains no elements. A set without elements is called the *empty set*. Note that by these definitions all empty sets are equal and that the empty set is a subset of any set. Sets may also contain other sets as elements.

We need one more definition before we can state some of Cantor's results. Two sets M and N are called *equivalent* if they can be put into one-to-one correspondence; that is, if it is possible by some rule to associate each element of M with exactly one element of N, and vice versa. For example, if M is the set of primes greater than 2 and less than 12, and N is the set of even numbers less than 10, there exists a one-to-one correspondence by the following pairing: $3 \leftrightarrow 2$, $5 \leftrightarrow 4$, $7 \leftrightarrow 6$, $11 \leftrightarrow 8$. Of course, there are other ways to make the pairing, but any one way satisfies the definition.

By making use of the concept of equivalence, we can define the concepts of finite and infinite sets. The essential idea is old; anticipations of it can be found in Plutarch (46–120), Proclus (410–485), Adam of Balsham (fl. 1130), Galileo (1564–1642), and Bernard Bolzano (1781–1848) [cf. Kneale 1962 440]. However, no one made anything of it until there appeared the work of Peirce, Dedekind, and Cantor in the nineteenth century. What was noted was that an infinite set is equivalent to a proper subset of itself, whereas a finite set is not. For example, the set of natural numbers is equivalent to a proper subset of itself. Various subsets can be chosen to prove this, as well as various ways of setting up the correspondence. Here are some possibilities:

a)   if the subset is the natural numbers greater than one:

$$1 \leftrightarrow 2, 2 \leftrightarrow 3, 3 \leftrightarrow 4, \ldots, n \leftrightarrow n + 1, \ldots;$$

b)   if the subset is the even numbers:

$$1 \leftrightarrow 2, 2 \leftrightarrow 4, 3 \leftrightarrow 6, \ldots, n \leftrightarrow 2n, \ldots;$$

c)   if the subset is the squares of natural numbers:

$$1 \leftrightarrow 1^2, 2 \leftrightarrow 2^2, 3 \leftrightarrow 3^2, \ldots, n \leftrightarrow n^2, \ldots.$$

Observations such as these become somewhat paradoxical if we reflect on the meaning of the statement that two sets have the same number of elements. Cantor suggested the extremely plausible definition that two sets have the same (cardinal) number of elements if they are equivalent. This follows ordinary usage very closely. For example, if in an auditorium every person occupies exactly one seat and if there are no empty seats, we say that there are the same number of seats as people. Whatever number we assign to the collection of seats we must also assign to the collection of people. From this it would follow, for example, that the set of even natural numbers contains just as many numbers as the set of natural numbers. This conclusion goes against common sense as well as traditional mathematical conceptions.

To see that the latter is true, we need only consider the fourth and fifth common notions in Euclid's *Elements*. They state, respectively, that "things which coincide with one another are equal to one another" and "the whole is greater than the part." Now if the common notions were meant to apply to any mathematical discipline, it seems fair to say that Cantor's definition of equivalence captures in the language of set theory the spirit of the fourth common notion. But, as we have just seen, this contradicts the fifth common notion, because in the case of infinite sets a part (say, the even numbers) is equivalent to the whole (say, the natural numbers). Thus what Cantor did was to show that the axiom *the whole is greater than the part* is false when applied to infinite sets. Of course, the fifth common notion could be saved if it meant only that the whole contains at least one element which the part does not. It seems unlikely, however, that this was all that was meant.

Further, it was apparently one of Zeno's paradoxes that "half a given time is equal to double that time" [cf. Aristotle 1941 336 (239b 36f)]. It may be that the sly Zeno was arguing that the whole is sometimes not greater than the part. Euclid's fifth common notion may have been added in order to deny such a possibility [cf. Szabó 1967 7]. Anyway, this addition makes his system inconsistent, since he was able to prove that there is an infinite number of prime numbers [cf. Euclid II 412 (IX, 20)], and it is simple to show that there are as many prime numbers as natural numbers. These conjectures are further complicated by the possibility that Euclid's fourth and fifth common notions may not be genuine [cf. Euclid 1926 I 224–226, 232]. In any case, what Cantor showed is that the notions of proper subset and non-equivalent were distinct with regard to infinite sets, and that traditionally these two notions were confused.

Problem 24. Bertrand Russell called the oddity of an infinite set being equivalent to a proper subset of itself the *paradox of Tristram Shandy*.

Tristram Shandy, as we know, employed two years in chronicling the first two days of his life, and lamented that, at this rate, material would accumulate faster than he could deal with it, so that, as years went by, he would be farther and farther from the end of his history. Now I maintain that, if he had lived forever, and had not wearied of his task, then, even if his life had continued as eventfully as it began, no part of his biography would have remained unwritten [1957 85–86].

Can you give an argument for Russell's conclusion?

Problem 25. There is a story that Hilbert [cf. Gamow 1947 17–18] used the following way to characterize the difference between the finite and the infinite.

If an innkeeper has a hotel with a finite number of rooms (however large) and every room is occupied, he must turn away new arrivals. However, this is not true in the case of an innkeeper who has a "fully occupied" hotel with an "infinite" number of rooms. Suppose you were the innkeeper of such a hotel and a new guest arrived. How could you accommodate him without putting guests who were formerly in different rooms in one room? The new guest, of course, must have a room to himself and everyone else simultaneously has a room to himself.

From all this it becomes natural to define a *finite set* as being one which cannot be made equivalent to a proper subset of itself and an *infinite set* as one which can. This conception, first proposed by Peirce in 1885 and Dedekind in 1888, has in fact been accepted. It was accepted because it corresponds with what we ordinarily mean by finite, which may be characterized as follows: Let N_n be the set of all natural numbers less than or equal to n (n may be any natural number). For example, N_3 is the set consisting of 1, 2, and 3. Now a set M is *finite in the ordinary sense* if there is some natural number n such that N_n and M are equivalent. M is *infinite in the ordinary sense* if there is no such n. It can be shown that the Peirce–Dedekind conception is equivalent to this ordinary one.

Once we have a definition of equivalence for sets it becomes natural to ask if all infinite sets are equivalent. A set which is equivalent to the set of all natural numbers is called a *denumerable set*. Are all infinite sets denumerable? Consider the set of positive rational numbers. At first it seems unlikely that it would be denumerable, for there is an infinite number of rationals between any two natural numbers. Further, in the ordering of the rationals in terms of size, there is no next-larger rational to any given one. Cantor, however, showed that the set is nevertheless denumerable, that is, he showed that there exists a one-to-one correspondence between the natural numbers and the positive rationals. The ordering that he used can be gleaned from the following diagram:

$$
\begin{array}{ccccccc}
\frac{1}{1} \to & \frac{2}{1} & \frac{3}{1} \to & \frac{4}{1} & \frac{5}{1} \to & \frac{6}{1} & \frac{7}{1} \to \cdots \\[4pt]
\frac{1}{2} & \frac{2}{2} & \frac{3}{2} & \frac{4}{2} & \frac{5}{2} & \frac{6}{2} & \frac{7}{2}\cdots \\[4pt]
\frac{1}{3} & \frac{2}{3} & \frac{3}{3} & \frac{4}{3} & \frac{5}{3} & \frac{6}{3} & \frac{7}{3}\cdots \\[4pt]
\frac{1}{4} & \frac{2}{4} & \frac{3}{4} & \frac{4}{4} & \frac{5}{4} & \frac{6}{4} & \frac{7}{4}\cdots \\[4pt]
\frac{1}{5} & \frac{2}{5} & \frac{3}{5} & \frac{4}{5} & \frac{5}{5} & \frac{6}{5} & \frac{7}{5}\cdots \\[4pt]
\frac{1}{6} & \frac{2}{6} & \frac{3}{6} & \frac{4}{6} & \frac{5}{6} & \frac{6}{6} & \frac{7}{6}\cdots \\[4pt]
\frac{1}{7} & \frac{2}{7} & \frac{3}{7} & \frac{4}{7} & \frac{5}{7} & \frac{6}{7} & \frac{7}{7}\cdots \\
\end{array}
$$

Consider the following sequence suggested by the diagram:

$$\tfrac{1}{1}, \tfrac{2}{1}, \tfrac{1}{2}, \tfrac{1}{3}, \tfrac{2}{2}, \tfrac{3}{1}, \tfrac{4}{1}, \tfrac{3}{2}, \tfrac{2}{3}, \tfrac{1}{4}, \tfrac{1}{5}, \tfrac{2}{4}, \tfrac{3}{3}, \tfrac{4}{2}, \tfrac{5}{1}, \tfrac{6}{1}, \tfrac{5}{2}, \tfrac{4}{3}, \tfrac{3}{4}, \ldots$$

Now if we reduce each fraction to its lowest form and cancel out any repetitions, we have a sequence of positive rationals that contains each positive rational exactly once. We can then exhibit the one-to-one correspondence of the natural numbers and the positive rational numbers as follows:

$$1 \leftrightarrow \tfrac{1}{1}, 2 \leftrightarrow \tfrac{2}{1}, 3 \leftrightarrow \tfrac{1}{2}, 4 \leftrightarrow \tfrac{1}{3}, 5 \leftrightarrow \tfrac{3}{1}, 6 \leftrightarrow \tfrac{4}{1}, 7 \leftrightarrow \tfrac{3}{2}, 8 \leftrightarrow \tfrac{2}{3}, 9 \leftrightarrow \tfrac{1}{4},$$

$$10 \leftrightarrow \tfrac{1}{5}, 11 \leftrightarrow \tfrac{5}{1}, 12 \leftrightarrow \tfrac{6}{1}, 13 \leftrightarrow \tfrac{5}{2}, 14 \leftrightarrow \tfrac{4}{3}, 15 \leftrightarrow \tfrac{3}{4}, \ldots$$

Thus the positive rational numbers are denumerable. In fact, it can be shown that the set of all positive and negative rational numbers is denumerable, and even that the set of all *algebraic numbers*—that is, roots of polynomials with integer coefficients—is denumerable.

Problem 26. Prove that the set of all positive and negative rational numbers is denumerable.

At this point one might conjecture that all infinite sets are equivalent. After all when one has a set which is "infinitely big," it doesn't seem that there can be any that are larger. It was one of Cantor's most important achievements to show that this is false. He did it by showing that the set of real numbers is not denumerable.

The proof is indirect. We begin by assuming that there exists a one-to-one correspondence between the natural numbers and the reals. Since a real number can be represented as a natural number followed by a decimal point and then a decimal expansion, we may represent this correspondence as follows:

$$1 \leftrightarrow N_1.a_{1,1}a_{1,2}a_{1,3}a_{1,4}\cdots$$

$$2 \leftrightarrow N_2.a_{2,1}a_{2,2}a_{2,3}a_{2,4}\cdots$$

$$3 \leftrightarrow N_3.a_{3,1}a_{3,2}a_{3,3}a_{3,4}\cdots$$

$$4 \leftrightarrow N_4.a_{4,1}a_{4,2}a_{4,3}a_{4,4}\cdots$$

$$5 \leftrightarrow N_5.a_{5,1}a_{5,2}a_{5,3}a_{5,4}\cdots$$

$$\vdots$$

In this array 'N_1', 'N_2', 'N_3', . . . are expressions which represent integers; the subscripts on them indicate the row in which the expression occurs. Each 'a' represents either 0 or a natural number less than 10. There are two subscripts following each 'a'. The first indicates the row in which it occurs, the second indicates the place it occupies in the decimal expansion. A specific example of such an array might be as follows:

$$1 \leftrightarrow \qquad 33.00003\overline{0}\ldots$$

$$2 \leftrightarrow \qquad 0.7133\overline{210}\ldots$$

$$3 \leftrightarrow -185{,}032.\overline{9876543210}\ldots$$

$$4 \leftrightarrow \qquad 3.14159\ldots = \pi$$

$$5 \leftrightarrow \qquad 1.57079\ldots = \frac{\pi}{2}$$

$$\vdots$$

Here, for example, $N_4 = 3$ and $a_{3,2} = 8$.

Now if we can construct a real number which is not in our assumed one-to-one correspondence, we shall have shown that the real numbers are not denumerable. We can do this by considering the following real number: $N_1.a_{1,1}a_{2,2}a_{3,3}a_{4,4}\ldots$, which has been constructed from digits along a diagonal line. For any given array, this is a fixed and well-defined real number. In our example above, it is 33.01759.... Now alter this number in the following way: If $a_{i,i} \neq 1$, let $b_i = 1$, and if $a_{i,i} = 1$, let $b_i = 2$. Thereby construct the following number: $N_1.b_1b_2b_3\ldots$. We shall call it the *diagonal number*. In our example the diagonal number is 33.12111.... This diagonal number cannot be in the original array. For if it were, it would correspond with some natural number n. But this number corresponding to n, $N_n.a_{n,1}a_{n,2}a_{n,3}a_{n,4}\ldots$, could not equal the constructed diagonal number. In particular $a_{n,n}$ would not equal b_n. We can only conclude that the real numbers are not denumerable.

Problem 27. Construct another real number not in the original list from another diagonal, thus giving another proof of the nondenumerability of the real numbers.

The argument which Cantor used to prove the nondenumerability of the reals is called the *diagonal argument*. It is important to note that by this argument the deficiency in the original list is necessary. It cannot be repaired, say, by adding the newly created number to the original list. For then the same process could be applied again, or (as indicated in Problem 27) it could be applied to other diagonals. Note that even in the construction along the original diagonal, an infinite number of new numbers are possible; the choice of 1 and 2 was arbitrary. Only 0 and 9 need to be avoided so as to get rid of ambiguities such as $1.\overline{0} = 0.\overline{9}$.

Problem 28. It might seem at first glance that there are only a finite number—say k—of possibilities along the original diagonal, contrary to what was just asserted. Give an argument which will prove that there are an infinite number. [*Hint:* Assume that there are k possibilities and construct one more.]

The diagonal argument has turned out to be very important. The above result was just the first use of this type of argument which has led to a number of important and sometimes startling results in mathematics and logic. Because the concept of real number has traditionally been associated with a point on a line, it might be useful to reproduce another proof which has a more geometric flavor [cf. Courant and Robbins 1953 82–83]. We can begin by first noticing that there is a one-to-one correspondence between the points of an infinite straight line and those of a finite straight line. This can easily be seen from Fig. 19.

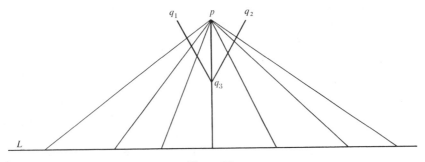

Figure 19

Given that L is a line and p a point off that line, construct a V of unit length (that is, "bend" it in half) such that the endpoints of the V (q_1 and q_2) are the same perpendicular distance from L as p is and the "bend" point (q_3) is on the perpendicular from p to L. Now there is a one-to-one correspondence between points on L and points on the unit interval (excluding q_1 and q_2). This correspondence can be found by drawing straight lines from p to L. There are also analytic proofs for this equivalence.

We can thus confine our proof to a unit interval; say, that between 0 and 1. Let's suppose that we have a one-to-one correspondence between the natural numbers and the points between 0 and 1; that is, $1 \leftrightarrow p_1$, $2 \leftrightarrow p_2$, $3 \leftrightarrow p_3, \ldots$. Now let us enclose point p_1 with an interval of length $1/10$, p_2 with an interval of length $1/10^2$, p_3 with an interval of length $1/10^3$, etc. If all the points in the interval were in our list, we would have covered all the points in the interval. But $1/9 = 1/10 + 1/10^2 + 1/10^3 + \cdots$. (We can see this by substituting $1/10$ in the equation in the answer to Problem 10 and multiplying both sides by $1/10$.) The result then is a covering of a unit interval with a set of intervals whose total length is $1/9$. The $1/9$ is arbitrary. We could make it as small as we please, for example, by starting with $1/10^2$

or $1/10^3$, etc. There is no acceptable way of getting around this conclusion except by denying our original assumption, namely, that the number of points between 0 and 1 is denumerable.

We have now established that there are at least two different sorts of infinite sets. Are there more? Before answering this question, we should inquire what Cantor meant by *cardinal number*. According to him, *the cardinal number of a set M* is

the general concept which, by means of our active faculty of thought, arises from the aggregate M when we make abstraction of the nature of its various elements m and of the order in which they are given [1895–1897 86].

In other words, the cardinal number of a set M is what M has in common with all sets equivalent to M. Cantor used the symbol '$\overline{\overline{M}}$' to signify the cardinal number of a set M. The double bar indicates a double abstraction: first, from the nature of the elements, second, from their order. If we have only made the first abstraction we have the concept of ordinal number of M for which Cantor uses the symbol '$\overline{M}$'. Note that the concept of cardinal number involves only correspondence. Ordering of a set—such as in counting—involves the notion of succession. Strictly speaking, for ordinal numbers we should use the expressions, 'first', 'second', 'third', etc. In short, cardinal numbers indicate *how many* members there are in a set, ordinal numbers the *position* of members in a sequence.

We can now return to the question of how many different types of infinite sets there are. Cantor gave the name of *aleph-null*—in symbols '$\aleph_0$'—to the cardinal number of the set of all natural numbers. He defined the concept *greater than* as follows: A set M is *greater than* a set N if and only if N is equivalent to a subset of M but M is not equivalent to any subset of N. If $\aleph$ is the cardinal number of the real numbers, we know that $\aleph$ is greater than $\aleph_0$. Cantor showed that $\aleph_0$ is the smallest infinite cardinal. Now the question is: Is there any greater cardinal than $\aleph$?

To answer this question Cantor proved a general proposition which is now known as *Cantor's theorem*: For any set M there exist sets larger than M; in particular the set of all subsets of M is larger than M.

We shall not prove this statement, but perhaps a word of explanation is in order. For any finite set with n elements, the number of subsets is 2^n. For example, the empty set has one subset, itself ($2^0 = 1$). A set with one element has two subsets ($2^1 = 2$), a set with two elements has four subsets ($2^2 = 4$), etc.

Problem 29. Consider the set of the first three natural numbers. Describe each of its eight subsets.

Cantor's theorem suggests why it is very natural to extend this symbolism to *transfinite cardinals*, as the cardinals of infinite sets are called. Thus what we have is an infinite hierarchy of transfinite cardinals:

$$\aleph_0, 2^{\aleph_0}, 2^{2^{\aleph_0}}, 2^{2^{2^{\aleph_0}}}, \ldots$$

The question immediately arises: Where does the cardinal number of the reals belong in this sequence? It can be shown that it equals $2^{\aleph_0}$. In other words, the set of all subsets of the natural numbers is equivalent to the real numbers. But then the question arises: Is there a transfinite cardinal which is greater than $\aleph_0$ but less than $2^{\aleph_0}$? Cantor conjectured that there was no such cardinal. It has become known as the *continuum problem*, since it asks where the cardinal number of the continuum (that is, the real number line) belongs in the sequence of transfinite cardinals. Cantor tried very hard to prove his conjecture; in fact, he had a mental breakdown which was no doubt partly caused by his extreme efforts. However, he failed, and—despite the effort of numerous great mathematicians—the question was not resolved until 1963. In that year Paul Cohen showed that Cantor's conjecture was independent of Cantor's other assumptions, and therefore could not be proved or refuted [see Church 1966 and Cohen and Hersh 1967 for popular accounts of this result].

What Cantor did then was to extend our idea of number beyond the finite, and thus we get the cardinal number series $1, 2, 3, \ldots, \aleph_0, \aleph_1, \aleph_2, \ldots$. Here the subscripts indicate the ordering by size of the transfinite cardinals; the subscripts thus refer to ordinal numbers. By Cohen's result, additional assumptions must be added to Cantor's to determine this ordering unambiguously. In any case, to the question as to how many transfinite numbers there are the answer is: There are an infinite number of them. Indeed the above statement—as well as the above description of set theory—is a simplification of what turned out to be an extremely rich and complicated theory.

Cantor also extended the concept of ordinal number into the infinite. He gave the name 'ω' (*omega*) to the first transfinite ordinal. The following series may give a rough indication of the richness of the theory:

$$1, 2, 3, \ldots,$$
$$\omega, \omega + 1, \omega + 2, \ldots,$$
$$\omega \cdot 2, (\omega \cdot 2) + 1, (\omega \cdot 2) + 2, \ldots,$$
$$\omega \cdot 3, (\omega \cdot 3) + 1, (\omega \cdot 3) + 2, \ldots,$$
$$\vdots$$
$$\omega^2, \omega^2 + 1, \omega^2 + 2, \ldots,$$
$$\omega^2 + \omega, \omega^2 + (\omega + 1), \omega^2 + (\omega + 2), \ldots,$$
$$\vdots$$

$$\omega^3, \omega^3 + 1, \omega^3 + 2, \ldots,$$

$$\vdots$$

$$\omega^\omega, \omega^\omega + 1, \omega^\omega + 2, \ldots,$$

$$\vdots$$

$$\omega^{\omega^\omega}, \omega^{\omega^\omega} + 1, \omega^{\omega^\omega} + 2, \ldots,$$

$$\vdots$$

$$\varepsilon_0, \varepsilon_0 + 1, \varepsilon_0 + 2, \ldots,$$

$$\vdots$$

The dots indicate omissions, not always omissions of the same structure. The third last ordinal listed is ε_0 (*epsilon zero*) and is the least ordinal which is greater than the infinite sequence

$$\omega, \omega^\omega, \omega^{\omega^\omega}, \ldots. \qquad \text{That is,} \qquad \varepsilon_0 = \omega^{\omega^{\omega^{\omega^{\cdot^{\cdot^{\cdot}}}}}}.$$

Of course, we have not here even defined addition, multiplication, or exponentiation for transfinite ordinals. This account is only meant to be suggestive. One can get a hint of the richness of the theory by considering that each of the signs for ordinals serves as a subscript in the series of alephs. In spite of the novelty of transfinite ordinals, note that ordinals such as ω or ε_0 are defined in a way analogous to real numbers. To see this, let v (nu) be a variable finite ordinal. According to Cantor,

ω is the least transfinite ordinal number which is greater than *all* finite numbers; exactly in the same way that $\sqrt{2}$ is the limit of certain variable, increasing, rational numbers, with this difference: the difference between $\sqrt{2}$ and these approximating fractions becomes as small as we wish, whereas $\omega - v$ is always equal to ω. But this difference in no way alters the fact that ω is to be regarded as definite and completed as $\sqrt{2}$, and in no way alters the fact that ω has no more trace of the numbers v which tend to it than $\sqrt{2}$ has of the approximating fractions. The transfinite numbers are in a sense *new irrationalities*, ... We can say that the transfinite numbers stand or fall with finite irrational numbers, in their inmost being they are alike, for both are definitely marked off modifications of the actual infinite [1895–1897 77].

It is for this reason that those few mathematicians who still reject Cantor's theory must also reject large parts of classical mathematics, especially those parts dealing with irrational numbers. But Cantor himself felt very strongly that there was no other choice than to accept the actual infinite as the way things are. As he put it:

In spite of the essential difference between the conceptions of the *potential* and the *actual* infinite, the former signifying a *variable* finite magnitude increasing beyond all finite limits, while the latter is a *fixed, constant* quantity lying beyond all finite magnitudes, it happens only too often that the one is mistaken for the other ... A certain *horror infiniti* has grown up in extended scientific circles, which finds its classic expression and support in a letter of Gauss, yet it seems to me that the consequent uncritical rejection of the legitimate actual infinite is no lesser violation of the nature of things, which must be taken as they are [cf. Benardete 1964 28].

Set theory, then, was important for mathematical logic in two ways. First, like non-Euclidean geometry, it provided a blow to the notion that our clear and distinct intuitions are criteria for truth. For what could be clearer than that the whole is greater than the part? Yet this was found to be false when applied to infinite collections. Even the notion of a completed infinite collection was counter-intuitive to many mathematicians. The "new irrationalities" were as difficult to accept as the old irrationalities had been.

Second, set theory also provided the prototype of a diagonal argument which was to become an important method of mathematical logic. Note that this kind of argument is also an impossibility proof, in the sense that it shows that a one-to-one correspondence between the natural numbers and the real numbers is impossible. If there had been as strong a prejudice in favor of one type of infinity as there had been in favor of Euclidean geometry, it might have taken years of trying to find a proof of the denumerability of the real numbers before anyone would have thought that such a proof might be impossible.

There was yet another way in which set theory was destined to become important in the development of mathematical logic. This was the discovery of logical contradictions within the theory. The discovery was so startling, and adequate means of dealing with it so elusive, that the impact of it has been compared with the unsettling effects of the discovery of the irrationals by the Pythagoreans. If, in a respectable mathematical theory, unexpected logical contradictions appear, then this occurrence naturally leads to new analyses of valid and invalid arguments, as well as new inquiry into the possibility of establishing safeguards to prevent their occurrence in other areas of mathematics. But what are these logical contradictions?

§12 PARADOXES

The paradox that is historically the most important, as well as logically the most fundamental, is known as *Russell's paradox*, after Bertrand Russell (1872–1970). Russell discovered this paradox in 1901. A popularization of it concerns a village in which the postman delivers mail to all and only those who do not get their own mail. The question arises: How does the postman get his mail? Suppose he gets his own mail. Then he does not get his own mail, since he delivers mail to exactly those who do not get their own mail.

Suppose he doesn't get his own mail. Then he does get his own mail, since he delivers mail to exactly those who do not get their own mail. In short, we have a contradiction.

The Russell paradox is similar. One notes that many sets do not contain themselves. Examples are the set of all books (which is not itself a book), the set of all primes less than ten, the set of all things which are not themselves sets, etc. We may call such sets *normal*. *Abnormal* sets are those sets which do contain themselves. Examples are the set of all sets, the set of all sets describable in fewer than 20 English words, the set of all sets which contain more than three members, etc. Now consider the set of all normal sets and ask whether it is normal or abnormal. If it is normal, then it doesn't contain itself, by definition of *normal*. However, if it doesn't contain itself, it does contain itself, since it is the set of all normal sets. If, on the other hand, it is abnormal, then it contains itself, by definition of *abnormal*. However, if it contains itself, it is excluded from itself, since it is the set of only normal sets. In short, we again have a contradiction.

It is important, however, to notice the difference between the popularization and the paradox itself. The popularization is not really a paradox because it may be thought of as a *reductio ad impossibile* proof that such a postman exists; or if he does, that he is an inhabitant of the village; or if he is, that he receives mail, etc. In short, there are many ways to escape the paradoxical consequences of the popularization. This is not the case with Russell's paradox. There all the assumptions appear to be *a priori* true and all the ways of escape appear to have logically unwanted consequences. There doesn't seem to be any way of consistently defining the difference between normal and abnormal sets. Yet definitions like these are necessary for classical mathematics. Russell's paradox is a *reductio ad impossibile* argument from *a priori* premises. Obviously there is a logical mistake, but the exact nature of the mistake is not clear. Traditional logical theory provided no clarification. The paradox thus provided a stimulant to logical inquiry to rid set theory of contradictions.

Nor was Russell's the only paradox that was discovered. Even before Russell, Cantor himself discovered several paradoxes, of which the following is known as *Cantor's paradox*: Consider the set of all sets. If we form the set of all subsets of this set, we know by Cantor's theorem that the cardinal number of this new set is larger than the original one. Yet our original set by definition includes *all* sets. Thus we have constructed a set larger than the set of all sets, which is paradoxical to say the least.

Russell's and Cantor's paradoxes are examples of what are known as *logical* or *syntactical* paradoxes. They involve only logical and mathematical notions. With the appearance of this kind of paradox—and there are a number of this sort not mentioned here—there also developed a renewed interest in another kind of paradox. This latter kind is now often known as an *epistemological* or *semantical* paradox. Such paradoxes make reference to language or thought or symbolism or denotation or truth; in short, they make some kind of reference to the relationship of language to the world.

The best-known example of an epistemological paradox is that of the Liar. As we have seen, this paradox goes back to antiquity, but interest in it is still strong in the present day. The paradox has many forms, the most common of which is: *This proposition is not true* [cf. §5B]. Here is a Dualist form of it, dating from medieval times:

Socrates says, "What Plato says is false." Plato says, "What Socrates says is true." Neither says anything else. Now: Is what Socrates says true?

Still another medieval version is: *God exists, so this argument is invalid.* The paradoxical quality of this argument can be seen by remembering that the premiss was thought to be a necessarily true statement. Now is the argument valid? That is, is the conclusion false?

These paradoxes—indeed, this type of paradox—might at first seem to have little to do with mathematics. As we shall see, however, they are quite relevant. An example of a semantical paradox which is more obviously relevant is the following one, due to Jules Richard [1905]. There are also numerous versions of this paradox, but we shall consider one which closely parallels Cantor's diagonal argument.

Consider an enumeration of all possible combinations of the 26 letters in the alphabet, the ten numerals '0', '1', '2', '3', '4', '5', '6', '7', '8', '9', and the standard punctuation marks. A blank space will be considered one of the possible "signs." Fix some standard "alphabetical" order for these signs. Then an enumeration of all possible sequences of symbols could easily be made.

Problem 30. Define one possible enumeration of these sequences of signs.

In this enumeration there will be many instances of nonsense, examples of expressions of a foreign language, etc. Delete all expressions which do not in English define or characterize a real number. Examples of sequences remaining might be: 'the even prime number', 'the square root of the ratio between the diameter and area of a circle', 'the number whose decimal expansion is all sevens', etc. Let us call the first definition remaining 'D_1', the second 'D_2', etc. Then we would have the following enumeration:

$$1 \leftrightarrow D_1 = N_1.a_{1,1}a_{1,2}a_{1,3}a_{1,4}\cdots$$

$$2 \leftrightarrow D_2 = N_2.a_{2,1}a_{2,2}a_{2,3}a_{2,4}\cdots$$

$$3 \leftrightarrow D_3 = N_3.a_{3,1}a_{3,2}a_{3,3}a_{3,4}\cdots$$

$$4 \leftrightarrow D_4 = N_4.a_{4,1}a_{4,2}a_{4,3}a_{4,4}\cdots$$

$$5 \leftrightarrow D_5 = N_5.a_{5,1}a_{5,2}a_{5,3}a_{5,4}\cdots$$

$$\vdots$$

Now consider the following phrase: 'the diagonal number of the above enumeration'. This phrase defines a specific real number. By Cantor's diagonal argument, it cannot be in the above enumeration. Yet, as an English phrase which defines a real number, it is in the above enumeration.

Problem 31. Once attention was focused on paradoxes, many were formulated. Formulate one concerning English adjectives which do or do not apply to themselves (called *autological* and *heterological* adjectives, respectively). [*Hint:* Keep in mind the Russell paradox.]

It seemed obvious to those who considered the paradoxes that the difficulty might lie in self-reference, since the paradoxes all seem to depend on this feature. However, the solution of banishing all instances of self-reference appeared much too stringent a cure. There are, for example, a number of quite harmless examples of it in everyday use, such as: 'All English sentences begin with a capital letter'. Further, and more important, there are a number of places in classical mathematics in which self-reference is essential. For example, it occurs in the calculus, where one must define a *least upper bound*.

We can sum up the general problem of finding out what is harmful in self-reference by considering the following variants of self-referring sentences:

1. This sentence is short.
2. This sentence is in German.
3. This sentence contains four words.
4. This sentence contains five words.
5. This sentence is not true.

It is clear that (1) is true, (2) false, (3) false, (4) true. All these sentences involve self-reference, but only the last leads to a contradiction. Traditional logical theory had no answer to the question of what it was about the last sentence that made it alone a self-contradictory example of self-reference.

§13 SUMMARY

A rebirth of logical inquiry took place in the nineteenth and the early twentieth century. The occasion for this rebirth was the development within mathematics of strange and surprising doctrines. Non-Euclidean geometry and set theory both contained numerous statements which are counter-intuitive. Are these statements really true? One's doubts about the soundness of reasoning in these areas increases with the appearance of the paradoxes. It seemed that the wages of the sin of biting from the tree of knowledge of infinity was self-contradiction. Yet mathematicians were reluctant to leave Cantor's Eden. "No one shall be able to drive us from the paradise that Cantor created for us," Hilbert once declared [1925 376]. The dogmas of the past were inadequate and a new approach was needed. This new approach was to use mathematical methods. It resulted in the new discipline: mathematical logic.

CHAPTER 3

MATHEMATICAL LOGIC

§14 INTRODUCTION

The new birth of logic was not due to one man or even to thinkers in one country. The contributors are numerous and their interrelations complex. Hence we must abandon the historical style of the previous chapters. There the simplicity of our story was somewhat forgivable because of the paucity of evidence or primary researchers. Now, however, the sources of evidence are rich, and we shall never reach our goal if we continue to indulge in the luxury of history.* Suffice it to say that the awakening of logic began with George Boole (1815–1864) and Augustus De Morgan (1806–1871). The most important early work, however, was the *Begriffsschrift* [1879] of Gottlob Frege (1848–1925). From then on the development was very rapid. The work of Schröder (1841–1902), Peirce (1839–1914), and Peano (1858–1932) paved the way for the great systematization of *Principia Mathematica* [1910–1913] due to Whitehead (1861–1947) and Russell. Subsequently, the new area of what was later called *metalogic* developed: the work of Löwenheim [1915], Skolem [1920 ff], Gödel [1930 ff], Tarski [1929 ff], Post [1936 ff], Church [1936 ff], Kleene [1936 ff], Turing [1936 ff], and Cohen [1963] virtually created a new universe. It is the developments in this area which have proved to be the most philosophically exciting, and which we shall describe in Chapter 4. Here we turn to a description of the logical systems on which the metalogical results are based.

There were essentially two methods of mathematics which transformed logic into a new discipline. The first was the algebraic method, in which relations between mathematical entities are reflected in the relationships which hold between the symbols for these entities. Thus new relationships between mathematical entities can be discovered by manipulation of the symbols in accordance with certain rules. The second was the axiomatic method. What is meant here by the phrase *axiomatic method* is not the method that Euclid used, but a revision of that method based on various insights and discoveries made since his time. To distinguish it from Euclid's, it is often called the *formal axiomatic method*.

---

* For those who want a brief account see Church 1962; the best longer route is Kneale 1962 and van Heijenoort 1967. The intermediate-length account in Edwards 1967 is also excellent.

In this chapter we shall first describe this method and then show in some detail how both it and the algebraic method can be applied to various areas of logic. We shall then be ready to understand some of the metatheoretical results of mathematical logic.

§15 FORMAL AXIOMATIC METHOD

Let us begin our discussion of the formal axiomatic method by considering deficiencies in the method used by Euclid. Most of these inadequacies were discovered more than 2000 years after Euclid. In view of the subtlety of the errors and the enormity of his achievement, we should not blame Euclid very much for not discovering them. On the contrary, no other textbook has been so slow in going out of date.

There are, in essence, three sorts of deficiencies: the first having to do with the nature of the definitions, the second with logical gaps in the proofs, and the third with the implicit rather than explicit treatment of logic.

It was apparently Blaise Pascal (1623–1662) who first noticed that the same argument that applies to propositions also applies to definitions; namely, just as (on pain of infinite regress) there are some propositions we can't prove, so there must also be some terms we can't define. By trying to define all his geometric terms Euclid was led into circularity; as, for example, in his definition of a *straight line* as "that which lies evenly with the points on itself." But the analogy goes further; for just as all propositions must be proved in terms of the unproved propositions, so all terms must be defined in terms of the undefined terms. The axioms then would contain only logical terms and undefined (mathematical) terms.

In one sense the axioms are not about anything, since no meaning can be assigned to the undefined terms. In another sense they are about any set of objects which makes the axioms true. The set of objects must be defined outside the formal system. We have seen examples of this when we were considering non-Euclidean geometry. Thus the undefined terms in the axioms might mean an indefinitely large number of different things. For this reason the postulates are often regarded as *implicit definitions* of the undefined terms. What this means is that the axioms determine to which sets of objects the undefined terms may apply.

The idea of an implicit definition was introduced by J. D. Gergonne. He gave the example of someone who knew the meanings of *triangle* and *quadrilateral* but not *diagonal*. If the person then read that "quadrilaterals have two diagonals, each dividing the quadrilateral into two triangles," he would know what a diagonal is, without having an explicit definition. Another analogy is a set of three algebraic equations with three unknowns, which implicitly define the unknowns.

It might be thought that this feature of the undefined terms in the axioms applying to many different things represents a defect of the formal axiomatic method. Instead it represents its great strength and versatility, for it allows the working out once and for all of a large body of theorems which are then true of any set of objects of which the axioms are true. This is not to suggest that this method exhausts mathematics for (as we shall see) some informal mathematics is necessary even to set up a formal system. Further, the very point of setting up such a system is that it can later be interpreted in various ways. Nevertheless the use of undefined terms has become of such central importance in mathematics that Henri Poincaré (1854–1912) even defined it in terms of this feature: "Mathematics is the art of giving the same name to different things."

A second way in which Euclid's *Elements* is defective is in the logical gaps which occur in numerous proofs. As we have seen, Euclid's very first theorem assumes without any justification that the constructed circles meet. Here is another benefit of undefined terms, for it makes less likely the introduction of an implicit assumption which is not justifiable on the basis of the axioms. Even in ancient times the use of a faulty diagram was seen as a fallacy. But now any essential reference to any diagrams or any other assignment of meanings of the undefined terms is seen as illegitimate. Kant's philosophy had made the appeal to diagrams—or at least constructions in the *a priori* intuition—an essential feature of geometric knowledge. This is now seen to be an inessential feature. Whenever a reference to diagrams (or any other assignment of meanings of undefined terms) has to be made, the axioms are deficient, and should be changed or enlarged to make such a reference unnecessary.

This is, in fact, what Hilbert did, for example, in introducing his axiom of continuity to get around the logical gap in the first proposition of Euclid. Nor was this the only logical gap. There were numerous others, such as his assumption that a line has two sides or that closed figures have insides and outsides. Apart from logical gaps, there were also examples of logical redundancy; that is, Euclid would give a geometric proof of a proposition that was logically equivalent (or followed from) an earlier proposition or propositions. For example, in Book I of the Elements, Proposition 27 is the contrapositive of Proposition 16, but is nevertheless given a separate geometric proof [cf. Euclid 1926 I 309]. Although this sort of deficiency is only one of elegance, it seems probable that Euclid would have avoided it had he known of it [cf. Problem 38].

The redundancy of some of his proofs was no doubt due to the third deficiency of Euclid: the implicitness of his logic. We have seen that Euclid's system is inconsistent, but if the logical rules of his system had been stated the inconsistency might have been recognized in antiquity. There are also other reasons why the implicitness of Euclid's logic represented a deficiency.

For even in ancient times there was controversy over the question of whether certain types of inference were correct. Since the same mathematical axioms give rise to different sets of theorems making use of different sets of logical rules, it is necessary to be explicit. It is also historically unfortunate that Euclid did not make explicit his logic. For if the logic of, say, *magnitudes being in the same ratio* had been investigated, it is very probable that the inadequacy of both Aristotelian and Stoic logic would have been recognized.

The formal axiomatic method attempts to correct these deficiencies of Euclid by making the basic terms undefined, completing the logical gaps, and stating the logical rules. In order to describe the method, we need to make clear the important distinction between a *language* and a *metalanguage*.

We can use language, of course, to talk of many things: shoes and ships and sealing wax and cabbages and kings. We can also use language to talk about another language. We can use English to study, say, Japanese grammar. In such a situation we shall say that English is the metalanguage and Japanese the object language; that is, English is used to talk about Japanese and Japanese is the object of talk in English. Now, strictly speaking, we don't make use of Japanese expressions in our metalanguage, we only talk about them, and thus use names of Japanese expressions in our meta-expressions. Similarly, were we to study English grammar in Japanese, English would be the object language, Japanese the metalanguage, and we would use Japanese names of English expressions. The distinction between language and metalanguage (and, of course metalanguage and metameta-language, etc.) is subtle, and it isn't usually important outside of logic and philosophy.

Yet in these areas it sometimes becomes crucial. For example, if we ignore the distinction, the following becomes a sound argument:

> Romeo loves Juliet.
>
> Juliet is one word.
>
> Therefore Romeo loves one word.

The trouble with the argument is the second premiss. Juliet is not a word, but rather a certain young girl who is loved by Romeo. If we wish to talk about Juliet's name we need the name of her name. Logicians have adopted the convention of using single quotation marks around a word to designate the name of that word. Thus:

> 'Juliet' is the name of Juliet,
>
> ' 'Juliet' ' is the name of 'Juliet', etc.

Problem 32. Let 'Mary' be a name of Juliet's name. Determine the truth or falsity of the following statements.

        a)   Mary begins with an 'M'.
        b)   Juliet begins with a 'J'.
        c)   Romeo loves 'Juliet'.
        d)   Romeo loves Mary.

It is quite important that a name be treated as a whole and not be broken up into parts. Otherwise we shall not be able to substitute equals for equals. Consider the following argument:

$$\text{Peggy} = \text{Peggy}$$
$$\text{egg} = \text{ovum}$$
$$\text{Therefore Peggy} = \text{Povumy}$$

The argument is invalid ('Povumy' might not be the name of Peggy) and to avoid this we must treat names as indivisible units. Here is a similarly invalid mathematical example:

$$\text{'1 + 1' contains a plus sign.}$$
$$1 + 1 = 2.$$
$$\text{Therefore '2' contains a plus sign.}$$

We may further clarify the language-metalanguage distinction by noting the difference between what Quine calls the *use* and *mention* of a term. In our argument about Romeo and Juliet, the first premiss uses the names 'Romeo' and 'Juliet' to mention two Shakespearean characters. It does not mention the names 'Romeo' or 'Juliet'. The second premiss uses the word 'Juliet' to mention itself. We shall henceforth not allow such an autonomous use of a word wherever it might lead to invalid arguments.

Problem 33. Determine which of the following are used, which mentioned, and which neither used nor mentioned in each of the sentences of Problem 32: (i) 'Juliet', (ii) ' '*M*' ', (iii) Juliet, (iv) 'Romeo'.

We may summarize the distinction between object language and metalanguage. In the object language we use the symbols, words, sentences, etc., of the object language, but do not mention them; in the metalanguage we use the symbols, words, sentences, etc., of the metalanguage to mention the expressions of the object language, but we make no use of the expressions of the object language. Instead we use the names of the symbols of the object language. The same distinction can be applied to other areas. For example, we have logic and metalogic, mathematics and metamathematics, etc. An example of a metageometric problem is the problem of showing that Euclid's fifth postulate is independent of his other postulates [cf. §9]. This problem is not about points, lines, triangles, etc. (as is the object language); but rather about the logical relationship of the fifth postulate to the other postulates.

Problem 34. A child asks his father, "Dad, how much are two and two?" Wearily and somewhat inattentively the father answers, "Four." Whereupon the delighted child declares, "You're wrong! It's twenty-two." Making use of the distinctions between use-mention and mathematics-metamathematics, clarify this misunderstanding.

We are now in a position to describe a formal axiomatic system. The basis of it—often called the *primitive basis*—consists of four things:

1. *A list of symbols.* This list must include all the symbols which are to be used in the system. Examples might be ' + ', ' = ', '0'. The list in most systems is potentially infinite and is then represented in the time-honored way by using three dots, for example: '0', '0'', '0''', '0''''', ...

2. *Formation rules.* The formation rules indicate the ways in which symbols may be legally combined to produce formulas. Thus, for example, '0 + 0 = 0' might be a formula, whereas ' + + + = +' might not be. The motivation here is that when the system is given an interpretation the formulas will express propositions, propositional functions, or related meaningful units.

3. *A list of initial formulas.* This list consists of the formulas with which we start to operate. It can be finite or infinite. The way in which an infinite list of initial formulas can be indicated will be shown when we come to the propositional calculus. It is intended that when the initial formulas are interpreted they will become true axioms for the system.

4. *A list of transformation rules.* These are the rules which, when applied to the axioms, produce new formulas called *theorems*. The guiding motive is that when the system becomes interpreted the rules will correspond with logical rules of inference.

A formal system has some analogy with a natural language. The symbols correspond to the letters of the alphabet, punctuation marks, numerals, and so forth. The formation rules correspond to the grammatical rules of a natural language. The transformation rules correspond to various operations any speaker can perform on the language. For example, if it is true that 'Romeo loves Juliet', it is also true that 'Juliet is loved by Romeo'. The initial formulas have less identifiable corresponding items in the natural language, although it might include such sentences as 'whatever is, is' or '*A* is *A*'. There are, of course, many important differences between natural languages and formal systems. Nevertheless the analogy is close. Indeed many linguists—such as Noam Chomsky—would argue that a real understanding of language requires a comprehension of the nature of formal systems. *Generative* or *transformational* grammar without an understanding of formal systems would be like physics without an understanding of calculus. When formal systems are interpreted they are sometimes called *artificial languages*.

When we apply our transformation rules to the initial formulas the result is a theorem. The exhibition of the application of the rules is a *proof*. More explicitly, a *proof* is a finite sequence of formulas, such that each formula is an initial formula or follows from an earlier formula by the application of a transformation rule. The last line of the proof is a *theorem*. We require that the transformation rules be such that it is merely a mechanical procedure to determine whether or not a given sequence of formulas is a proof. Note that this requirement is important; it ultimately derives from the idea of the Pre-Socratics that there is no royal road to knowledge. If knowledge is claimed, and a proof is given as evidence, this proof must be open to inspection by all. This requirement distinguishes logical proof from some theological "proofs" (of God's existence) where "faith" or "grace" is needed to "see" the so-called proofs. What is required then is that no ingenuity or special insight be needed; in other words, that it be mechanical. We shall say more about this requirement in Chapters 4 and 5.

An analogy which is often used to introduce the formal axiomatic method is that of a board game such as chess [cf. Nagel and Newman 1958 34–36]. Let us exploit the analogy. (1) The list of symbols in the formal axiomatic method corresponds to the 32 red squares, 32 black squares, 16 red chess pieces and 16 white chess pieces. (2) The formation rules correspond (for the board) to the arrangement of the 64 squares into a square array in which all squares (we assume they are of the same size) are juxtaposed with at least two other squares and no square shares a border with another square of the same color. There would be further ("formation") rules for the allowable positions of the pieces on the board, such as, no piece may occupy more than one square, or no square contain more than one piece, and so forth. (3) The initial formulas correspond to the beginning array of pieces on the board. Note that just as different mathematical systems may have different axioms, so different initial positions would lead to different games. (4) The transformation rules correspond to the rules for the manipulation of pieces on the board. These rules imply, for example, that on the initial play no piece may be moved except a knight or a pawn.

The theorems correspond to the configurations of pieces on the board after the initial configuration. In both cases there is some order of development. One cannot move a rook without first moving a pawn. Similarly, in order to prove some theorems, one must first prove other theorems.

Finally, we may note that the analogy extends to interpretation. Normally we do not give any "interpretation" to the game of chess. However, one could. For example, a spy might use it to signal messages, different messages depending on the way the game is played. Of course, chess can have many different "interpretations."

The case is similar for a formal system. We consider a set of objects which are defined outside the formal system in question. We give an *interpretation* to the system by assigning a set of meanings to the primitive symbols in such a way that the initial formulas (now axioms) become true of the set of objects. This set of objects is then called a *model* for the system.

Before turning to an extended example of the formal axiomatic method, let us say something about the motivation for using such a method. Sometimes even those quite competent in mathematics (not to mention those of lesser expertise) find the formality forbidding, anti-intuitive, and somehow artificial. These difficulties may be admitted, and yet no better way has been found to solve two recurrent problems in mathematics: First, how do you avoid paradox? Or this might be put in another way: How do you know when something is proved? We have seen that Euclid probably had some such motive involved in his axiomatization. This avoided the criticism (noted by Aristotle) that either nothing can be proved or everything can. Let us suppose that someone raises the question of proof with regard to some particular proposition. We can answer this question by showing that such-and-such a formula is a theorem in such-and-such a formal system with such-and-such initial formulas and such-and-such undefined terms using such-and-such transformation rules; and that upon interpretation in such-and-such a specified way, the formula expresses the proposition in question, the initial formulas become axioms accepted by the individual, and the transformation rules become rules of inference similarly accepted by the individual. The proposition is then proved. Thus all questions of proof are reduced to questions of the acceptability of interpretations of formal systems. This, of course, does not settle all arguments; for example, some mathematicians will, and some will not, accept a *reductio ad impossibile* proof. Yet, at the very least, the axiomatic method clarifies the problem by making precise the exact point at which there is disagreement. In most cases it settles the question because of the power of the method to reduce complexities to acceptable simplicities.

Second, how do we know that two mathematical structures are the same? We have seen that Descartes and Fermat discovered the similarity between plane Euclidean geometry and part of algebra. This was done without any axiomatization of algebra. But the similarity was shown by rather casual methods. For example, Descartes said, in effect, that if you give me any statement of plane Euclidean geometry I will show you how to translate it into algebra, and if you give me any statement in a certain area of algebra I will show you how to translate it into plane Euclidean geometry. But suppose someone said that he granted that Descartes could do this for any proposition that he (the questioner) was likely to think of, but what he really wanted from Descartes was a proof that it was possible to make such a translation for any and all of the nondenumerable infinitude of possible statements in the two areas of mathematics. It is here that the formal axiomatic method becomes useful. For this latter question can be answered in the following way : Construct a formal axiomatic system such that under one interpretation it becomes plane Euclidean geometry and under another it becomes a certain fragment of algebra.

Thus the artificiality of the formal axiomatic method has the same kind of justification as the stilted legal language of lawyers : to clarify issues and to protect against certain kinds of doubts and objections. It may happen that a formal system is developed first (perhaps out of idle curiosity) and at some later time an interpretation found. However, it is more common that an already familiar body of knowledge is formalized, so that there is an intended interpretation which guides the choice of undefined terms, initial formulas, etc. Of course no use of that interpretation may be made in the development of the system. We shall now illustrate this process by first developing several areas of logic informally and then formalizing each area. Having done this, we shall find that many more advantages than those we have mentioned will accrue to us as a result. We shall start with the most elementary part of logic.

§16 PRIMARY LOGIC: THE PROPOSITIONAL CALCULUS

This area of logic is called *primary* because other parts of logic presuppose it. For example, Aristotle's theory of reduction depends on a calculus of propositions which he never made explicit. Since primary logic has its historical roots in the Stoic logicians, it is perhaps appropriate that we begin with an example from Chrysippus.

> Either the first or the second or the third.
>
> Not the first.
>
> Not the second.
>
> Therefore the third.

What has been presented here is not an argument but an argument form, a kind of logical skeleton. The occurrence of the phrases 'the first', 'the second', 'the third', is not necessary in order that this argument form be presented. By using '*A*', '*B*', and '*C*' we can present the same form. Thus :

> Either *A* or *B* or *C*.
>
> Not *A*.
>
> Not *B*.
>
> Therefore *C*.

We may turn an argument form into an argument by substituting sentences for the dummy letters. Our only requirement is that we substitute the same sentence throughout for the same dummy letter. For example:

> Either the animal went this way or that way or the other way.
>
> Not this way.
>
> Not that way.
>
> Therefore the other way.

Chrysippus said that even dogs are capable of this form of argument, for he had seen a dog chasing after an animal come to a threefold division in the path. The dog sniffed at the two paths that the animal hadn't taken and without sniffing raced off down the third path.

We shall henceforth use capital letters 'A', 'B', 'C', 'A_1', 'B_1', etc., as letters which will stand for fixed propositions, and call them *propositional constants*. This is done by analogy with mathematics, in which, for example, in the equation '$y = ax + b$' the letters 'a' and 'b' are numerical constants because they stand for fixed numbers. Also we shall often not make any concession to idiomatic English (contrary to the practice of the above example), and rather insist that a complete proposition be substituted for each propositional constant (our first example of logical legalese). Our revised practice would give us:

Either the animal went this way or the animal went that way or the animal went the other way.

Not the animal went this way.

Not the animal went that way.

Therefore the animal went the other way.

From our present logical point of view we shall treat propositions as unanalyzed, thus making no attempt to discern their logical structure. We shall be concerned only with the relations between propositions, and then only insofar as those relations concern truth or falsehood. Let it again be noted that we are not concerned with the truth or falsehood of the propositions themselves, but only with the way the truth or falsehood of a given proposition relates to the truth or falsehood of another proposition.

From our unanalyzed propositions we may build compound ones. The first premiss above is an example of such. Let's consider it further. It presumably means that exactly one of the three simple propositions contained within it is true. But there are other propositions of similar form which only mean that at least one of the component propositions is true. For example, a father who said to his son, "Either you wash the car or you cut the grass or you clean the cellar before you go out" would have no right to complain if his son went out after doing all three chores. We shall use the symbol

'∨'—suggested by the Latin word '*vel*'—to indicate the sense of the word 'or' in 'Either *A* or *B* or both'. A statement in this form is called a *disjunction* (or *alternation*). The sense of 'or' in 'Either *A* or *B*, but not both' we shall treat below. Using the now-familiar device of a truth table—a device whose use in modern times dates from the independent work of Post and Wittgenstein, but whose historical origins go back to the Stoics—we may define '∨' exactly.

p	q	$p \lor q$
T	T	T
T	F	T
F	T	T
F	F	F

Truth tables make use of *propositional variables*, whose domain of definition is a set of truth values for propositions. We use the lower-case letters 'p', 'q', 'r', 'p_1', 'q_1', etc., for propositional variables, just as 'x' and 'y' are numerical variables in '$y = ax + b$'. The left-hand side of the truth table indicates all the possible combinations of truth and falsehood (symbolized by 'T' and 'F', respectively). The right-hand side indicates when the compound form is true and when false. Thus 'Socrates is alive ∨ Plato is alive' is false, since both components are false (corresponding to the fourth line of the truth table), while 'Socrates is alive ∨ Plato is dead' is true, since the second component is true (corresponding to the third line of the truth table).

In a similar way we may use the symbol '⁻' for the word 'not'. Using a truth table, we would have:

p	$\bar{p}$
T	F
F	T

This table tells us that we are to consider '$\bar{p}$' false when 'p' is true and '$\bar{p}$' as true when 'p' is false. '$\bar{p}$' is called the *negation of p* and is read 'not p'. The idiomatic awkwardness is justified by the elimination of the ambiguity which results in English by allowing the 'not' to come at various places in the sentences. For example, 'All men are not good' could mean either 'there are no good men' or 'not all men are good'.

⁻ and ∨ are examples of truth functions; the former is a *singulary* function (since it operates on one proposition), the latter a *binary* function. A *truth* function then is a function of one or more propositional variables such that the assignment of a set of truth values (that is, either truth or falsehood) to each propositional variable determines a unique assignment of a truth value for the function. There are four possible singulary truth functions and sixteen possible binary ones. It is not necessary to consider ternary, etc., truth functions because all truth functions of more than two variables may be expressed in terms of binary truth functions. Consider the possibilities for singulary and binary functions:

Singulary truth functions

p	$f_1^1(p)$	$f_2^1(p)$	$f_3^1(p)$	$f_4^1(p)$
T	T	T	F	F
F	T	F	T	F

Binary truth functions

p	q	$f_1^2(p,q)$	$f_2^2(p,q)$	$f_3^2(p,q)$	$f_4^2(p,q)$	$f_5^2(p,q)$
T	T	T	T	T	T	T
T	F	T	T	T	T	F
F	T	T	T	F	F	T
F	F	T	F	T	F	T

$f_6^2(p,q)$	$f_7^2(p,q)$	$f_8^2(p,q)$	$f_9^2(p,q)$	$f_{10}^2(p,q)$	$f_{11}^2(p,q)$
T	T	T	F	F	F
F	F	F	T	T	T
T	F	F	T	T	F
F	T	F	T	F	T

$f_{12}^2(p,q)$	$f_{13}^2(p,q)$	$f_{14}^2(p,q)$	$f_{15}^2(p,q)$	$f_{16}^2(p,q)$
F	F	F	F	F
T	F	F	F	F
F	T	T	F	F
F	T	F	T	F

Not all these functions turn out to be useful enough to merit a special symbol. Among singulary functions, only f_3^1 is so deserving, and we have given it the symbol '‾'.*

Problem 35. One of the truth functions which is useful is 'and', which is symbolized by a dot '·'. By taking examples of compound statements connected by '·', determine what truth function best represents it. For example, 'Socrates is alive · Plato is alive' is false and so the fourth line in the truth table for '·' must be 'F'. Determine the other three lines and thus the truth function.

---

* The superscript on the function symbol refers to the number of propositional variables for that function. '$\bar{p}$' is used instead of '$-(p)$' because it simplifies the writing of some formulas. Similarly, since $\vee = f_2^2$, we could write '$\vee(p, q)$'. Some books on logic either do this or something analogous, though we shall not follow the practice.

Now that we have several truth functions, we can combine them in numerous ways. Consider the following truth table:

(1)	(2)	(3)	(4)	(5)	(6)	(7)
p	q	$\bar{p}$	$\bar{p} \vee q$	$\bar{q}$	$p \cdot \bar{q}$	$\overline{(p \cdot \bar{q})}$
T	T	F	T	F	F	T
T	F	F	F	T	T	F
F	T	T	T	F	F	T
F	F	T	T	T	F	T

In the table column (3) is gotten from (1), (4) from (3) and (2), (5) from (2), (6) from (1) and (5), and (7) from (6). In the sequel we shall be interested in just (4) and (7). Columns (3), (5), and (6) were just subsidiary and were written only to explain how we arrived at columns (4) and (7).

Now we come to the most controversial question in the theory of truth functions, namely, how to represent the locution 'if...then ____'. The controversy is not over which binary truth function best represents it, but whether 'if...then ____' is a truth-functional connective at all. For example, the truth or falsehood of 'Socrates believes that p' does not solely depend on the truth or falsehood of 'p', and therefore the expression is not truth-functional. Could it be that 'if...then ____' is also not truth-functional? To answer this question we shall first indicate some of the reasons why f_5^2 was chosen from the 16 binary truth functions to represent 'if...then ____', and after that, why it is adequate for our purposes.

An 'if...then ____' proposition is called a *conditional*, the proposition replacing '...' is called the *antecedent*, and that replacing the '____' is called the *consequent*. Hence, in 'if it isn't raining, then he is at the game', the antecedent is 'it isn't raining' and the consequent is 'he is at the game'. Now it is clear if 'it isn't raining' is true and 'he is at the game' is false, then the whole conditional is false. This suggests that the conditional is equivalent to denying that the antecedent is true and the consequent is false, which in turn suggests that the conditional be expressed by $\overline{(p \cdot \bar{q})}$ and $f_5^2(p, q)$. Following a long-standing convention, we shall use '$\supset$' as the special symbol for the conditional. '$p \supset q$' may be read as either 'if p, then q' or 'p only if q'. Note that the latter reading preserves the meaning of the former and is somewhat more convenient.

There are in essence three different kinds of objections to the truth-functional conditional; we shall briefly consider each. The first is that the truth-functional conditional does not require that the antecedent be relevant to the consequent. For example, 'if Socrates is dead, then there is no highest prime number' is true under the interpretation of 'if...then ____' as f_5^2. Further, 'if Socrates isn't dead, then there is a highest prime number' is also true! Hence our logical analysis appears to be going against common usage.

Yet we sometimes use the 'if . . . then _____' locution without presupposing any connection between the antecedent and the consequent. For example, 'if Mohammed is a Christian, then I'm a monkey's uncle', or again 'if the sun rises tomorrow, Mohammed will win'. The first sentence is a means of denying the antecedent by having an obviously false consequent follow from it. A similar device is sometimes used in mathematical logic by allowing 'f' to stand for some standard false sentence. Then '$\bar{p}$' can be defined as '$p \supset f$'. Note that this definition requires that when 'p' is true, '$p \supset f$' is false, and when 'p' is false, '$p \supset f$' is true. Analogous remarks could be made about a standard true sentence—t—with regard to our second example; that is, 't $\supset p$' is an alternative way of asserting 'p'. Our second example is thus equivalent to 'Mohammed will win'. Thus it is not true that there is always a logical connection between the antecedent and consequent in an 'if . . . then _____' statement in ordinary English.

The second objection follows from the first, namely, that if we allow a lack of relevance between antecedent and consequent, then any 'if . . . then _____' statement becomes true which has a false antecedent or a true consequent. Surely this is paradoxical. Yet if it is paradoxical, it is paradoxical in the sense of contrary to what was expected, not paradoxical in the sense of an antinomy. We *usually* don't make conditional statements unless the truth or falsehood of both antecedent and consequent are unknown to us. What we are doing in our artificial language is to be explicit where our natural language is silent. This explicitness, it will be recalled, was part of the very *raison d'être* of an artificial language.

The third objection is that the truth-functional interpretation of 'if . . . then _____' takes no cognizance of the subjunctive mood. Indeed the statement above that "we *usually* don't make conditional statements unless the truth or falsehood of both antecedent and consequent are unknown to us" applies only to the indicative mood. To see this, consider the following example: 'No cube of sugar was put in this glass of pure room-temperature water in the last hour'. Let's suppose that this is true. It would nevertheless be false that 'if a cube of sugar had been put in this glass of pure room-temperature water, then the sugar would not dissolve'. The only adequate answer to this objection is to admit the point and reduce the claims made for '$\supset$' in the following way: '. . . $\supset$ _____' will be taken as a sufficiently close approximation of the indicative 'if . . . then _____' as to be very useful for logical purposes. It does not approximate the subjunctive conditional; indeed, no other truth function would either. The analysis of such conditionals—often called *contrary-to-fact* conditionals—is more complex, and whether there now exists any adequate treatment is a moot point. In any case, the ultimate justification for the translation of the indicative 'if . . . then _____' as '. . . $\supset$ _____' is pragmatic; it has proved very effective in logical analysis.

There is yet another truth function which is so useful as to deserve a special name. We have translated 'p only if q' as 'p ⊃ q'. Now 'p if q' is certainly equivalent to 'if q, then p'; that is, 'q only if p'. Thus '(p if q) and (p only if q)' may be translated as '(q ⊃ p)·(p ⊃ q)'. When this state of affairs holds, we shall write 'p ≡ q', which may be read 'p if and only if q'. ≡ is the same as f_7^2 and is called the *biconditional*.

Problem 36. It is sometimes useful to have special symbols for the binary truth functions that represent the *negations* of statements containing exactly one binary connective. The special symbol used for this is the old symbol with a vertical line drawn through it (on the analogy of '=' and '≠'). Thus '(... ∨ ____)' is represented by '... ⩔ ____', '(... · ____)' by '... | ____', '(... ⊃ ____)' by '... ⊅ ____', and '(... ≡ ____)' by '... ≢ ____'. Determine which of the 16 binary truth functions are respectively represented by '⩔', '|', '⊅', '≢'.

The purpose of our introduction of special symbols was to facilitate logical analysis. We can now turn to some examples to show in detail that this purpose is fulfilled. In particular we shall show how truth tables may be used to determine the validity or invalidity of arguments. Let's consider the dog argument referred to at the beginning of this section. It can be represented as follows:*

$$(A \lor B) \lor C$$
$$\bar{A}$$
$$\bar{B}$$
$$\therefore \quad C$$

As in Aristotelian logic, a *valid argument* is an argument in which it is impossible for all the premisses to be true and the conclusion false. We can test the above argument by showing that it is impossible for any argument of the same form to have true premisses and a false conclusion. The *form of an argument* is the result of replacing all propositional constants in that argument by propositional variables, where the same propositional constant is replaced throughout the argument by the same propositional variable, and different propositional constants are replaced by different propositional variables. Now the form of the dog argument is

$$(p \lor q) \lor r$$
$$\bar{p}$$
$$\bar{q}$$
$$\therefore \quad r$$

---

* We use '∴' as an abbreviation for 'therefore'.

We can test it by a truth table (Table 7). In this table, only the seventh line represents a possibility in which all the premises are true. In that line the conclusion is also true, so we have a valid argument.

Table 7

			First premiss	Second premiss	Third premiss	Conclusion
p	q	r	$(p \lor q) \lor r$	$\bar{p}$	$\bar{q}$	r
T	T	T	T	F	F	T
T	T	F	T	F	F	F
T	F	T	T	F	T	T
T	F	F	T	F	T	F
F	T	T	T	T	F	T
F	T	F	T	T	F	F
F	F	T	T	T	T	T
F	F	F	F	T	T	F

Problem 37. The careful reader might object to the translation of the first premiss as '$(A \lor B) \lor C$', since this means 'Either (A or B or both) or C or both', while the original English, 'Either the animal went this way or the animal went that way or the animal went the other way', means that the animal went exactly one of the ways. To translate this latter interpretation, we need to capture the sense of the expression: 'Either p or q, but not both'. This may be done, for example, by '$(p \lor q) \cdot \overline{(pq)}$'. On the basis of this hint, see if you can translate the premiss under this interpretation and show that the argument is still valid by the truth table.

To each valid argument there corresponds a true conditional statement in which the antecedent is a conjunction of the premises and the consequent is the conclusion. Thus the following is true: 'If (either the animal went this way or the animal went that way or the animal went the other way) and (not the animal went this way) and (not the animal went that way), then the animal went the other way'. To capture the notion of *logically true* for the propositional calculus, we shall define a propositional form to be a *tautology* if and only if its truth table contains only 'T's. In the propositional calculus all logical truths are tautologies and all tautologies are logical truths. A proposition will also be said to be a *tautology* if its propositional form is a tautology. For example, the statement 'either Socrates is dead or not Socrates is dead' is a tautology, since '$p \lor \bar{p}$' contains only 'T's in its truth table. We are now in a position to give an equivalent definition of a *valid argument*: An argument in the propositional calculus is valid if and only if its corresponding conditional proposition is a tautology. To see that this definition is indeed equivalent to our earlier one, consider the dog argument. The truth table of the form of its corresponding conditional would be like Table 8. Observe that the antecedent of the statement form will have a 'T' in its truth table only if all the premises do. Thus by the truth table for the conditional, the whole truth function will be false exactly when the premises are all true and the conclusion is false.

Table 8

p	q	r	$((((p \lor q) \lor r) \cdot (\bar{p})) \cdot (\bar{q})) \supset r$
T	T	T	T
T	T	F	T
T	F	T	T
T	F	F	T
F	T	T	T
F	T	F	T
F	F	T	T
F	F	F	T

Problem 38. The truth-table technique is adequate to deal with many rather complicated arguments, but it gets to be rather lengthy as the number of propositional variables increases.* Both these points can be illustrated by the following example of logical redundancy from Euclid. In Book I from Proposition 5 and Proposition 18 (which is proved independently of 6) it is possible logically to deduce Propositions 6 and 19; that is, no use is made of any further geometric argument to get Propositions 6 and 19 [cf. Euclid 1926 I 284–285]. The form of the argument is as follows:

$$((p(\bar{q}\bar{r})) \lor (\bar{p}(q\bar{r}))) \lor (\bar{p}(\bar{q}r))$$

$$((p_1(\bar{q}_1\bar{r}_1)) \lor (\bar{p}_1(q_1\bar{r}_1))) \lor (\bar{p}_1(\bar{q}_1r_1))$$

$$p \supset p_1$$

$$q \supset q_1$$

$$r \supset r_1$$

$$\therefore \quad ((p_1 \supset p) \cdot (q_1 \supset q)) \cdot (r_1 \supset r)$$

Show by means of a truth table that the argument form is valid.

---

* Given that n is the number of propositional variables in the truth table, then 2^n is the number of lines in it.

Not only long problems, but also relatively short but tricky problems may be clarified by truth-functional analysis. Consider the following example:

Suppose that Socrates is in such a condition that he does not wish to visit Plato, unless Plato wishes to visit him; and that Plato is in such a condition that he does not wish to visit Socrates, if Socrates wishes to visit him, but wishes to visit Socrates if Socrates does not wish to visit him. Does Socrates wish to visit Plato or not? [cf. *Journal of Symbolic Logic* VI 35].

Letting 'A' stand for 'Socrates wishes to visit Plato' and 'B' stand for 'Plato wishes to visit Socrates' we have:

$$\bar{A} \vee B$$

$$(A \supset \bar{B}) \cdot (\bar{A} \supset B)$$

Note that in ordinary English the antecedent is sometimes stated after the consequent, as in the second premiss. Now we want to know which of the following is a tautology:

$$((\bar{A} \vee B) \cdot ((A \supset \bar{B}) \cdot (\bar{A} \supset B))) \supset A$$

or

$$((\bar{A} \vee B) \cdot ((A \supset \bar{B}) \cdot (\bar{A} \supset B))) \supset \bar{A}.$$

Table 9 is a truth table for the forms of each of these statements.

Table 9

p	q	$((\bar{p} \vee q) \cdot ((p \supset \bar{q}) \cdot (\bar{p} \supset q))) \supset p$	$((\bar{p} \vee q) \cdot ((p \supset \bar{q}) \cdot (\bar{p} \supset q))) \supset \bar{p}$
T	T	T	T
T	F	T	T
F	T	F	T
F	F	T	T

Since only the right-hand column has all 'T's, it follows that Socrates does not wish to visit Plato. If neither had all 'T's, it would mean that the premisses were not strong enough to decide the issue in question. On the other hand, suppose that both columns were all 'T's. This would mean that the premisses were contradictory. Hence they could not all be true and would never describe any reality. An example of this situation is the Protagoras–Euathlus argument [cf. §3]. If we let 'A' stand for 'Euathlus wins this case', 'B' stand for 'Euathlus wins his first case', and 'C' stand for 'Euathlus ought to pay Protagoras', then the following represents a somewhat simplified version of their arguments:

Protagoras' argument	Euathlus' argument
$A \supset B$	$A \supset \bar{C}$
$B \supset C$	$\bar{A} \supset \bar{B}$
$\bar{A} \supset C$	$\bar{B} \supset \bar{C}$
$A \lor \bar{A}$	$A \lor \bar{A}$
$\therefore \quad C$	$\therefore \quad \bar{C}$

A truth-table check of the corresponding conditional of the forms of both these arguments reveals that they are both tautologies, in spite of the fact that they come to contradictory conclusions. This implies that the premises taken together are contradictory and thus at least one is false. In this case it appears that both '$B \supset C$' and '$\bar{B} \supset \bar{C}$' are false, since the validity of the arguments presumably depends on the judgment of the court.

Problem 39. In Cervantes' story about the man and the bridge [cf. Problem 3 (§3)], let 'A', 'B', 'C', 'D' respectively stand for 'he crosses the bridge', 'he is hanged on the gallows', 'the oath to which he swears is true', 'the law is obeyed'. The conditions in the story can then be expressed by the three formulas: '$C \equiv (AB)$', 'A', '$D \supset (B \equiv (A\bar{C}))$'. What would be wrong with having '$D \supset (A\bar{C} \supset B)$' instead of the third formula?

Problem 40. Read Appendix A, which contains the well-known "Barber Shop" argument of Lewis Carroll. Letting 'A', 'B', 'C' respectively stand for 'Allen is out', 'Brown is in', 'Carr is out', determine by a truth table whether C can be true under the conditions stated by Carroll.

The use of special symbols and truth functions is useful not only in determining validity but also in making clear the structure of various types of propositions and arguments. For example, consider the following sequence of argument forms:

(1)	(2)	(3)	(4)	(5)
$p \supset r$	$p \supset r$	$p \supset r$	$p \supset p$	
$q \supset s$	$q \supset r$	$\bar{p} \supset r$	$\bar{p} \supset p$	$\bar{p} \supset p$
$p \lor q$	$p \lor q$	$p \lor \bar{p}$	$p \lor \bar{p}$	
$\therefore \quad r \lor s$	$\therefore \quad r$	$\therefore \quad r$	$\therefore \quad p$	$\therefore \quad p$

Argument form (1) is called a *complex constructive dilemma* and is often used in debate. The debater tries to get his opponent to admit the truth of the three premises, and then his opponent is stuck with a Hobson's choice of unpalatable alternatives. Argument form (2) is called a *simple constructive dilemma*, and is a special case of (1) in which the consequents of the first and second propositional forms are identical. (Note that the conclusion reduces to r.) Argument form (3) is called a *simple analytic dilemma* because the third premiss is a tautology and thus is analytic. It is a special case of (2). Argument form (4) is the *consequentia mirabilis* used by Saccheri [cf. §6]. Argument form (5) is a simplified form of (4), since the first and third premisses of (4) are not necessary to obtain the conclusion [as they are in (1), (2), and (3)]. Without special notation the similarities and differences among the argument forms (1) through (5) would take some time in explaining. Here they leap to the eye.

A number of propositional and argument forms are useful enough to deserve special names. Table 10 presents a sample of these forms and their names.

Table 10

Propositional forms	
Law of non-contradiction	$\overline{(p \cdot \bar{p})}$
Law of excluded middle	$p \vee \bar{p}$
Law of double negation	$p \equiv \bar{\bar{p}}$
Law of exportation	$((p \cdot q) \supset r) \equiv (p \supset (q \supset r))$
Laws of De Morgan	$\overline{(p \vee q)} \equiv (\bar{p} \cdot \bar{q})$
	$\overline{(p \cdot q)} \equiv (\bar{p} \vee \bar{q})$
Laws of tautology	$p \equiv (p \vee p)$
	$p \equiv (p \cdot p)$
Laws of association	$(p \vee (q \vee r)) \equiv ((p \vee q) \vee r)$
	$(p \cdot (q \cdot r)) \equiv ((p \cdot q) \cdot r)$
Laws of distribution	$(p \cdot (q \vee r)) \equiv ((p \cdot q) \vee (p \cdot r))$
	$(p \vee (q \cdot r)) \equiv ((p \vee q) \cdot (p \vee r))$

Argument forms	
Modus ponens	$p \supset q$
	p
	$\therefore\ q$
Modus tollens	$p \supset q$
	$\bar{q}$
	$\therefore\ \bar{p}$
Hypothetical syllogism	$p \supset q$
	$q \supset r$
	$\therefore\ p \supset r$
Disjunctive syllogism	$p \vee q$
	$\bar{p}$
	$\therefore\ q$
Conjunction	p
	q
	$\therefore\ p \cdot q$
	p
Addition	$\therefore\ p \vee q$

A word of caution: Do not see too much in a name. The names given in Table 10 are the common ones; but, for example, the laws of De Morgan were not first discovered by Augustus De Morgan, but were known in scholastic times and probably even in antiquity. And in any case the laws he stated were the corresponding laws in set theory. Similarly, 'tautology' in 'laws of tautology' does not have any particular connection with what we defined 'tautology' to be above.

One further point about Table 10 needs explanation. If the corresponding conditional of each of the named argument forms were composed, it would be a tautology. In such a circumstance, we shall say that the antecedent *implies* the consequent. For example, '$((p \supset q) \cdot p)$' implies 'q'. Similarly, if the biconditional is a tautology, we shall say that one component is *equivalent* to the other component. For example, '$\overline{(p \vee q)}$' is equivalent to '$\overline{p} \cdot \overline{q}$'. 'Implies' and 'equivalent' here mean 'logically implies' and 'logically equivalent'. As we shall have no further occasion to speak of any other kind of implication or equivalence, we shall drop the adverb.

There are differences between the notion of implication and that of the truth-functional conditional. In particular, implication is a concept from the metalanguage and connects names of propositional variables or propositional constants. The truth-functional conditional is from the object language and connects propositional variables or propositional constants. From this point of view it would be wrong to read '$\supset$' as 'implies'. Similar remarks may be made about '$\equiv$' and 'equivalence'. Further note that the concepts of true and false also belong to the metalanguage. Thus we have 'Socrates is mortal' is true, and not 'Socrates is mortal is true'.

Problem 41. By the above conventions, which of the following are correct and which incorrect?

1) $(p \cdot \bar{p})$ is a contradiction.
2) '$p \vee \bar{p}$' is a tautology.
3) p implies $p \vee q$.
4) 'p' $\supset$ 'q'.

We have now informally developed a propositional calculus. It is just one formulation of the propositional calculus; there are many other equivalent ones. Of course, the propositional calculus could be (and has been) developed much further, for example, by presenting means of shortening proofs of validity. We shall not do this, but instead will turn to the problem of presenting a formal axiomatic system for this theory. We shall use the letter '$\mathscr{P}$'—suggesting *primary logic* or *propositional calculus*—for the system. Our aim at formalization will be achieved if the informal theory presented above is an interpretation of the formal system.

Primitive Basis for $\mathscr{P}$

1) List of symbols

    a) Two logical symbols:

$$\supset \qquad \overline{}$$

    b) Two parentheses:

$$(\qquad)$$

    c) An infinite list of propositional constants:

$$A\ B\ C\ A_1\ B_1\ C_1\ A_2\ B_2\ C_2 \cdots$$

    d) An infinite list of propositional variables:

$$p\ q\ r\ p_1\ q_1\ r_1\ p_2\ q_2\ r_2 \cdots$$

2) Formation rules

    a) A constant standing alone is a formula.
    b) A variable standing alone is a formula.
    c) If **P** and **Q** are formulas, then so is **(P ⊃ Q)**.
    d) If **P** is a formula, then so is **(P̄)**.

3) List of initial formula schemata

    a) **(P ⊃ (Q ⊃ P))**
    b) **((P ⊃ (Q ⊃ R)) ⊃ ((P ⊃ Q) ⊃ (P ⊃ R)))**
    c) **(((P̄) ⊃ (Q̄)) ⊃ (Q ⊃ P))**

4) Transformation rule

    a) From **(P ⊃ Q)** and **P**, infer **Q**.

In (1) we have eliminated the single quotation marks for the sake of clarity. Here and henceforth we shall use the colon followed by a new line of centered text to be equivalent to single quotation marks. Variables and constants in the object language are printed in sans serif type to distinguish them from variables and constants in the metalanguage. The letters '**P**', '**Q**', and '**R**' are variables in the metalanguage whose domain of definition comprises sequences of symbols in the object language (a sequence will be understood as one or more symbols in some order). Except in the statement of the formation rules, these sequences will always be taken to be formulas. A particular concatenation of names in the metalanguage will refer to the corresponding concatenation of symbols in the object language. For example, '(p ⊃ p)' means the formula formed by juxtaposing '(', 'p', '⊃', 'p', and ')', in that order. The schemata in (3) each stand for an infinite number of formulas. The following are samples of (3(a)):

$$(p \supset (q \supset p))$$

$$(p \supset (p \supset p))$$

$$((p \supset q) \supset (r \supset (p \supset q)))$$

Thus the system has an infinite number of initial formulas, but all of them have one of three possible structures. There is no need to distinguish logical symbols and parentheses by use of different type because they never appear alone. For example, a '$\supset$' appearing in the context of italic lightface letters means that it belongs to the metalanguage. If it appears in the context of sans serif letters, it belongs to the object language. If it appears in the context of boldface letters, it belongs to the metalanguage, and is a name of the same symbol in the object language. Hence, contrary to first appearances, no formulas or even symbols from the object language appear in the statement of the primitive basis; at most, names of such occur. The order in which the propositional constants and variables are named is called their *alphabetic order*.

Problem 42. It is very important to understand that in the object language everything proceeds by rule, but that in the metalanguage we must proceed without rules. Read Appendix B, and on the basis of distinctions and conventions so far introduced, see if you can extricate Achilles from the logical morass in which the Tortoise put him.

It may be helpful to see a sample proof. Consider the following list of formulas:

1) $((p \supset ((q \supset p) \supset p)) \supset ((p \supset (q \supset p)) \supset (p \supset p)))$
2) $(p \supset ((q \supset p) \supset p))$
3) $((p \supset (q \supset p)) \supset (p \supset p))$
4) $(p \supset (q \supset p))$
5) $(p \supset p)$

Note that each line is either an initial formula or follows from earlier lines by our transformation rule. Lines (1), (2), and (4) are initial formulas. Line (3) is derived from (1) and (2) by the transformation rule. In this case, **P** is '$(p \supset ((q \supset p) \supset p))$' and **Q** is '$((p \supset (q \supset p)) \supset (p \supset p))$'. Line (5) is similarly gotten from (3) and (4). Thus this list of formulas is a proof of '$(p \supset p)$'. This may be expressed in the metalanguage by saying that '$(p \supset p)$' is a theorem. Since this predicate frequently occurs, a special metalinguistic symbol is used: '$\vdash$'. We shall write ' $\vdash$'$(p \supset p)$' ', which will be read ' '$(p \supset p)$' is a theorem'. Note that this proof does not prove, for example, '$(q \supset q)$'. We may, however, write out a proof schema in the metalanguage which will guarantee that any conditional formula is a theorem if the antecedent and consequent are identical:

$$((P \supset ((Q \supset P) \supset P)) \supset ((P \supset (Q \supset P)) \supset (P \supset P)))$$

$$(P \supset ((Q \supset P) \supset P))$$

$$((P \supset (Q \supset P)) \supset (P \supset P))$$

$$(P \supset (Q \supset P))$$

$$(P \supset P)$$

This is called a *proof schema* because it represents in the metalanguage an infinite number of proofs in the object language. We may thus write ' $\vdash(P \supset P)$ '.

§17 GENERAL LOGIC: THE PREDICATE CALCULUS

There are many different types of arguments which cannot be accounted for by the propositional calculus. To see this, consider the following argument:

> If Socrates is here, then he is not in Rhodes.
>
> Socrates is here.
>
> Therefore it is not the case that Socrates is in Rhodes.

This is a perfectly valid argument, as we can easily check. But compare the following ancient Stoic example (called *The Nobody* [cf. Mates 1953 85]) which is superficially similar to the above:

> If someone is here, then he is not in Rhodes.
>
> Someone is here.
>
> Therefore it is not the case that someone is in Rhodes.

This argument is invalid; its plausibility depends on an ambiguity. To see exactly what kind of ambiguity is involved, we might try to put both the above arguments into the *modus ponens* form of argument which they resemble. In the first argument, the 'he' in the consequent of the conditional premiss refers to Socrates. Thus it might be written:

> If Socrates is here, then it is not the case that Socrates is in Rhodes.
>
> Socrates is here.
>
> Therefore it is not the case that Socrates is in Rhodes.

This is an instance of *modus ponens* and is therefore valid.

> If we try a similar transformation on the second argument we get:

> If someone is here, then it is not the case that someone is in Rhodes.
>
> Someone is here.
>
> Therefore it is not the case that someone is in Rhodes.

There seem to be two plausible ways of interpreting these English words. On the one hand, we might take the word 'someone' in the consequent of the first premiss and in the conclusion to refer to a person different from the person referred to by the word 'someone' in the antecedent of the first premiss and in the second premiss. Interpreted in this way, the argument is valid, but the first premiss is obviously false. On the other hand, we might take all occurrences of 'someone' to refer to the same person. In this case the argument might be written:

If someone is here, then it is not the case that that someone is in Rhodes.

Someone is here.

Therefore it is not the case that that someone is in Rhodes.

This argument is valid, but it lacks the paradoxical quality of the original. That quality depended on interpreting the 'he' in the consequent of the conditional premiss as 'that someone' and the 'someone' in the conclusion not as 'that someone' but as just 'someone'.

What this example suggests is that we need a completely unambiguous way of making singular, particular, and general references and that this should contain within itself a means of making cross references. Actually English itself contains such devices. For example, words such as 'everything', 'something', 'nothing', 'someone', 'Socrates', etc., enable us to make singular, particular, or general references, and the use of pronouns allows us to make the cross references. Both the references and cross references can also be made unambiguous by using an extended description. Yet the description is likely to become very lengthy, so lengthy as to impair our understanding. We might compare this case to the lengthiness of legal documents, which comes about very often because of the necessity of avoiding ambiguity. But lengthiness itself can be a factor in diminishing perspicuity. Thus we would want our representation of arguments as compact as possible, provided, of course, that it is not so compact as to also hinder clarity.

Our purpose in this section is to extend our propositional calculus so as to include arguments whose validity depends not only on the external relationship of propositions but also on the internal construction of the propositions themselves. Simple illustrations of such arguments would include Aristotelian syllogisms (for example, if 'all mammals are animals' and 'all men are mammals', then 'all men are animals') or Galen's non-Aristotelian and non-Stoic arguments (for example, if 'Sophroniscus is father to Socrates', then 'Socrates is son to Sophroniscus'). We also want our new logic to have a notation which fits in with the propositional calculus, and, further, we want to keep the algebraic feature that manipulation of symbols according to certain rules leads to the expression of new logical relationships.

Our new expanded logic is sometimes called the *first-order* (or *restricted*) *calculus of propositional functions*, sometimes the *first-order* (or *restricted*) *functional calculus*, and sometimes the *first-order* (or *restricted*) *predicate calculus*. We shall use the latter name, leaving off the adjectives *first-order* or *restricted*. The significance of these terms will be explained in the next section.

The central notion to be explained is that of a propositional function (or *predicate*, as it will also be called). Let there be fixed some nonempty *domain of discourse*, that is, some set of objects which our logic will be about. Examples might include the set of physical objects, the set of living animals, or the set of natural numbers. The members of the domain of discourse will be called *individuals*. An *n-place propositional function* (or *n-place predicate*) is a function of n individual variables, where the domain of definition is the domain of discourse and the domain of values is a set of propositions. Hence, when each variable in a propositional function (predicate) has assigned to it an individual, the result is a proposition.

For example, suppose that our domain of discourse is the natural numbers. Then x is a prime, $x < y$, $(x < y) \lor (y < x)$ are examples of propositional functions (predicates). As soon as individuals are assigned to x and y, the result is a proposition. If 7 is assigned to x and 5 is assigned to y, then our first example of a propositional function (predicate) becomes a true proposition, the second a false proposition, and the third a true proposition. A one-place propositional function or predicate (such as our first example) is a *property*, a two-place propositional function or predicate a *binary relation*, a three-place propositional function or predicate a *ternary relation*, and so on. For simplicity we shall henceforth drop the term *propositional function* and use the shorter one *predicate*.

We shall use 'a', 'b', 'c', 'a_1', 'b_1', etc., to represent individual constants, which in a given context stand for fixed (though perhaps unknown) individuals. We shall represent individual variables by 'x', 'y', 'z', 'x_1', 'y_1', etc. We shall use 'A^n', 'B^n', 'C^n', 'A_1^n', 'B_1^n', etc., to represent *n*-place predicate constants, which in a given context stand for fixed (though perhaps unknown) predicates. In this section we shall not have need for predicate variables. The following are examples of predicates: $A^1(x)$, $B_1^3(x, y, z)$, $C_3^4(z, x_1, y_2, x_2)$. The superscript indicates the number of variables which follow the predicate letter, and will be omitted where no ambiguity will result. Sometimes, when a predicate has a well-known name, we shall use it in the normal way rather than in the way we just indicated. For example, if *less than* is represented by 'A', then $< = A$ and we could write either '$A(x, y)$' or '$<(x, y)$' for 'x is less than y', but we shall choose '$x < y$'.

If all individual variables in a predicate are replaced by individual constants, the result is a *substitution instance*. For example, $7 < 5$, $5 < 5$, $3 < 4$, are substitution instances of $x < y$. Predicates are neither true nor false, but when we take substitution instances of them they become propositions which are true or false. They become propositions about specific individuals.

However, sometimes we wish to express particular or universal propositions (as in the case of the syllogism). We may do this by making use of quantifiers, that is, operators which make particular or universal propositions out of predicates. We shall use two operators: '$\exists$', which is called the *existential quantifier*, and '$\forall$', which is called the *universal quantifier*. The use of quantifiers is most easily explained by examples. Therefore let us consider Table 11.

The formula '$(\exists x)A(x)$' is read 'there exists an x, A of x' or 'for some x, A of x'. It means that there is at least one individual in the domain of discourse which has the property represented by 'A'. The formula '$(\forall x)A(x)$' is read 'for all x, A of x' or 'given any x, A of x'. It means that every individual in the domain of discourse has the property represented by 'A'. The prefixing of the predicate $A(x)$ by either of the two quantifiers has the effect of turning it into a particular or universal proposition, that is, something either true or false. Hence (1(a)) is true and (1(b)) false. A quantifier is always preceded by a left-hand parenthesis and followed by the variable which it quantifies, which is itself followed by a right-hand parenthesis. This whole expression should then be followed by a pair of parentheses which indicate its *scope*; that is, all similar variables occurring within the parentheses refer back to the quantifier (here in part is where the cross references come in). Thus (1(a)) should have been written '$(\exists x)(A(x))$' and (3(b)) should be written '$(\forall x)((\forall y)((\forall z)((\exists x_1)(A(x_1, x, y, z)))))$'. However, in practice these parentheses are left out whenever no ambiguity results.

Consider now (1(c)). Here the x in $B(x)$ is not in the scope of the quantifier and is not quantified. The resulting expression is thus a predicate. In (1(d)), y is within the scope of the quantifier but, since the quantified variable is x, it is unaffected by it. The result is another predicate. A variable which occurs within a quantifier or within the scope of a quantifier containing the same variable is called a *bound* (or *apparent*) variable. Otherwise it is called a *free* (or *real*) variable. For example, in (1(c)) the first two occurrences of x are bound, the third is free.

The order in which the quantifiers occur is important. Notice the difference in the translations of (2(b)) and (2(c)). Propositions (4(a)) and (4(c)) are **A** and **O** propositions in our new notation. To be more precise, '$(\forall x)(A(x) \supset B(x))$' (which may be read 'for all x, if A of x then B of x') expresses the gist of the traditional **A** proposition, but is neutral on the question of whether anything exists which has the property A. What it expresses may be precisely rendered: 'If there are any athletes, then they are all brash'. The traditional **A** proposition was used in a context in which it was presupposed that all predicates had at least one exemplification. Whether ordinary English

Table 11*

Propositions	Predicates and Constants	Translations
1) (a) $(\exists x)A(x)$	$A(x)$: x is divisible by 2	There exists a natural number divisible by 2.
(b) $(\forall x)A(x)$	$B(x)$: x is divisible by 3	Every natural number is divisible by 2.
(c) $(\exists x)A(x)\cdot B(x)$		There exists a natural number divisible by 2 and x is divisible by 3.
(d) $(\exists x)(A(x)\cdot B(y))$		There exists a natural number divisible by 2 and y is divisible by 3.
2) (a) $(\exists x)(\exists y)A(x,y)$	$A(x,y)$: x is greater than y	Some natural number is greater than some natural number.
(b) $(\exists x)(\forall y)A(x,y)$		Some natural number is greater than all natural numbers.
(c) $(\forall x)(\exists y)A(x,y)$		Given any natural number, it is greater than some natural number.
(d) $(\forall x)(\forall y)A(x,y)$		Every natural number is greater than every natural number.
3) (a) $(\forall y)(\forall z)(\exists x)A(x,y,z)$	$A(x,y,z)$: x is between y and z	Given any two rational numbers, there exists a rational number between them.
(b) $(\forall x)(\forall y)(\forall z)(\exists x_1)A(x_1,x,y,z)$	$A(x_1,x,y,z)$: x_1 is the sum of x and y and z	Given any three real numbers, there exists a real number which is their sum.
4) (a) $(\forall x)(A(x)\supset B(x))$	$A(x)$: x is an athlete	All athletes are brash.
(b) $(\forall x)(A(x)\cdot \overline{B(x)})$	$B(x)$: x is brash	All human beings are brash athletes.
(c) $(\exists x)(A(x)\cdot \overline{B(x)})$		There exists an athlete who is not brash.
(d) $(\exists x)(A(x)\supset B(x))$		There exists a human being who, if he is an athlete, then he is not brash.
5) (a) $((\forall x)(A(x)\supset B(x))\cdot(\forall x)(C(x)\supset A(x)))\supset(\forall x)(C(x)\supset B(x))$	$A(x)$: x is a mammal $B(x)$: x is an animal $C(x)$: x is a man	If all mammals are animals and all men are mammals, then all men are animals.
(b) $A(a,b)\supset B(b,a)$	$A(x,y)$: x is father to y $B(x,y)$: x is son to y a: Sophroniscus b: Socrates	If Sophroniscus is father to Socrates, then Socrates is son to Sophroniscus.
(c) $(\forall x)(A(x)\supset B(x))\supset$ $(\forall x)((\exists y)(A(y)\cdot A(x,y))\supset$ $(\exists y)(B(y)\cdot A(x,y)))$	$A(x)$: x is a horse $B(x)$: x is an animal $A(x,y)$: x is the head of y	If all horses are animals, then the head of a horse is a head of an animal.

* The domain of discourse for (1) and (2) is the natural numbers; for (3(a)) the rational numbers; for (3(b)) the real numbers; for (4) human beings; for (5(a)) living things; for (5(b)) ancient Greeks; for (5(c)) living things and parts of living things.

requires this is a moot point. In any case, we could make our logic equivalent (on this issue) to the traditional one by adding the assumptions $(\exists x)A(x)$, $(\exists x)B(x)$, $(\exists x)C(x)$, etc., for each property. Here we shall not make this rather strong set of assumptions, but only the assumption that the domain of discourse is not empty.

Finally, the formulas (5(a)), (5(b)), and (5(c)) represent arguments. Note that the set of objects in, for example, (5(c)) is rather vaguely defined. Some degree of vagueness is allowable in the specification of the universe of discourse, so long as the predicates defined for the logic in question are incapable of defining borderline cases. Otherwise the predicates themselves would be undecidable for some given individual, something we will try to avoid. However, it is not clear that we can always be successful in this regard [cf. § 27]. Further, we require that our predicate logic make these arguments valid. Otherwise it would not serve our purpose of capturing an intuitive notion (for example, that of *valid*) in a formal system, and to that degree it would be inadequate [cf. § 3].

Problem 43. Translate into symbolic notation the **E** and **I** propositions for the predicates of (4).

As a first step toward defining validity for the predicate calculus, we might raise the question whether the technique of a truth table might be extended from the propositional calculus to the predicate calculus. The answer is yes. But the technique, while possible, does not give us a mechanical test of validity as was the case in the propositional calculus. Nevertheless, truth tables will enable us to define what we mean by validity in the predicate calculus.

It must be kept in mind that *qua* logicians we can make no assumptions about either the nature of the predicates or the character of the domain. All we assume is that each substitution instance of any predicate is a proposition which is either true or false. Of course, as logicians we don't claim to know which. However, for the purpose of introducing the general concept of validity, it will be useful to take a concrete example.

Consider the argument about not being in Rhodes. Assume that the domain of discourse consists just of Socrates and Plato, and that we let '$A^1(x)$' stand for 'x is here' and '$B^1(x)$' stand for 'x is in Rhodes'. Consider what the universal quantifier means in such a context. $(\forall x)B^1(x)$ tells us that everything in the domain of discourse is in Rhodes, in other words, that Socrates is in Rhodes and that Plato is in Rhodes. Thus in this case our universally quantified predicate, $(\forall x)B^1(x)$ means $B^1(\text{Socrates}) \cdot B^1(\text{Plato})$. In general, a universally quantified predicate with a domain of discourse of n objects is equivalent to a conjunction of n propositions.

Our existential quantifier admits of a similar translation. In our example, $(\exists x)B^1(x)$ means that either Socrates is in Rhodes or Plato is in Rhodes;

that is, it means B^1(Socrates) $\lor$ B^1(Plato). If the domain of discourse were n objects, we would have a disjunction of n propositions. It should now be noticed that the respective translations of the universal quantifier into a conjunction of propositions and the existential quantifier into a disjunction of propositions allows us to reintroduce the symbolism and techniques of the propositional calculus. Thus we might let 'A' stand for 'Socrates is here', 'B' for 'Socrates is in Rhodes', 'C' for 'Plato is here', 'A_1' for 'Plato is in Rhodes'. Our analysis of the argument may be illustrated as follows:

English language

If someone is here, then he is not in Rhodes.

Someone is here.

Therefore it is not the case that someone is in Rhodes.

Language of the predicate calculus

$$(\forall x)(A^1(x) \supset \overline{B^1(x)})$$

$$(\exists x)A^1(x)$$

$$\therefore \ \overline{(\exists x)B^1(x)}$$

$$(A^1(\text{Socrates}) \supset \overline{B^1(\text{Socrates})}) \cdot (A^1(\text{Plato}) \supset \overline{B^1(\text{Plato})})$$

$$A^1(\text{Socrates}) \lor A^1(\text{Plato})$$

$$\therefore \ \overline{(B^1(\text{Socrates}) \lor B^1(\text{Plato}))}$$

Language of the propositional calculus

$$(A \supset \overline{B}) \cdot (C \supset \overline{A}_1)$$

$$A \lor C$$

$$\therefore \ \overline{(B \lor A_1)}$$

Problem 44. Write out the formulas in the predicate and propositional calculi, respectively, for the case in which the domain of discourse is just one element (say Socrates), and for three elements (say Socrates, Plato, and Protagoras). Then test the validity for both the case of one and two elements by writing out the truth table for the corresponding argument form.

We would be tempted to view the notation of the predicate calculus as a mere convenience which is theoretically dispensable were it not for the case of infinite (both denumerable and nondenumerable) domains of discourse. Here the truth table itself becomes infinite and our reasoning about it must be of a generalized sort. As we shall see in the next chapter, there is no mechanical technique for determining the truth table of an arbitrary formula in the predicate calculus [cf. §21 and §23C]. However, there are some techniques which work for certain infinite sets of formulas.

For example, consider the universally quantified formula '$(\forall x)(P(x) \supset P(x))$' for some arbitrary nonempty domain D. Given that a is some arbitrary element of that domain, '$P(a) \supset P(a)$' is one of the conjuncts in the conjunction of sentences which results when we eliminate the quantifier. Since a is an arbitrary element, every conjunct is true, and thus the formula '$(\forall x)(P(x) \supset P(x))$' has only T's in its truth table. In such a case we shall call it *valid in* that domain. (If a formula contains at least one 'T' in its truth table, it is called *satisfiable in* that domain.)

A formula may be valid in one domain but not in another. For example, the formula corresponding to our argument about not being in Rhodes is valid in the domain of one object, but not valid in the domain of two or three objects; that is, it is 1-valid but not 2-valid or 3-valid. On the other hand, since D was an arbitrary domain, the formula '$(\forall x)(P(x) \supset P(x))$' is valid in every domain, in which case we shall call it *valid*. (If a formula is *satisfiable in* some domain, it is called *satisfiable*.) Hence *logically true* in the predicate calculus means *valid*, just as *logically true* in propositional calculus means a *tautology*. This is why we require a valid formula to be valid in every domain. We wish our logic to apply to a domain of discourse regardless of what the "facts" about that domain are.

Our definition of validity stems, therefore, from the old Leibnizian requirement that logic be true in all "possible worlds," the difference being, of course, that we substitute the more precise terminology of set theory for the relatively vague talk of "possible worlds." It turns out that a formula which is valid in one domain is valid in all smaller domains and that a formula invalid in one domain is invalid in all larger domains. Only the cardinal number of a domain is important with regard to the question of validity in that domain (remember that a cardinal number is a double abstraction, first from the nature of the elements and second from their order [cf. § 11]). Thus logic and set theory are intimately related, and there are some interesting results about domains of discourse. For example, there exist formulas which are valid in every finite domain but in no infinite one. However, be warned that here we are using set-theoretical concepts to define validity; that is, we are using concepts from a more-powerful theory to define concepts of a less-powerful theory. The fallacy of begging the question becomes an issue at this point. However, we shall postpone discussion of it [cf. § 30].

It must be further emphasized that this account is a rough characterization. Note that the concept of an infinite truth table hasn't been defined, but only characterized in a more or less vague way. Indeed it is hard to imagine what a nondenumerable truth table would be like. A similar statement applies to the formulas of infinite size which may appear in the table. An exact explanation of all this is rather long-winded, and it would not significantly advance our purpose. Instead let us consider the following chart, which presents typical valid formulas of the predicate calculus.

Examples of Formulas Valid in the Predicate Calculus

A

1) $(\forall x)A(x) \equiv \overline{(\exists x)\overline{A(x)}}$

2) $\overline{(\forall x)A(x)} \equiv (\exists x)\overline{A(x)}$

3) $(\forall x)\overline{A(x)} \equiv \overline{(\exists x)A(x)}$

4) $\overline{(\forall x)\overline{A(x)}} \equiv (\exists x)A(x)$

B

1) $(\forall x)A(x) \equiv (\forall y)A(y)$

2) $(\exists x)A(x) \equiv (\exists y)A(y)$

3) $(\forall x)(\forall y)A(x, y) \equiv (\forall y)(\forall x)A(x, y)$

4) $(\exists x)(\exists y)A(x, y) \supset (\exists y)(\exists x)A(x, y)$

5) $(\forall x)A(x) \supset (\exists x)A(x)$

6) $(\exists x)(\forall y)A(x, y) \supset (\forall y)(\exists x)A(x, y)$

C

1) $(\forall x)(A(x) \cdot B(x)) \equiv ((\forall x)A(x) \cdot (\forall x)B(x))$

2) $(\exists x)(A(x) \cdot B(x)) \supset ((\exists x)A(x) \cdot (\exists x)B(x))$

3) $((\forall x)A(x) \vee (\forall x)B(x)) \supset (\forall x)(A(x) \vee B(x))$

4) $((\exists x)A(x) \vee (\exists x)B(x)) \equiv (\exists x)(A(x) \vee B(x))$

5) $(\forall x)(A(x) \supset B(x)) \supset ((\forall x)A(x) \supset (\forall x)B(x))$

6) $((\exists x)A(x) \supset (\exists x)B(x)) \supset (\exists x)(A(x) \supset B(x))$

D

1) $(\forall x)((\forall x)A(x) \supset A(x))$

2) $(\exists x)((\exists x)A(x) \supset A(x))$

3) $(\forall x)(A(x) \supset (\exists x)A(x))$

4) $(\exists x)(A(x) \supset (\forall x)A(x))$

$$E^*$$

1) $(\forall x)D \equiv D$

2) $(\exists x)D \equiv D$

3) $(\forall x)(A(x) \cdot D) \equiv ((\forall x)A(x) \cdot D)$

4) $(\exists x)(A(x) \cdot D) \equiv ((\exists x)A(x) \cdot D)$

5) $(\forall x)(A(x) \vee D) \equiv ((\forall x)A(x) \vee D)$

6) $(\exists x)(A(x) \vee D) \equiv ((\exists x)A(x) \vee D)$

7) $(\forall x)(D \supset A(x)) \equiv (D \supset (\forall x)A(x))$

8) $(\exists x)(D \supset A(x)) \equiv (D \supset (\exists x)A(x))$

9) $(\forall x)(A(x) \supset D) \equiv ((\exists x)A(x) \supset D)$

10) $(\exists x)(A(x) \supset D) \equiv ((\forall x)A(x) \supset D)$

Group A in our list of valid formulas indicates the relationship between the existential and universal quantifier. Generalizing from the propositional calculus, we shall say that the antecedent implies the consequent of a valid conditional, and that the formulas flanking a valid biconditional are equivalent. Note that the biconditionals in Group A become quite intuitive when we take specific instances. For example, to say 'not everything is a mammal' is equivalent to saying 'there is something which is not a mammal' (A(2)). In Group B further quantifier relationships are given. Formulas (1) and (2) show that validity is preserved through an alphabetical change of bound variable, while (3) and (4) allow us to vary the order of quantifiers, provided they are all of the same type. That (5) is valid is due to our assumption that the domain of discourse is nonempty.

Problem 45. Find an example of a predicate and a domain in which the converse of (B(6))—that is, the formula

$$(\forall y)(\exists x)A(x, y) \supset (\exists x)(\forall y)A(x, y)$$

—is false.

Group C illustrates some of the relationships of quantifiers with '·', '∨', and '⊃'. In each case the converse of the conditional formulas does not hold. Note that the formulas in Group D are not conditionals (that is, compound sentences), but rather single complex sentences. Whenever a propositional connective occurs within·the scope of a quantifier, the complexity becomes an internal property of the proposition or propositional function.

---

* In these formulas we assume that 'D' is a formula in which 'x' is not free.

For example, $(\forall x)A(x) \supset (\forall x)B(x)$ is a compound proposition, while $(\forall x)(A(x) \supset B(x))$ is a noncompound proposition. They express different things. If our domain of discourse is living things and if '$A(x)$' stands for 'x is human' and '$B(x)$' stands for 'x is a fish', then '$(\forall x)(A(x) \supset B(x))$' means 'all humans are fish', which is a false sentence. On the other hand, since '$(\forall x)A(x)$' is false, it follows that '$(\forall x)A(x) \supset (\forall x)B(x)$' is true. Consequently the formula

$$((\forall x)A(x) \supset (\forall x)B(x)) \supset (\forall x)(A(x) \supset B(x))$$

is not valid. The converse of that formula is valid (see formula C(5)).

In Group E it is assumed that 'D' is a formula in which 'x' is not free. It could either be a formula expressing a predicate (such as '$A(y)$', '$(\exists y)B(y, z)$', etc.) or a proposition (such as 'A', '$(\forall x)(\forall y)(\exists z)P(x, y, z)$', etc.) or a propositional variable or form (such as 'p', 'q', '$p \lor q$', etc.).

We are now ready to state the primitive basis for an axiomatic formulation of the predicate calculus. Our aim again is to arrange the basis so that the above informal theory is an interpretation of the formal system. We shall use the letter '$\mathscr{L}$'—suggesting *Logic*—as a name for the system.

Primitive Basis for $\mathscr{L}$

1) List of symbols:

a) Three logical symbols:

$$\supset \quad ^- \quad \forall$$

b) Two parentheses and a comma:

$$(\quad) \quad ,$$

c) An infinite list of individual constants:

$$a \ b \ c \ a_1 \ b_1 \ c_1 \ a_2 \ b_2 \ c_2 \cdots$$

d) An infinite list of individual variables:

$$x \ y \ z \ x_1 \ y_1 \ z_1 \ x_2 \ y_2 \ z_2 \cdots$$

e) An infinite list of propositional constants:

$$A \ B \ C \ A_1 \ B_1 \ C_1 \ A_2 \ B_2 \ C_2 \cdots$$

f) An infinite list of propositional variables:

$$p \ q \ r \ p_1 \ q_1 \ r_1 \ p_2 \ q_2 \ r_2 \cdots$$

g) An infinite list of singular predicate constants:

$$A^1 \ B^1 \ C^1 \ A^1_1 \ B^1_1 \ C^1_1 \ A^1_2 \ B^1_2 \ C^1_2 \cdots$$

h) An infinite list of binary predicate constants:

$$A^2 \quad B^2 \quad C^2 \quad A_1^2 \quad B_1^2 \quad C_1^2 \quad A_2^2 \quad B_2^2 \quad C_2^2 \quad \ldots$$

.
.
.

2) Formation rules:
   a) A propositional constant or variable standing alone is a formula.
   b) If P^n is a predicate constant, and if $x_1, x_2, \ldots, x_n$ are individual variables or constants or some combination of both, then $P^n(x_1, x_2, \ldots, x_n)$ is a formula.
   c) If P and Q are formulas, then so is $(P \supset Q)$.
   d) If P is a formula, then so is $(\bar{P})$.
   e) If P is a formula and x an individual variable, then $(\forall x)(P)$ is a formula.

3) List of initial formula schemata:
   a) $(P \supset (Q \supset P))$
   b) $((P \supset (Q \supset R)) \supset ((P \supset Q) \supset (P \supset R)))$
   c) $(((\bar{P}) \supset (\bar{Q})) \supset (Q \supset P))$
   d) $((\forall x)(P \supset Q) \supset (P \supset (\forall x)Q))$, where x is not free in P.
   e) $((\forall x)P \supset Q)$, where Q is the result of replacing all free occurrences of x in P by some individual variable or constant x_1. If x_1 is a variable, it must be free in Q at exactly those places at which x is free in P.

4) Transformation rules:
   a) From $(P \supset Q)$ and P, infer Q.
   b) From P, infer $(\forall x)(P)$.

As in the primitive basis for $\mathscr{P}$, variables and constants in the object language are represented in sans serif type and their order in each list is called the *alphabetic order*. The letters 'P', 'Q', and 'R' are variables in the metalanguage whose domain of definition comprises sequences of symbols in the object language. Except in the statement of the formation rules, these sequences will always be taken to be formulas. The letter 'x' (with or without a subscript) is a variable of the metalanguage whose domain of definition is the set of individual constants and individual variables of the object language. As before, there is no need to distinguish the logical symbols, parentheses, and the comma by different type because they never appear alone. Note that the system $\mathscr{L}$ includes the system $\mathscr{P}$. The initial formula schema (3(d)) may be compared with formula $(E(7))$ in the list of formulas valid in the predicate calculus. The restriction on x is necessary, for otherwise it would license invalid inference. For example, we can prove $(P \supset P)$ as we did in $\mathscr{P}$. By (4(b)), called the *rule of generalization*, we get $(\forall x)(P \supset P)$, and thus $(P \supset (\forall x)P)$ by (4(a)) applied to (3(d)) and $(\forall x)(P \supset P)$. We can then get $((\forall x)\bar{P} \supset (\forall x)P)$, which is clearly false for some domains and predicates.

We shall not carry out the derivation here.* Similarly, without the restriction in (3(e)), we can derive

$$(\exists y)P^2(y, y) \qquad \text{from} \qquad (\forall x)(\exists y)P^2(x, y).$$

This would be invalid.

Problem 46.   Find a domain and a predicate which makes an instance of $((\exists x)P \supset (\forall x)P)$ false. Find a domain and a predicate which makes an instance of $(\forall x)(\exists y)P^2(x, y)$ to be true and makes an instance of $(\exists y)P^2(y, y)$ to be false, thus showing that an inference from the former to the latter is invalid.

§18 SET-THEORETIC LOGIC: HIGHER-ORDER PREDICATE CALCULI

Once the notion of quantification is understood, it is natural to introduce predicate variables and to extend its use to them. For this purpose we shall use the letters 'P', 'Q', 'R', 'P_1', 'Q_1', etc., to represent predicate variables. We are now able to express such sentences as 'No property applies to everything' in the following way: '$\overline{(\exists P)(\forall x)P(x)}$'. This may be read, 'it is not the case that there exists a P such that for all x P of x'. Here $P(x)$ is thought of as a propositional function of two variables, each one having its own domain of discourse. The domain for x is the set of individuals which we may specify in any given case. But what is the domain for P? An obvious answer is that it is a set of properties which we may specify in any given case. There are difficulties in this obvious answer. The central one is: How do we know when properties are identical and when they are different?

For example, in the domain of natural numbers, suppose that '$A(x)$' means 'x is less than 3 and even' and '$B(x)$' means 'x is less than 50,000 and an even factor of 110,158'. It happens that these two predicates are true only of the number 2. Does this mean that $A(x)$ and $B(x)$ are identical? The answer that they are different raises the very difficult problem of specifying exactly when properties are identical. Certainly a *necessary* condition that two properties be the same is that they apply to the same sets of objects.

The question of what would constitute a sufficient condition over and above the necessary one is a question to which there is at present no universally accepted answer. The one which commands the greatest following takes the necessary condition as a sufficient one also. That is, two properties are identical if and only if they apply to the same set of objects. This last statement is a version of the *axiom of extensionality*, and it is embodied in most systems of higher-order logic.

---

* But cf. formulas (E(9)) and (A(4)) in our list of formulas valid in the predicate calculus. In view of (A(4)), we shall introduce an abbreviation in the metalanguage; that is, '$(\exists x)P$' will denote the same set of formulas as '$\overline{(\forall x)\overline{P}}$'.

The acceptance of this axiom allows us to treat the notions of *property* and *set* as equivalent from a logical point of view. Thus we shall consider the two following sentences to mean the same thing: 'No property applies to everything' and 'No set contains everything'. Or again, 'Roses are red' means the same as 'Roses are members of the set of all red things'. Or we can introduce the symbolism of set theory into the symbolism of quantification. Suppose we let '$x \in M$' symbolize 'x is a member of M', and let M be a variable whose domain of definition is some set of sets. Then '$\overline{(\exists M)(\forall x)(x \in M)}$' means the same as '$\overline{(\exists P)(\forall x)P(x)}$'. Hence we can now specify the domain of discourse for predicate variables: It consists of sets of individuals. If the number of sets of individuals is finite, we may eliminate quantifiers with predicate variables, in the same way that we eliminated quantifiers with individual variables when the domain of discourse was finite. As before, however, quantification is necessary because of the possibility of infinite domains, that is, because of the possibility of domains of discourse which contain an infinite number of sets of individuals.

A logic which contains quantification over only one domain—the individuals—is called a *first-order* (or *restricted*) predicate calculus. In this book we shall in general continue to leave off the adjectival phrase *first-order* (or *restricted*) when we are referring to the first-order predicate calculus.

A logic which contains quantification with two types of variables—individual variables and predicate variables for individuals—is called a *second-order predicate calculus*. Just as a first-order predicate calculus may contain predicate constants (that is, 'A', 'B', 'C', etc.) for properties of individuals, so a second-order predicate calculus may contain predicate constants for properties of properties of individuals (for symbols, we shall use '^2A', '^2B', '^2C', '$^2A_1$', etc.). An example of the use of such a predicate constant may be seen in the translation of the following sentence: 'Natural rubber has no good property which synthetic rubber lacks'. Let '$A(x)$' stand for 'x is synthetic rubber', '$B(x)$' stand for 'x is natural rubber', and '$^2A(P)$' stand for 'P is a good property'. Then one possible translation might be

$$\overline{(\exists P)(^2A(P) \cdot (\forall x)(\forall y)((A(x) \cdot B(y)) \supset (P(y) \cdot \overline{P(x)})))}.$$

This may be read as 'it is not the case that there exists a P such that P is a good property, and given any x and any y, if x is synthetic rubber and y is natural rubber, then y is a P and x is not a P'. As before, we shall leave off superscripts if no ambiguity will result.

Problem 47. Let our domain for individual variables be the nonnegative integers, and the domain for predicate variables be sets of nonnegative integers. Let '0' stand for 0, '0'' for 1, '0''' for 2, etc. What does the following formula of the second-order predicate calculus represent:

$$(\forall P)((P(0) \cdot (\forall x)(P(x) \supset P(x'))) \supset (\forall x)P(x))?$$

Ordinary discourse admits of statements which may require a third-order predicate calculus for a translation, as, for example, in the following: 'Of all the Christian virtues, none is more misunderstood than hope'. Here we might have four domains of discourse: the first for individuals (say, human beings), the second for properties of individuals (say, those mentioned in some specific dictionary), the third for properties of properties of individuals (say, such ones as follow: virtues, faults, common, rare, etc.) and finally the fourth for properties of properties of properties of individuals (say, such ones as follows: Christian, Moslem, Hindu, Stoic, etc.). In spite of the complexity of even ordinary sentences, our logic soon outruns that complexity as far as orders are concerned. There are fourth-, fifth-, . . . , nth-order logics. And, as mathematicians are wont to do, the process is carried into the transfinite: ω-order logic has received extensive investigation.

From what has been said, it is not surprising to learn that the higher-order predicate calculi are each isomorphic with various fragments of set theory. In fact, it seems likely that the theory as originally proposed by Cantor was meant to embody somewhat the same theoretical structure as is found in ω-order logic. In the latter, of course, there are individuals, properties of individuals, properties of properties of individuals, ... This immediately raises the question of paradoxes. Is it, for example, possible to get an analog of Russell's paradox in higher-order logic? Consider trying to exploit the isomorphism:

Language of sets	*Language of properties*
1. Let normal sets be sets which do not contain themselves, for example, the set of all books.	1. Let impredicable properties be properties which do not apply to themselves, for example, the property of being a book.
2. Let abnormal sets be sets which are not normal, for example, the set of all sets.	2. Let predicable properties be properties which are not impredicable, for example, the property of being a property.
3. Consider the set of all normal sets: Is it normal or abnormal?	3. Consider the property of being impredicable: Is it impredicable or predicable?
4. Suppose it is normal. Then it doesn't contain itself, and thus is a member of the set of all normal sets. This makes it abnormal because it contains itself.	4. Suppose it is impredicable. Then it doesn't apply to itself, and thus is an impredicable property. This makes it predicable because it applies to itself.

5. Suppose it is abnormal. Then it contains itself, but this makes it normal since it is the set of all normal sets.

5. Suppose it is predicable. Then it applies to itself, but this makes it impredicable since it is the property of being impredicable.

It is part of the purpose of the distinction among types of predicates to make impossible the derivation of this kind of paradox. Implicit in the distinction is the requirement that any given predicate of type $n > 1$, say ^nA, applies only to predicates of type $n - 1$, say ^{n-1}A. Then '$^3A(^2A)$' is a formula which represents a predicate, but '$^2A(^3A)$' and '$^3A(^1A)$' are not. The notion of being impredicable cannot be defined. Similarly, the analog of *set of sets* cannot be defined, and so Cantor's paradox cannot be derived. Other conventions—notably the distinction between language and meta-language—give protection against the Liar and Richard paradoxes. But how do we know that set theory does not harbor some other contradiction? This is a question which we shall postpone [cf. §29]. Another doubt which may arise concerns the stringency of the requirement of types. For example, is the word 'Christian' used in two senses in the following sentence: 'A Christian man is a man who embodies Christian virtues'? The division of predicates into types would require that these senses be different.

In spite of this and other artificialities, set theory remains important because it is thought that all or at least most of mathematics can be derived within a suitably rich system of set theory. The stronger claim is somewhat doubtful; for example, it is far from clear that the theory of categories can be derived. On the other hand, the weaker claim is plausible. In particular, the notions of natural number, integer, rational number, real number, and complex number can be defined. Further, so can operations such as addition, multiplication, etc. For this reason we shall consider higher-order predicate calculi to be "mathematics" and not "logic." As we shall see in the next chapter, there is a more profound reason for drawing the distinction between logic and mathematics at this point [cf. §24]. Henceforth the terms *logic* or *general logic* without qualification will mean the *first-order predicate calculus*. Higher-order predicate calculi will often be referred to as indis-criminately as *set theory*.

Second-order logic enables us to define the binary predicate of identity between two individuals. Following common practice, the equality sign ('$=$') will be used for this predicate, although in our symbolism it should be a capital letter from the beginning of the alphabet. The definition of identity (which stems primarily from Leibniz, but also is found in Aristotle [cf. Kneale 1962 42]) can be put in symbolic notation in the following way:

$$(\forall x)(\forall y)((x = y) \equiv (\forall P)(P(x) \equiv P(y))).$$

Since a biconditional is a conjunction of two conditionals, this formula may be broken up into two parts in order to understand it better. The first part is

$$(x = y) \supset (\forall P)(P(x) \equiv P(y)).$$

This states that if x and y are identical, then they share all properties. This is known as the principle of the *indiscernibility of identicals* [cf. Kneale 1962 604 and Quine 1961 139]. The second part is

$$(\forall P)(P(x) \equiv P(y)) \supset (x = y).$$

This states that if x and y share all properties, then they are identical. This is known as the principle of the *identity of indiscernibles*. The combination of these two principles gives us the definition of identity which we need. Note that we needed a second-order logic to define identity between two individuals. A similar definition in third-order logic would give us a definition of identity of two properties of individuals; that is,

$$(\forall P)(\forall Q)((P = Q) \equiv (\forall^2 P)(^2 P(P) \equiv {}^2 P(Q))).$$

Such definitions could be extended into higher-order logics. For each order they respectively serve as an axiom of extensionality.

Problem 48. Consider the following three sentences:

1) Bacon is Shakespeare.

2) Bacon is corrupt.

3) Man is corrupt.

The word 'is' is used in three different senses above. See if you can identify the senses and determine how each is represented in our symbolism.

We shall now present some typically valid formulas of higher-order predicate calculi:

Examples of Formulas Valid in Higher-Order Predicate Calculi

A

1) $(x = y) \supset (\forall P)(P(x) \equiv P(y))$

2) $(\forall P)(P(x) \equiv P(y)) \supset (x = y)$

3) $(\forall P)((P(0) \cdot (\forall x)(P(x) \supset P(x'))) \supset (\forall x)P(x))$

B

1) $(\forall P)(\forall x)P(x) \equiv (\forall x)(\forall P)P(x)$

2) $(\exists P)(\exists x)P(x) \equiv (\exists x)(\exists P)P(x)$

3) $(\exists P)(\forall x)P(x) \supset (\forall x)(\exists P)P(x)$

4) $(\exists x)(\forall P)P(x) \supset (\forall P)(\exists x)P(x)$

C

1) $$x = x$$

2) $$(x = y) \equiv (y = x)$$

3) $$((x = y) \cdot (y = z)) \supset (x = z)$$

4) $$(\mathbf{x} = \mathbf{y}) \supset (\mathbf{P(x)} \supset \mathbf{P(y)})$$

D

1) $$(\mathbf{P(0)} \cdot (\forall \mathbf{x})(\mathbf{P(x)} \supset \mathbf{P(x')})) \supset (\forall \mathbf{x})\mathbf{P(x)}$$

2) $$(x' = y') \supset (x = y)$$

3) $$\overline{(x' = 0)}$$

E

1) $$(x + 0) = x$$

2) $$(x + y') = (x + y)'$$

3) $$(x \cdot 0) = 0$$

4) $$(x \cdot y') = ((x \cdot y) + x)$$

The formulas $(A(1))$ and $(A(2))$ express the two principles that make up our definition of identity. Note that we have left off the universal quantifiers for 'x' and 'y'. They are inessential. A formula with free variables is always interpreted in the same way as its universal closure. The *universal closure* of some formula **P** means the formula

$$(\forall \mathbf{x_n})(\forall \mathbf{x_{n-1}}) \ldots (\forall \mathbf{x_1})\mathbf{P},$$

where $x_1, \ldots, x_n$ are the free variables of **P** in alphabetic order. Similar definitions would apply for predicate variables. Part of the purpose of our rule of generalization was to ensure this interpretation of free variables. For clarity, universal quantifiers for individuals have often been omitted elsewhere.

Group (B) indicates some of the relationships which hold between quantifiers. These should be compared with those given for general logic [cf. (B) in §17].

The first three formulas in group (C) are the basic principles of identity which can be derived from $(A(1))$ and $(A(2))$. Note that they contain no quantification over predicate variables. Formula $(C(4))$ is a schema and represents infinitely many formulas of some object language.

Let the domain of discourse be as in Problem 47, let '0' stand for 0, and interpret '′' to mean 'successor'. Even so, formula $(D(1))$ is not quite the schema for mathematical induction; the full statement is given in $(A(3))$.

Mathematical induction cannot in fact be stated in the symbolism of general logic; (D(1)) is the closest one can come to it. This fact, as we shall see, is quite important, but the explanation of it will be postponed [cf. §24]. Formula (D(2)) states that no two distinct numbers have the same successor. Formula (D(3)) states that 0 is not the successor to any number.

Group (E) gives the formulas that characterize addition and multiplication. The concepts of equality, successor, plus, and times are all capable of being defined within set theory. Hence this increases the likelihood that most of mathematics can be derived within it.

It is, however, precisely set theory's power which provides part of the motivation for studying systems which are stronger than the first-order predicate calculus but weaker than set theory. In particular, the following system, which we shall denote by '$\mathscr{A}$', is a widely studied one. Take the system $\mathscr{L}$ and change it in the following ways [cf. §17]:

1. Replace the stated infinite list of individual constants by one individual constant, namely, '0'.

2. Eliminate the lists of propositional constants and variables as well as all predicate constants with the exception of one binary predicate constant '=' (equals).

3. Add three function symbols: '′' (successor), '+' (plus), '·' (times).

4. Eliminate reference to propositional constants or variables in the formation rules.

5. Add the following formation rule: if **u** and **v** are terms, then (**u**) = (**v**) is a formula. (A *term* is either an individual constant or a variable standing alone, or is one of the forms (**u**)′, (**u**) + (**v**), (**u**) · (**v**), where **u** and **v** are terms.)

6. Finally, add the formulas in Groups (C), (D), and (E) as initial formulas. Of course, we would have to eliminate abbreviations. For example, formula (E(1)) would become '((x) + (0)) = (x)'.

'$\mathscr{A}$' is used for the system, since under one interpretation it is ordinary elementary arithmetic. In §23 we shall study the metatheory of $\mathscr{A}$. A system that is still weaker—but stronger than $\mathscr{L}$—would be formed by adding the formulas in Group (C) to the initial formulas of $\mathscr{L}$ and making the other obvious necessary changes. Such a system is called the *first-order predicate calculus with equality* and is denoted by '$\mathscr{L}^=$'. Elementary arithmetic and general logic with equality are both systems interesting in their own right, but part of the motivation for studying them is fear that stronger systems—such as set theory—may be inconsistent.

The statement of the primitive basis for higher-order predicate calculi is a long and involved one, and we shall avoid it. Instead we shall informally present the axioms of a widely used set theory, that of Fraenkel.

Axioms for Set Theory

1. *Axiom of Extensionality.* A set is determined by the totality of its elements. In other words, two sets are equal ($=$) if, and only if, they contain the same elements [Fraenkel 1953 21].

2. *Axiom of Subsets.* Given a set S and a property π meaningful for the elements of S, there exists the set containing those elements of S, and only those, which possess the property π [Fraenkel 1953 22].

3. *Axiom of Pairing.* For any two different sets S' and S'' there exists a set S that contains just S' and S'' [cf. Fraenkel and Bar-Hillel 1958 34].

4. *Axiom of Sum-Set.* Given a set S whose elements are again sets, there exists the set containing just the elements of the elements of S. It is called the *sum set* of S [cf. Fraenkel 1953 28–29].

5. *Axiom of Infinity.* There exists at least one infinite set: the set of all natural numbers [Fraenkel 1953 42].

6. *Axiom of Power Set.* To any given set S there exists the set whose elements are all the subsets of S. It is called the *power set* of S [cf. Fraenkel 1953 97].

7. *Axiom of Choice.* If S is a nonempty set of sets, and if there are no two distinct elements of S which themselves have an element in common, then there exists a set S' which consists solely of one and only one element from each set in S. [cf. Kleene 1967 190].

The intention is that these principles be sufficient for the derivation of mathematics. Whether or not this is fulfilled is a controversial point. It must be emphasized that there are a fair number of different systems intended to serve the same function. These systems are not all equivalent, but the differences between them—however vital they are for some questions—are not important for this introduction.

CHAPTER 4

THE METATHEORY OF
MATHEMATICAL LOGIC

§19 INTRODUCTION

There are many different questions one can raise about the systems described in the preceding chapter. These questions can be of different sorts. They can be philosophical: For example, one might ask what is the nature of the objects that are assumed to exist in the various domains of discourse. They can be esthetic: For example, one might ask whether this or that formulation of the propositional calculus is more elegant. They can be pragmatic: For example, one might ask whether a given system is consistent.

All these kinds of questions (and this list is far from complete) are metalogical questions in the sense that they concern logical systems. We shall postpone any consideration of philosophical, esthetic, or pragmatic questions until Chapter 5. Here we shall be concerned with logical questions of metalogic, in particular, with questions concerning consistency, independence, categoricalness, completeness, and decision procedure. We shall discuss each in turn.

The general idea of consistency is that a statement (or system of statements) is *consistent* if it is possible for it to be true. This is sometimes called a *semantical definition of consistency*, since it makes reference to truth. *Semantics*, as we shall understand the term, is a study of interpretations of languages, including formal languages (cf. the use of the term in §12). One way to prove the consistency of a statement, or system of statements, is to show that it is in fact true. As we have seen, this method was used in the nineteenth century, but it does not work for many interesting systems. The idea arose that perhaps this difficulty could be avoided if the definition of consistency was changed from a semantical form to a syntactical form. *Syntactics*, as we shall understand the term, is a study of the relations among symbols in a given language, including formal languages (cf. the use of the term in §12). We can now present a *syntactical definition of consistency*: A formal system is *consistent* if there is no formula such that both it and its negation are theorems. Of course, the negation might be represented in different ways in different systems. For example, in one system the negation of the formula 'A' might be '$\overline{A}$', in another '$A \supset \mathfrak{f}$' (cf. §16). As we shall see, a new syntactical method of proving consistency—apparently first proposed by Hilbert—will enable us to get around some of the limitations of the nineteenth century methods.

As in the case of defining consistency, explaining the idea of the independence of an axiom may make a reference to truth. That is, an axiom is considered *independent* if it is possible for it to be either true or false, under the assumption that all the other axioms are true. Since we are dealing with formal systems, we shall again substitute a definition that makes no reference to the semantical property of truth: An initial formula of a formal system will be called *independent* if it is impossible to prove either it or its negation from the other initial formulas. As can easily be seen, independence, although a logical feature of an axiom, is related to esthetic questions. An elegant logical or mathematical system should not (at least according to one criterion of elegance) make use of logically unnecessary axioms. When all the initial formulas of a system are independent, we shall call the system itself *independent*; otherwise, it is called *redundant*.

Our third topic is *categoricalness*. This notion is not as simple to explain as consistency or independence. We shall do so by considering two games. In the first game—which we may call ACE—there are two players. The game begins in a situation in which there are 9 playing cards lying face up on a table between the two players. The cards are the ace, 2, . . . , 9 of some suit. The players take turns picking a card, each player trying to be the first to obtain three cards whose sum is 15. In the second game—called HOT— there are also two players. Between them on the table are 9 plain white cards, with one word printed on each. No word is repeated. The words are: 'WASP', 'BRIM', 'WOES', 'HEAR', 'FROM', 'TIED', 'SHIP', 'TANK', and 'HOT'. The players take turns picking a card, each player trying to be the first to hold three cards that have the same letter. These games are related by what mathematicians call an *isomorphism*. For an isomorphism to hold, there must be a one-to-one correspondence between elements, and the elements must have the same structural relationships. To see an example, consider the following correspondence: $1 \leftrightarrow$ BRIM, $2 \leftrightarrow$ HOT, $3 \leftrightarrow$ WASP, $4 \leftrightarrow$ WOES, $5 \leftrightarrow$ HEAR, $6 \leftrightarrow$ TIED, $7 \leftrightarrow$ TANK, $8 \leftrightarrow$ SHIP, $9 \leftrightarrow$ FORM. Now note that there are 8 letters that occur in three words, as indicated in the chart at the top of the facing page, in which the corresponding numbers are put in parentheses.

A quick check shows that the numbers in each column add up to 15. A further check shows that there aren't any other triplets whose sum is 15. Thus the two games are identical in structure. One can see this very clearly by considering the following possibility. Suppose that the words of HOT were not written on plain white cards, but on playing cards, according to the above correspondence; that is, 'BRIM' on the ace, etc. Then there would be no way for an observer to tell which game two players were playing. Or to put the same point even more sharply, if neither player spoke (or otherwise gave his game away), then it would be possible for the players to play in-

R	I	H	O
BR<u>I</u>M(1)	BR<u>I</u>M(1)	<u>H</u>OT(2)	<u>H</u>OT(2)
H<u>E</u>AR(5)	TI<u>E</u>D(6)	H<u>E</u>AR(5)	W<u>O</u>ES(4)
FO<u>R</u>M(9)	SH<u>I</u>P(8)	S<u>H</u>IP(8)	F<u>O</u>RM(9)

T	A	S	E
HO<u>T</u>(2)	W<u>A</u>SP(3)	WA<u>S</u>P(3)	WO<u>E</u>S(4)
<u>T</u>IED(6)	H<u>E</u>AR(5)	WO<u>E</u>S(4)	H<u>E</u>AR(5)
<u>T</u>ANK(7)	T<u>A</u>NK(7)	<u>S</u>HIP(8)	TI<u>E</u>D(6)

definitely with one thinking he is playing ACE and the other thinking he is playing HOT. From a logical point of view, the games are the same; that is, there is no difference between them except in the use of symbols or notation.

Problem 49. Can you think of another common game (not a card game) which is isomorphic to these two games? If you can, see if you can prove that it is isomorphic by use of a diagram. If you cannot, check the answer given in the footnote on the next page and then try to prove the isomorphism.

We shall recall that a *model* of a formal system is a set of objects defined outside that system such that it is possible to find an interpretation of the system in that set of objects. Two models will be called *isomorphic* if there is a one-to-one correspondence between the elements of the models that pre-serves truth for all propositions in the system. What is meant by the latter phrase is this: If a statement of the formal system is true (or false) when interpreted in one model, it must also be true (or false) when interpreted in the other model under the systematic change indicated by the one-to-one correspondence. For example, as we have seen, a formal system may have plane Euclidean geometry be one interpretation, and a fragment of algebra be another (cf. §15). The corresponding models are then isomorphic, since any true (or false) statement of the former changes systematically into a true (or false) statement of the latter.

We are now ready to define *categorical*. A formal system is *categorical* if all its models are isomorphic. To understand the significance of a system being categorical, recall that the purpose of setting up a formal system is to define implicitly the undefined terms such that some informal theory is axiomatized (cf. §15). This informal theory is then called the *intended inter-pretation*. The proof that a system is categorical is an indication of success, for this proof would show that all interpretations of the system were like the intended one except in matters of terminology and notation, which from a strictly logical point of view are irrelevant. Later we shall see examples of categorical systems (cf. §20 and §21).

For the notion of completeness, we shall need to define two different senses of the word: *expressive completeness* and *deductive completeness*. The general idea of expressive completeness is that a formal system be able to "express" all the statements, whether true or false. I have put the word 'express' in double quotes because formulas do not express statements at all (since they are uninterpreted). To define the concept, we must make reference to an interpretation: A formal system is called *expressively complete* if under the intended interpretation it is possible to express all the sentences (both true and false) of the informal theory. Expressive completeness is thus related to primitive symbols and formation rules, and is a semantical concept.

Deductive completeness is related to axioms and rules of inference. A formal system is *deductively complete* if under the intended interpretation there is no truth which is not also a theorem. As in the case of consistency and independence, we would like to substitute a syntactical definition for the semantical one. However, it turns out that the syntactical definitions of *deductive completeness* vary much more from system to system than do the definitions of *consistency* or *independence*. Hence it is not possible to give a general definition. However, the unifying idea, for all the definitions we shall give, is that in a deductively complete system the set of theorems cannot be enlarged without the undesirable consequence that the intended interpretation would no longer be an interpretation at all. The problem is to find a syntactical counterpart for this core idea in each system considered.

Related to deductive completeness is the concept of correctness. A system is *correct* if under the intended interpretation every theorem is a truth. It is essential that a system be correct to be useful; it is only desirable that it be deductively complete. Like deductive completeness, syntactical versions of correctness vary from system to system and a common definition cannot be given. However, the semantical definition, which guides the syntactical one, is clear enough.

Finally we come to the notion of a decision procedure. Any effective finite method by which it can be determined whether or not an arbitrary formula of a formal system is a theorem is called a *decision procedure*. By *effective finite method* is meant the same thing that used to be called an *algorithm*. As an example we might consider the problem of determining whether the first of two arbitrarily given natural numbers evenly divides the second. Note that this is an infinite set of problems. Nevertheless, anyone familiar with long division has an algorithm by which (barring death, loss of interest, etc.) he can solve any problem given to him in a finite amount of time without any new insights (that is, no insights beyond what he learned in the long-division procedure).

Footnote to Problem 49: Ticktacktoe.

The existence of a decision procedure decreases the theoretical interest we might have in an area. Although only a finite number of long divisions have actually been carried out (and an infinite number are untried), no one has a theoretical interest in the subject any more. The problems left over are practical ones, such as the question of inventing better short cuts. The existence of a decision procedure for a formal system in a sense resolves the theoretical problems for those areas which are models for the system. Any system for which there is a decision procedure is called *decidable*.

This completes our list of metalogical or metamathematical properties of formal systems. However, let it not be thought that this represents a complete list of properties logicians are interested in. All these concepts— consistency, independence, categoricalness, completeness, and decision procedure—represent general terms which include a number of kinds. For example, we have defined what is called *simple consistency*. But, as we shall see, there are other kinds of consistency (cf. §23). Nor are these types of questions the only metalogical or metamathematical sort that can be asked. Note that, except for expressive completeness, the concepts are all about the axioms or the consequences of the axioms.

However, we might ask some similar questions about other aspects of the system. For instance, we might ask if the primitive terms are *independent*, that is, whether or not any one of them can be defined in terms of the others. There is even an analog of the notion of completeness which can be applied to primitive terms. Or again, the notion of independence can be applied to the rules of inference. If a system has two rules of inference but the set of theorems is unchanged by the removal of one of them, there is a fault (at least from one esthetic point of view) in the system. Some other desirable properties could even be added to the ones here presented.

Now let us imagine that we have a formal system which is consistent, categorical, and expressively and deductively complete. Let us further assume that there is a decision procedure for it and that the primitive terms, axioms, and rules of inference are each independent. For good measure, let us assume that this formal system has the desirable but unspecified properties mentioned in the preceding paragraph. Let us suppose that the metalogical methods which are used to prove these things are of such simplicity that only the self-defeating philosophical sceptic has any doubts. Finally, let us suppose that we try to devise such a formal system (or series of formal systems) to encompass all of mathematics. A proposal similar to this was in fact made by Hilbert early in this century. It became known as *Hilbert's program*. Note that if such a program were to be carried out it would rid mathematics once and for all of the paradoxes and thus show that it has a firm foundation. It would certainly go down as one of the monumental achievements of human reason. However, it would have a further, less-desirable effect: It could then be truly said that mankind had conquered that area of knowledge and that the inquiring mind should then turn to other areas of inquiry. In other words, it would make mathematics theoretically uninteresting.

What was the outcome of this program? The answer is that it failed. However, the failure produced results which in a sense are more exciting than success would have been. In particular, it was shown that it is not possible to complete Hilbert's program by any means which we have at our disposal. This statement is vague and will be clarified later, but the upshot is clear: The results of mathematical logic leave mathematics as fascinating as ever.

Now we shall investigate some metalogical properties—consistency, independence, categoricalness, completeness, and decidability—of the logical and mathematical systems of the preceding chapters. For some of the stated results we shall give a sketch of the proof. These sketches are meant to serve two purposes: (1) To give some idea of how a particular result was achieved to the reader who doesn't want to take time to examine a complete proof. (2) To serve as an aid in understanding the complete proof to the reader who does wish to take the time.

§20 THE METATHEORY OF THE PROPOSITIONAL CALCULUS

The consistency of the propositional calculus is established by making use of the fact that $(P \supset (\bar{P} \supset Q))$ are theorems (although we shall not prove this fact here). If P could be proved and then $\bar{P}$, it would follow, by two applications of *modus ponens* [see (4(a)) of §16], that any formula is a theorem. Thus, if there is at least one formula which is not a theorem, then $\mathscr{P}$ is consistent. If we could show that all theorems have a certain property which some formula lacked, consistency would be established. We could prove that all theorems have a certain property by showing that each initial formula possesses it, and that the transformation rule preserves it; that is, if the premisses of the rule have the property, the consequence does also. A property that is preserved by the transformation rules of a given system is called a *hereditary property* for that system.

All this may perhaps be made clear by an analogy: Assume (as in a now-old-fashioned theology) that Adam and Eve are literally the first two humans (initial formulas) and that a property which they share—original sin (property of the initial formulas)—is hereditary and that thus all their descendants (theorems) share it. The existence of a human without that property—Jesus (a formula)—saves humanity (the system) from damnation (inconsistency). This is so because we know that if all humans (formulas) are descended (inferred) from Adam and Eve (initial formulas), then mankind (the system) is doomed (inconsistent).

The property we shall use is that of being a tautology. We have to show three things:

1) Each initial formula is a tautology.

2) The property of being a tautology is hereditary.

3) There exists a formula which is not a tautology.

The job of checking (1) is a simple mechanical one, as we have seen. Finding a formula that is not a tautology in order to satisfy (3) is also easy. For example, 'p' or '(p ⊃ q)' are non-tautologous formulas. The real work of the proof concerns (2). It may be restated in the following way: Every theorem is a tautology. A *correct* formal system is one in which only truths are provable. Since in the propositional calculus *truths* means *logical truths*, which means *tautologies*, we may again restate (2) to say: The propositional calculus is correct. Note that we have substituted a syntactical counterpart (tautology) for a semantical term (truth).

To prove that $\mathscr{P}$ is correct, we shall show that if the premisses of *modus ponens* are tautologies, the conclusion is also. This is easily done by a *reductio ad impossibile* argument. Suppose that the premisses are tautologies— that is, **P** and **(P ⊃ Q)**—and that the conclusion—**Q**—is not. Then the conclusion must have an 'F' in its truth table. But if **P** has only 'T's in its truth table, then the table for **(P ⊃ Q)** will contain an 'F', in the line or lines in which **Q** has an 'F'.

Turning now to the proofs of independence, we see that our proof of consistency has done half the work. For if an initial formula is independent, neither it nor its negation is provable as a theorem from the other initial formulas, and if the system is consistent, the negation of the initial formula will not be provable. Hence it remains to prove that the initial formula itself is not derivable from the other initial formulas. The proof here is exactly the same as for consistency, except that we redefine *tautology* (in a new but related sense, to be indicated). To prove the independence of any initial formula, we have to show that:

1′) Every other initial formula is a tautology (in the new sense).

2′) The property of being a tautology is hereditary.

3′) The initial formula in question is not a tautology.

What is this "new sense"? In order to answer this question we must remember that, just as there are different systems of geometry, so there are different systems of logic. One of these systems is called a *three-valued logic*, in which, instead of just truth and falsehood for the values of our propositions, we have truth, falsehood, and undetermined. Such a system was probably first considered in order to take care of the philosophical problem of future contingents.

This problem goes at least as far back as Aristotle. One aspect of the problem is that it seems that the acceptance of a two-valued logic commits us to fatalism. To take Aristotle's example, if every proposition is either true or false, it must be true now that there will be a sea battle tomorrow or it must be false now that there will be a sea battle tomorrow. Consequently, all our piety and wit and tears will not be able now to cancel the truths that must be. In short, everything is determined. However, some philosophers have thought that if we reject a two-valued logic for a three-valued logic, then this fatalistic argument is defeated. In such a logic the proposition that there will be a sea battle tomorrow is undetermined—it is neither true nor false.

The controversy concerning future contingents is a tangled one, and fortunately we do not have to decide it. The main point is that it is quite possible to extend the 2-valued propositional calculus to a 3-valued, 4-valued, ..., n-valued, or even an infinite-valued calculus. Such systems may have applications. For example, it has been proposed that a 3-valued logic is necessary in order to adequately take care of the complexities of quantum mechanics. Or, again, it has been suggested that an infinite-valued calculus is necessary to properly interpret probability theory. We shall not be concerned with such interpretations here [for further discussion see Rescher 1969 and the references therein]. We shall be concerned only that a 3-valued propositional calculus is possible, not that it has an interpretation which might be useful for some purpose or other.

Taking three truth values—true, false, and undetermined—we could define our truth functions as in Table 12.

Table 12

p	$\bar{p}$		p	q	$p \supset q$
T	T		T	T	T
U	U		T	U	F
F	U		T	F	F
			U	T	F
			U	U	F
			U	F	T
			F	T	T
			F	U	T
			F	F	T

By these definitions of '¯' and '⊃', the first initial formula schema is not a tautology. For we would have the results shown in Table 13.

However, the second and third initial formula schemata become tautologies (that is, they contain only 'T's in their 3-valued truth tables) and this property is hereditary. Thus independence is established. Elaborations of

Table 13

P	Q	$(P \supset (Q \supset P))$
T	T	T
T	U	F
T	F	T
U	T	T
U	U	T
U	F	F
F	T	T
F	U	T
F	F	T

this technique will establish independence for the other axioms. It should be emphasized that the definitions of the truth functions given by truth tables to establish independence do not in general have intuitive plausibility. All that is necessary is that the definitions satisfy the conditions (1′), (2′), and (3′).

Problem 50. To establish the independence of some initial formulas, it is not necessary to use a 3-valued system, but only to give a different 2-valued interpretation to the truth-functional connectives. Show, in particular, that it is possible to establish the independence of the third axiom in this way.

We now turn to completeness. A propositional calculus is *expressively complete* if all truth functions can be defined in it. Is this true for $\mathscr{P}$? The answer is yes for binary truth functions, as can be established by the following definitions:

$$f_1^2(P, Q) \quad \text{is defined as} \quad (\bar{P} \supset \bar{P})$$

$$f_2^2(P, Q) \quad \text{is defined as} \quad (\bar{P} \supset Q)$$

$$f_3^2(P, Q) \quad \text{is defined as} \quad (\bar{P} \supset \bar{Q})$$

$$f_4^2(P, Q) \quad \text{is defined as} \quad P$$

$$f_5^2(P, Q) \quad \text{is defined as} \quad (P \supset Q)$$

$$f_6^2(P, Q) \quad \text{is defined as} \quad Q$$

$$f_7^2(P, Q) \quad \text{is defined as} \quad ((P \supset Q) \supset \overline{(Q \supset P)})$$

$$f_8^2(P, Q) \quad \text{is defined as} \quad \overline{(P \supset Q)}$$

$$f_9^2(P, Q) \quad \text{is defined as} \quad (P \supset \bar{Q})$$

$$f_{10}^2(P, Q) \quad \text{is defined as} \quad ((P \supset Q) \supset \overline{(Q \supset \bar{P})})$$

$f_{11}^2(\mathbf{P}, \mathbf{Q})$	is defined as	$\bar{\mathbf{Q}}$
$f_{12}^2(\mathbf{P}, \mathbf{Q})$	is defined as	$\overline{(\mathbf{P} \supset \mathbf{Q})}$
$f_{13}^2(\mathbf{P}, \mathbf{Q})$	is defined as	$\bar{\mathbf{P}}$
$f_{14}^2(\mathbf{P}, \mathbf{Q})$	is defined as	$\overline{(\bar{\mathbf{P}} \supset \bar{\mathbf{Q}})}$
$f_{15}^2(\mathbf{P}, \mathbf{Q})$	is defined as	$\overline{(\bar{\mathbf{P}} \supset \mathbf{Q})}$
$f_{16}^2(\mathbf{P}, \mathbf{Q})$	is defined as	$\overline{(\mathbf{P} \supset \mathbf{P})}$

Note that the expression on the right gives the required truth table for the expression on the left. A generalization of this technique will establish expressive completeness for an arbitrary truth function of an arbitrary number of variables. In fact, there are 7 pairs of truth functions, each pair of which is a sufficient basis to form an expressively complete propositional calculus. They are:

1) $^-$ and $\cdot$
2) $^-$ and $\vee$
3) $^-$ and $\supset$
4) $^-$ and $\not\supset$
5) $\supset$ and $\not\equiv$
6) $\equiv$ and $\not\supset$
7) $\supset$ and $\not\supset$

If we include the propositional constants t or f there are two more:

8) $\supset$ and f
9) $\not\supset$ and t

It is also interesting to note that there are just two binary truth functions which alone are sufficient:

$$1) \quad | \qquad 2) \quad \psi$$

For example, in reference to the system based on '$|$', '$(\mathbf{P} \supset \mathbf{Q})$' could be defined as '$(\mathbf{P}|(\mathbf{Q}|\mathbf{Q}))$'.

A propositional calculus is *deductively complete* if all tautologies (that is, logical truths in the system) are theorems. The central idea in the proof of deductive completeness is the translation of a formula into an equivalent one in disjunctive normal form.

A proposition or propositional form is said to be in *disjunctive normal form* if:

1) it contains no propositional connectives except $\cdot$, $\vee$, $^-$,

2) $^-$ applies only to propositional variables and/or constants,

3) $\cdot$ does not conjoin any disjunctions.

For example, the formula

$$(pqr) \vee (pq\bar{r}) \vee (p\bar{q}r) \vee (p\bar{q}\bar{r}) \vee (\bar{p}qr) \vee (\bar{p}q\bar{r}) \vee (\bar{p}\bar{q}r) \vee (\bar{p}\bar{q}\bar{r})$$

is in disjunctive normal form (some parentheses have been omitted for clarity); whereas '$(p \supset p)$' is not (since it contains '$\supset$'), '$\overline{(p \vee q)}$' is not (since '$^-$' applies to a disjunction), and '$(p \vee q)\cdot(q \vee p)$' is not (since '$\cdot$' here conjoins a pair of disjunctions). Note that writing a truth table for the above formula with eight disjuncts is easy, since each disjunct is true in exactly one line and no other disjunct is true in that line. Hence it can easily be seen that any tautology has an equivalent one in disjunctive normal form containing the same number of variables as the original tautology and the same number of disjuncts as there are lines in the truth table of the original tautology.

Let 'A' stand for some arbitrary tautology of $\mathscr{P}$. It can be shown (in an elementary but tedious proof) that there exists an equivalent formula in disjunctive normal form which is provable if and only if A is provable. Let 'B' be the name of the equivalent formula. Let the constants or variables in B be represented by $p_1, p_2, \ldots, p_n$. Then it can be shown that B is equivalent to $(p_1 \vee \bar{p}_1)\cdot C$, where C is a formula which does not contain p_1 or $\bar{p}_1$. To get this result, we first factor out p_1 and $\bar{p}_1$ from B by our distribution laws to get $p_1 C \vee \bar{p}_1 C$.* Then by another application of the distribution laws, we get $(p_1 \vee \bar{p}_1)\cdot C$.

Problem 51. If, in a similar way, we applied our distribution laws to the formula with eight disjuncts named in the paragraph before the last, what would we successively get?

We are now ready to apply mathematical induction. Suppose that A contains just one variable. Then its disjunctive normal form is $p_1 \vee \bar{p}_1$, and this is provable (although again the proof is somewhat tedious). Now suppose that all tautologies in disjunctive normal form are provable if they have k variables. Consider A as containing $k + 1$ variables. We can prove it if we can prove that $(p_{k+1} \vee \bar{p}_{k+1})\cdot C$, where C contains exactly k variables. But now C is provable (by our assumption), $p_{k+1} \vee \bar{p}_{k+1}$ is provable (by

---

* We assume, here and henceforth, that all our examples of valid formulas and argument forms in §16 hold for $\mathscr{P}$. Of course, in a full treatment, this would have to be proved.

the same argument whereby $\mathbf{p_1} \vee \bar{\mathbf{p}}_1$ was proved), so $(\mathbf{p_{k+1}} \vee \bar{\mathbf{p}}_{k+1}) \cdot \mathbf{C}$ is provable by conjunction. We thus have the deductive completeness of the propositional calculus.

It is easy to show that $\mathscr{P}$ is not categorical. For to be categorical, all models of a system must be isomorphic. As we have seen, it is quite possible for there to be a 2-valued, 3-valued,..., n-valued propositional calculus. Even within the 2-valued calculus all models are not isomorphic, because of the existence of contingent propositional forms. By definition, these can be true under one assignment of truth values and false under another. Any formula which is not a tautology or the negation of a tautology (since this formula will be F in all lines of the truth table) would serve as an example: 'p' or '(p ⊃ q)' or '($\bar{\text{A}}$)', etc.

It may be thought that the failure of categoricalness is a failure of ingenuity. However, this is not the case. The propositional calculus can be made categorical by leaving out all propositional variables and constants and replacing them by just two constants: t and f. We did not do this because we intended to embody the propositional calculus in the stronger systems (which, like $\mathscr{L}$, take into account the structure of propositions). For example, we wish to be able to express

$$(2 + 2 = 5) \vee (x^{n+2} + y^{n+2} \neq z^{n+2})$$

when we don't know whether both or just one of the disjuncts are false (cf. §25). In the propositional calculus this would be expressed as $(\mathbf{A} \vee \mathbf{B})$, and thus would not be provable, since its truth table contains an 'F'. In the categorical version just mentioned, the formula would be represented by a 't' or 'f' according as

$$x^{n+2} + y^{n+2} \neq z^{n+2}$$

is or is not true. But we have no way of knowing whether such a symbolization is correct. Instead of allowing uncertainty in the application of our semantical rules, we enrich our symbolism to allow for the expression of contingent statements. Of course, we hope that the larger theory (in which we embody the propositional calculus) will prove to be categorical so that every proposition that was contingent on the basis of the propositional calculus alone will be resolved.

Finally, it turns out that the propositional calculus is decidable. That is, there is a decision procedure by which we can decide, in a finite number of steps, whether any arbitrary formula is a theorem. Further, the procedure will enable us to actually produce a proof of the theorem. Our truth-table technique provides us with a method by which we can decide validity (that is, tautologousness) in a finite number of steps. Because a formula is valid if and only if it is provable (correctness and deductive completeness), the same procedure works for provability. If we wish actually to find the proof, the unabridged proof of completeness gives the detailed instructions for proving any formula. The proofs so generated are often intolerably long (although always finite) for even relatively short formulas. We are not here concerned with pragmatic considerations, but only with theoretical possibility. Many attempts have been made to significantly shorten proofs. Some of these have been more or less successful, but the usual price of success has been the sacrifice of independence for either the axioms or the rules of inference.

We may now summarize our results. The propositional calculus $\mathscr{P}$ is consistent, correct, independent, expressively and deductively complete, and decidable. It is not categorical, but may be made categorical if we so desire.

§21 THE METATHEORY OF THE PREDICATE CALCULUS

We can prove the consistency of the predicate calculus by reducing it to the propositional calculus. This may be done by the semantical expedient of assuming that there is just one element (or, of course, any finite number of elements) in the domain of discourse. Consistency is then easily established. Note that we are assuming the principle that if a formal system is true of something, then it is consistent.

Problem 52. How do we know that the predicate calculus might not prove inconsistent for some other domain? In particular, how do we know that the assumption of an infinite domain will not lead to inconsistency?

However, we intend whenever possible to have a syntactical proof, that is, one which makes no reference to an interpretation. Here is a sketch of such a proof. Consider any formula of $\mathscr{L}$, say

$$(\forall x)((\forall y)((\forall z)((A^2(x, y) \supset (A^1(x) \supset A^1(z)))))).$$

First, erase all quantifiers, then all parentheses which are required by their presence, and then all individual variables and the parentheses and commas required by their presence; in our case we are left with

$$(A^2 \supset (A^1 \supset A^1)).$$

Now replace each predicate letter with a propositional variable, replacing different predicate letters by different propositional variables and the identical predicate letters by identical propositional variables. Further, we shall always work from left to right, replacing predicate letters one by one, and using the first unused propositional variables in alphabetic order; we are now left with '$(p \supset (q \supset q))$'. A systematic comparison of these instructions with the formation rules of $\mathscr{P}$ and $\mathscr{L}$ reveal that we shall always be left with a formula of the propositional calculus. We shall call this the *associated formula of the propositional calculus*. Note that it is unique.

Now we prove consistency of $\mathscr{L}$ in the same way that we proved the consistency of $\mathscr{P}$: by showing that every initial formula of $\mathscr{L}$ has a tautology as its associated formula of the propositional calculus, and that this property is preserved under the transformation rules. We then exhibit a formula (such as our example) which does not have this property. Consistency is thus established.

Problem 53. Consider the initial formula schemata (3(d)) and (3(e)) in the primitive basis of $\mathscr{L}$ (§17). Find the associated formula schemata of $\mathscr{P}$. Then show that in $\mathscr{L}$ the property of having the associated formula be a tautology is preserved under transformation rule (4(b)).

We obtain the independence of the initial formulas of $\mathscr{L}$ by means similar to those for $\mathscr{P}$. Since the first three initial formulas of $\mathscr{L}$ are the same as those for $\mathscr{P}$, independence in $\mathscr{L}$ is obtained by showing that the two additional initial schemata do not disturb the independence established for $\mathscr{P}$. We may do this in the following way: First, we interpret the initial formula schemata (3(d)) and (3(e)) of the primitive basis of $\mathscr{L}$ as their respective associated formulas of the propositional calculus. Second, we show that the first three initial formula schemata remain independent. That is, we reduce the problem to showing that the first three of the following five schemata are each independent:

$$(P \supset (Q \supset P))$$

$$((P \supset (Q \supset R)) \supset ((P \supset Q) \supset (P \supset R)))$$

$$(((\bar{P}) \supset (\bar{Q})) \supset (Q \supset P))$$

$$((P \supset Q) \supset (P \supset Q))$$

$$(P \supset P)$$

The proof here proceeds in a way that is analogous to the way it did for the propositional calculus. We demonstrate the independence of (3(d)) by interpreting every formula or part of a formula which has the form $(\forall x)P$ (where P is any formula not containing any '$\supset$'s and x is any individual variable) as $(\overline{R \supset R})$. The first three initial formula schemata remain unchanged, but (3(d)) and (3(e)) become, respectively,

$$((\forall x)(P \supset Q) \supset (P \supset (\overline{R \supset R}))) \quad \text{and} \quad ((\overline{R \supset R}) \supset Q).$$

All initial formulas are valid except (3(d)) and the transformation rules preserve this property. We can prove the initial formula schema (3(e)) independent by using $(R \supset R)$ in place of $(\overline{R \supset R})$.

The deductive completeness of $\mathscr{L}$ is the first proof in the metatheory of general logic which is not elementary. One's first instinct is to continue the kind of proof that was used in proving the consistency of $\mathscr{L}$. That is, it was shown that every provable formula in $\mathscr{L}$ has a (provable) tautology as its associated formula of the propositional calculus. If we could show that valid formulas in $\mathscr{L}$ correspond to tautologies, and invalid ones do not, then we could show that $\mathscr{L}$ is complete.

However, while it is true that every valid formula has a tautology as its associated propositional formula, it is also true that some invalid formulas do. For example, consider the formulas

$$(\exists x)P(x) \supset (\forall x)P(x).$$

By using the technique of Problem 44 (§17) it is easy to see that these formulas are valid in the domain of one individual but invalid in larger domains, and therefore invalid. Yet their associated propositional formula is the tautology '$(p \supset p)$'. Reflection on this example gives the incompleteness of the predicate calculus in a strong sense not defined above. A formal system will be said to be *complete in the strong sense* if there is no formula in the system that would become an independent initial formula if it were added to the initial formulas. The propositional calculus as presented above—that is, the system $\mathscr{P}$—is not complete in the strong sense. A propositional variable—say 'p'—could be added as an initial formula and the system would not become redundant or inconsistent. However, we could easily make the system complete in the strong sense by adding a substitution transformation rule which would allow one to infer the result of substituting any formula for a variable throughout another formula. For example, from '$(p \supset (q \supset p))$' we could infer

$$((p_1 \supset p_1) \supset (q \supset (p_1 \supset p_1))).$$

Curiously, the addition of this powerful rule would not enlarge the set of theorems; but, without reducing the number of theorems, it would allow us to reduce the infinite number of initial formulas of $\mathscr{P}$ to three initial formulas:

$$(p \supset (q \supset p))$$

$$((p \supset (q \supset r)) \supset ((p \supset q) \supset (p \supset r)))$$

$$((\bar{p} \supset \bar{q}) \supset (q \supset p))$$

For the predicate calculus, however, introducing a rule of substitution—for example, for individual variables—will not make it complete in the strong sense. To make it complete in the strong sense, we would have to add initial formulas in a way that would rule out some interpretations which we wish to have. To see this, consider the following list of schemata, each valid in the domain of n individuals but invalid in the domain of $n + 1$ individuals:

1) $(\exists x)P(x) \supset (\forall x)P(x)$

2) $(\exists x)(\exists y)((P(x) \cdot P(y)) \cdot (Q(x) \equiv \overline{Q(y)})) \supset (\forall x)P(x)$

3) $(\exists x)(\exists y)(\exists z)((P(x) \cdot P(y) \cdot P(z)) \cdot (P(x) \equiv \overline{P(y)}) \cdot$

$\quad\quad\quad\quad\quad (Q(x) \equiv \overline{Q(z)}) \cdot (R(y) \equiv \overline{R(z)})) \supset (\forall x)P(x)$

$$\vdots$$

(1) is valid in the domain of 1 individual but invalid in any larger domain. (2) is valid in domains ≤ 2 but invalid in any larger domain. (3) is valid in domains ≤ 3 but invalid in any larger domain, and so forth. Since we wish our logical system to apply to *any* size domain, we do not want to be able to prove any of the above formulas or their negations.

Problem 54. Give an argument which will at least make it plausible that the above formulas have the stated properties.

There are even formulas which are valid in every finite domain but in no infinite one. An illustration might make plausible the existence of such formulas. Suppose that our domain is that of the natural numbers. Following the usual interpretation of '$<$' as *less than*, we would certainly be able to define some of its properties in the following way:

a) $(\forall x)\overline{(x < x)}$

b) $(\forall x)(\exists y)(x < y)$

c) $(\forall x)(\forall y)(\forall z)(((x < y) \cdot (y < z)) \supset (x < z))$

Each is true in the domain of natural numbers. However, taken together, they are not true in any finite domain of natural numbers. To see this, suppose we start with a finite domain consisting of just the first natural number. Then eliminating quantifiers (a) would be '$\overline{(1 < 1)}$' and (b) would be '$(1 < 1)$'; that is, a contradiction. Thus we need at least two individuals, say, the first two natural numbers. Then (a) becomes '$\overline{(1 < 1)} \cdot \overline{(2 < 2)}$'. (b) first reduces to

$$(\exists y)(1 < y) \cdot (\exists y)(2 < y)$$

and then becomes

$$((1 < 1) \lor (1 < 2)) \cdot ((2 < 1) \lor (2 < 2)).$$

This is false when '$<$' is interpreted as 'less than'; but (a) and (b) so reduced are at least consistent. We can see this by interpreting '$<$' as *less than or greater than.* But a step-by-step elimination of quantifiers for (c) gives us first

$$(\forall y)(\forall z)(((1 < y) \cdot (y < z)) \supset (1 < z)) \cdot$$
$$(\forall y)(\forall z)(((2 < y) \cdot (y < z)) \supset (2 < z)),$$

then second

$$(\forall z)(((1 < 1) \cdot (1 < z)) \supset (1 < z)) \cdot$$
$$(\forall z)(((1 < 2) \cdot (2 < z)) \supset (1 < z)) \cdot$$
$$(\forall z)(((2 < 1) \cdot (1 < z)) \supset (2 < z)) \cdot$$
$$(\forall z)(((2 < 2) \cdot (2 < z)) \supset (2 < z)),$$

and finally,

$$(((1 < 1) \cdot (1 < 1)) \supset (1 < 1)) \cdot$$
$$(((1 < 1) \cdot (1 < 2)) \supset (1 < 2)) \cdot$$
$$(((1 < 2) \cdot (2 < 1)) \supset (1 < 1)) \cdot$$
$$(((1 < 2) \cdot (2 < 2)) \supset (1 < 2)) \cdot$$
$$(((2 < 1) \cdot (1 < 1)) \supset (2 < 1)) \cdot$$
$$(((2 < 1) \cdot (1 < 2)) \supset (2 < 2)) \cdot$$
$$(((2 < 2) \cdot (2 < 1)) \supset (2 < 1)) \cdot$$
$$(((2 < 2) \cdot (2 < 2)) \supset (2 < 2)).$$

Now this formula is false under the interpretation of '<' as *less than or greater than* (cf. the sixth conjunct). Hence it is inconsistent with (a) and (b). This example has been written out at length to make plausible the impossibility of satisfying (a), (b), and (c) in a finite domain: Formula (a) guarantees that an individual doesn't have the relation to itself, formula (b) that for any individual there always is another individual with that relation to it, and formula (c) rules out the possibility of a "circle" (as above in our "less than or greater than" interpretation).

Now if we conjoin (a), (b), and (c) in a single formula, we get

d)
$$((\forall x)\overline{(x < x)} \cdot (\forall x)(\exists y)(x < y) \cdot \quad \text{See (b) on p. 144}$$
$$(\forall x)(\forall y)(\forall z)(((x < y) \cdot (y < z)) \supset (x < z))).$$

Thus (d) is not satisfiable in any finite domain, but it is satisfiable in an infinite domain. However, it follows by definition that the negation of a formula which is not satisfiable is valid. Hence the negation of (d) is valid in every finite domain but in no infinite one.

One might think that we could determine the cardinal number of the domain of discourse (or at least an upper bound on that domain) by adding suitable initial formulas to $\mathscr{L}$; that is, we could, for any cardinal number, devise a logical system that could be applicable only to domains not larger than that number. This is indeed true for all cardinals less than $\aleph_0$, that is, for all finite cardinal numbers. All we need do is to decide on a finite cardinal, and then add the corresponding formula schema in the infinite list described above to the initial formula schemata. For example, if we want a domain of at most 3 members, we would add (3) to the list. However, when we come to transfinite cardinals, the situation changes. Suppose that we add an initial formula schema to those of $\mathscr{L}$, which will ensure a domain at least equal to $\aleph_0$. On the one hand, it turns out that any such formal system will have models corresponding to any transfinite cardinal. That is, we can no longer put an upper bound on the domain of discourse by suitable initial formula schemata. On the other hand, it turns out that we can't put a lower bound on the domain either! That is, if we try to add an initial formula schema which will ensure that the domain be larger than $\aleph_0$—say, if we try to axiomatize the real numbers—then there will always be an interpretation of the system which makes the theorems true in the domain of natural numbers (that is, a domain equal to $\aleph_0$). Of this we shall say more later.

To return to the question of deductive completeness: We have seen that there are formulas which are valid in domains only up to some specific finite size and that none of these formulas is provable (or refutable) by the axioms. But the question arises: Suppose that there is a formula which is valid in *every* domain. Is it provable? The answer is yes, and thus $\mathscr{L}$ is complete in the sense that every valid formula is provable. The proof of this—due originally to Kurt Gödel—is complicated, and we shall present only a generalized sketch.

Suppose that some specific formula **A** is valid. We must show that **A** is provable in $\mathscr{L}$. The first step is similar to the first step in the proof of completeness of the propositional calculus, namely, to reduce the problem to formulas in a specifiable form. It can be shown that a formula is provable if and only if its Skolem normal form is provable. A formula is in *Skolem normal form* if it has the following structure:

$$\overbrace{\underbrace{(\exists\ )(\exists\ )(\exists\ )\ldots(\exists\ )}_{m \text{ distinct quantifiers}}\underbrace{(\forall\ )(\forall\ )(\forall\ )\ldots(\forall\ )}_{n \text{ distinct quantifiers}}}^{m + n \text{ distinct quantifiers}}\mathbf{P}$$

Here $m \geq 1$, $n \geq 0$, and **P** contains no quantifiers but has $n + m$ free variables (that is, those in the indicated quantifiers). The following illustrate Skolem normal form:

$$(\exists x)(\exists y)(\forall z)\mathbf{A}^3(x, y, z), \qquad (\exists x)(\forall y)(\mathbf{A}^2(x, y) \supset \mathbf{B}^2(y, x)), \qquad (\exists x)\mathbf{A}^1(x).$$

The general idea of the proof is as follows: Suppose that some formula in Skolem normal form is valid. Then if we leave off the existential quantifiers, there is some way of replacing the free variables so created so that the result is a substitution instance of a tautology. An illustration may serve to make this clear. Consider the formula

$$(\exists x)(\exists y)(\exists z)(\forall x_1)(A^2(x, y) \supset A^2(z, x_1)).$$

It is provable and thus is valid. Arrange in order all possible triplets (corresponding to three existential quantifiers) of variables. One such ordering might be:

$$x, x, x$$
$$x, x, y$$
$$x, x, z$$
$$x, y, x$$
$$x, y, y$$
$$x, y, z$$
$$x, z, x$$
$$x, z, y$$
$$x, z, z$$
$$y, x, x$$
$$y, x, y$$
$$\vdots$$

Consider the following infinite sequence of formulas:

$$(A^2(x, x) \supset A^2(x, y))$$

$$(A^2(x, x) \supset A^2(x, y)) \vee (A^2(x, x) \supset A^2(y, z))$$

$$(A^2(x, x) \supset A^2(x, y)) \vee (A^2(x, x) \supset A^2(y, z)) \vee (A^2(x, x) \supset A^2(z, x_1))$$

$$(A^2(x, x) \supset A^2(x, y)) \vee (A^2(x, x) \supset A^2(y, z)) \vee (A^2(x, x) \supset A^2(z, x_1)) \vee$$
$$(A^2(x, y) \supset A^2(x, y_1))$$
$$\vdots$$

Note how the free variables corresponding to the existential quantifiers follow the ordering indicated in the infinite sequence of triplets, and that if any formula in the sequence of formulas is valid, all subsequent ones are also. Finally, note that in each formula the variable corresponding to the universal quantifier is different from all those occurring earlier in the list of formulas (or in the same formula). Subject to this condition, we have used alphabetic ordering for the variables.

Now the key to the proof is this: A formula in Skolem normal form is valid if and only if one of the formulas in its corresponding infinite list is a substitution instance of a tautology. No proof of this statement will be given here. But note that in our example the fourth formula is a substitution instance of a tautology, namely,

$$(p \supset q) \lor (p \supset r) \lor (p \supset p_1) \lor (q \supset q_1).$$

Returning to the general case, we consider two possibilities. Suppose that in the systematic listing of formulas there is one which is a substitution instance of a tautology. Given this fact, it is possible to give complete instructions—again, these instructions are long and involved, but they always work—on how to find a proof of the formula in Skolem normal form. Thus we have a complete proof procedure; that is, if a formula is valid, there is a proof of it in $\mathscr{L}$. The procedure goes as follows:

i) We are given a *valid* formula **A**.

ii) We find its Skolem normal form **B**.

iii) We then make a listing of all possibilities (analogous to our above example) until we come to an instance of a tautology.

iv) Using the instructions given in the completeness proof for the propositional calculus, we prove *that* tautology.

v) Then, using the instructions alluded to earlier in the paragraph, we prove **B**.

vi) Finally, using the instructions given in the exposition of Skolem normal form, we prove **A**.

But let's consider the other possibility. Suppose that in our systematic listing of formulas there is no instance of a tautology. It can be shown that there exists one infinite assignment of truth values which makes all these formulas false and hence the formula is not valid. For consider the corresponding formula of the propositional calculus. To the variables **p, q, r**, etc., we assign 'T' or 'F' according as that assignment does or does not occur in an infinite number of falsifying assignments. Let us call this falsifying assignment of truth values *the falsifying assignment*. Then it will be impossible for any formula in our sequence to be satisfied by the falsifying assignment.

However, our proof yields more than we set out to prove, for a simple argument now suffices to show that if a formula is invalid, it is invalid in the domain of natural numbers. For consider any invalid formula **C**. Find the systematic list of formulas corresponding to it. Give each formula in the list the following interpretation: To each variable assign the natural number which indicates its place in the alphabetic numbering (for example, to z assign 3). Assign to each predicate an interpretation which makes it true or false (of the natural numbers assigned to its variables) according as *the* falsifying assignment for it is true or false. There will then be an interpretation of C which makes it false in the domain of natural numbers. Hence any valid formula or set of valid formulas which has an infinite model also has the natural numbers as a model.

We are ready to draw some conclusions from this proof. What we have shown is the following: *If* a formula is valid, then we have a step-by-step procedure for finding a proof of it. If a formula is not valid, then there exists a falsifying assignment for it and so it isn't a theorem.

Problem 55. How could we show it isn't a theorem?

We may conclude that a formula is valid if and only if there is a proof of it. However, there is an important difference in the situation with regard to validity and invalidity. If a formula is valid, we have a step-by-step procedure for finding a proof of it. If it is invalid, all we know is that there exists a falsifying assignment; we are not given instructions for finding that assignment. How do we know, for example, whether **p** is assigned 'T' or 'F' in an infinite number of falsifying assignments? Thus this proof yields no decision procedure. To see this, suppose that we are given an arbitrary formula. We might decide to prove it valid. But suppose that we go through a large number of lines (say, a million) and we come to no tautologous formula. We might then decide that it is invalid. But then suppose we laboriously check each domain and it is valid in the domain of 1 individual, 2 individuals, 3 individuals, and so on up to, say, a million individuals. We can't conclude that it is valid because it might be invalid in the domain of a million and one individuals. In fact, we know that for any given natural number n, there are formulas valid in any domain less than n and invalid in any domain greater than n (cf. Problem 54). And as we have seen, there are formulas valid in every finite domain but invalid in any infinite domain. Hence our proof yields no algorithm for deciding theoremhood.

Problem 56. A careless reader might object: "There is a possibility you have overlooked. Instead of working on just one formula we should always work on two: the formula and its negation. One of these has to be valid (they can't both be) and so if we try to prove each valid by the means given in the completeness proof we are guaranteed an answer in a finite number of steps. Thus we do have a decision procedure." Find the error in the above argument.

Were it not for the existence of those formulas which are valid in every finite domain but invalid in any infinite one, we would gain a decision procedure by the completeness proof. For consider the class of formulas which are either valid or, if invalid, invalid in some finite domain. Then the procedure suggested (that is, in the last paragraph) would be guaranteed to end in a finite number of steps. We would either have a proof of validity or a demonstration of invalidity in a finite domain.

Hence the class of formulas which are valid in every finite domain but invalid in an infinite domain is an interesting one from a theoretical point of view. It becomes even more interesting when we consider the last part of the completeness proof, according to which a formula which is invalid in any domain is invalid in the domain of natural numbers. Our decision problem reduces to the following: Find a procedure by which we can distinguish formulas which are valid in every finite domain but invalid in the domain of natural numbers, from formulas which are valid in every domain (that is, just plain valid). Suppose that we found such a procedure. Then the decision problem for $\mathcal{L}$ would be solved. For if we were given an arbitrary formula of $\mathcal{L}$, we could carry out three different operations on it:

i–vi) Try to prove it valid by the instructions given in the proof of completeness.

vii)   Try to prove it invalid by finding a finite domain in which it is invalid and then using the truth-table technique.

viii)   Use the above hypothesized procedure to tell us whether it is a formula valid in all finite domains but in no infinite one.

If we first did the first step of (i–vi), then the first step of (vii), then the first step of (viii), then the second step of (i–vi), then the second step of (vii), and so on, we would have a guarantee that our procedure would terminate. It turns out that the deficiency in the completeness proof (in that it does not give us a decision procedure) cannot be corrected, because there is no decision procedure for $\mathcal{L}$ and thus no procedure such as (viii).

It is important to note that the lack of a decision procedure for $\mathcal{L}$ is not due to a deficiency in ingenuity; there just isn't one to be had. The situation is analogous to the question of a proof of Euclid's fifth postulate from the remaining four. It isn't that the Greeks didn't have enough ingenuity to prove it; we now possess a proof that such a proof is impossible. We now also have a proof that a proof of the existence of a decision procedure for the predicate calculus is impossible. It was first discovered by Alonzo Church in 1936, and hence is called *Church's theorem*.

We shall postpone consideration of Church's theorem until later, and instead turn our attention to the question of categoricalness. We have seen that $\mathcal{L}$ is not categorical, since it allows for models with different cardinal numbers. Since validity or invalidity in a domain depends only on the cardinal number of that domain, it might seem that we could make $\mathcal{L}$ categorical by adding to the initial formulas at least one initial formula which would require the cardinal number of all interpretations to be some fixed number. For example, we could add the formulas $(\exists x)P(x) \supset (\forall x)P(x)$. Our system would then apparently be categorical, since every model contains exactly 1 element. However, we have no means of doing likewise in cases in which the domain is larger than 1. For example, (2) on page 144 is valid in the domain of either 1 or 2 elements. Hence if we added (2) to the initial formulas of $\mathscr{L}$, the system would not be categorical. Within the expressive power of $\mathscr{L}$, there is no way to get around the difficulty. However, if we consider the system $\mathscr{L}^=$, then we can get formulas which are valid only in domains of some specifiable number of elements. Consider the following list of schemata:

1') $(\forall x)(\forall y)(x = y)$

2') $(\exists x)(\exists y)((x \neq y) \cdot (\forall z)((z \neq x) \supset (z = y)))$

3') $(\exists x)(\exists y)(\exists z)((x \neq y) \cdot (y \neq z) \cdot (x \neq z) \cdot (\forall x_1)(((x_1 \neq x) \cdot$

$(x_1 \neq y)) \supset (x_1 = z)))$

$$\vdots$$

(1') states that there is at most one individual. The initial formulas guarantee at least one individual. Thus adding (1') to the initial formulas of $\mathscr{L}^=$ would give us a 1-element domain. If instead of (1') we added (2'), we would have a 2-element system. (2') states that there are exactly 2 elements, since the left conjunct within the schema guarantees at least 2 elements, while the right conjunct guarantees at most 2. Similarly, (3') guarantees exactly 3 elements, etc. The result of adding any one of these schemata to the initial schemata of $\mathscr{L}^=$ would seem to produce a categorical system. However, just as for $\mathscr{P}$ we do not want to make the system categorical, so here we do not want such a system. For otherwise it would apply only to some given finite domain and we want our logic to apply to any size domain. Nevertheless, the hope is that when $\mathscr{L}$ is taken as the logical basis and applied to some mathematical or other axiomatic system, the resulting applied system can be made categorical.

Problem 57. In this paragraph we have assumed that if we can show a one-to-one correspondence in all models, the resulting version of the predicate calculus is categorical. Yet in order to be categorical we require that all models be isomorphic, that is, that any two models possess a one-to-one correspondence *and* that any formula of the system which is true when interpreted in one model must also be true when interpreted in the other model. But the latter requirement is necessary and the additions suggested do not produce categorical systems. Can you see why?

§22 THE THEORY OF RECURSIVE FUNCTIONS

We now turn to a development which at first sight appears to have nothing to do with formal systems. In fact, imagine that you have tired of the complexities of metatheory and for diversion have turned to another book. The author of this different book is familiar with the paradoxes of the infinite and the complexities introduced into mathematical logic in order to get rid of them. However, he feels that both the paradoxes and complexities are unnecessary and that they are caused by the use of quantifiers which range over infinite domains. Such an author was Thoralf Skolem, who after studying *Principia Mathematica* wrote a paper with an alternative answer to the problems of the infinite. The paper—written in 1919 and published in 1923—is called "The foundations of elementary arithmetic established by means of the recursive mode of thought, without use of apparent variables ranging over infinite domains." The paper marked the beginning of the development of an area of mathematics whose importance is comparable to geometry or algebra. It is today called the *theory of recursive functions*.

To understand the title of Skolem's paper, some background is needed. What Skolem calls the *recursive mode of thought* is exemplified in the use of mathematical induction, which goes back at least as far as Fermat (1601–1665) and Pascal (1623–1662), but it wasn't until the nineteenth century that the principle itself was studied, in particular by Frege, Dedekind, and Peano. All of them studied it with a view toward understanding the nature of number. Skolem's motivation was different; he thought he could use it as part of the *foundation* for arithmetic. This foundation was intended to be so certain and the inferences from it so indubitable, that not only wouldn't the paradoxes arise, but we would also be completely convinced that they couldn't arise. His motivation then was somewhat similar to Euclid's; that is, he wanted to start from statements indubitably true and proceed in such a way that doubts could never set in. His success has been such that not only has everyone accepted his work (even mathematicians and philosophers who have doubts about some parts of classical mathematics), but also his theory has become a paradigm of indubitable reasoning.

Skolem proceeds informally; in fact, he does not even use the axiomatic method. He assumes that the following notions are already understood: natural number, successor of x, substitution of equals for equals, and the recursive mode of thought. From these he attempts to define all arithmetic notions. By *natural number* Skolem meant $1, 2, 3, \ldots$, that is, the numbers used in counting. By *successor of* x he meant the number coming after x, denoted by 'x''. The *substitution of equals for equals* is an obvious rule, since here two expressions flanking an equals sign means that they denote the same object. Thus if $a = b$ and $b = c$, then $a = c$, because this means that if a denotes the same object as b, and b denotes the same object as c, then a denotes the same object as c. In other words, we have transitivity.

The expression *recursive mode of thought* refers to two things: First, all definitions are definitions by recursion and second, all proofs are proofs by mathematical induction. These may be made clear by an example.* Suppose that we wish to define the function of addition; that is, we wish to define a function f which will correspond to what we normally mean by addition. It can be done by the following schema for addition:

$$f(x, 0) = x \qquad\qquad x + 0 = x$$

$$f(x, y') = (f(x, y))' \qquad x + y' = (x + y)'$$

The right-hand pair of equations is the same as the left-hand one, except that it uses more familiar notation.

The addition schema should be thought of as a set of rules which tells us how to compute the value of $+$ for any given pair of numbers. Consider $7 + 5$. It could be computed by the following two lists of 21 equations:

$0''''''' + 0 = 0'''''''$	$7 + 0 = 7$
$0''''''' + 0' = (0''''''' + 0)'$	$7 + 1 = (7 + 0)'$
$0''''''' + 0' = (0''''''')'$	$7 + 1 = 7'$
$(0''''''')' = 0''''''''$	$7' = 8$
$0''''''' + 0' = 0''''''''$	$7 + 1 = 8$
$0''''''' + 0'' = (0''''''' + 0')'$	$7 + 2 = (7 + 1)'$
$0''''''' + 0'' = (0'''''''')'$	$7 + 2 = 8'$
$(0'''''''')' = 0'''''''''$	$8' = 9$
$0''''''' + 0'' = 0'''''''''$	$7 + 2 = 9$
$0''''''' + 0''' = (0''''''' + 0'')'$	$7 + 3 = (7 + 2)'$
$0''''''' + 0''' = (0''''''''')'$	$7 + 3 = 9'$
$(0''''''''')' = 0''''''''''$	$9' = 10$
$0''''''' + 0''' = 0''''''''''$	$7 + 3 = 10$
$0''''''' + 0'''' = (0''''''' + 0''')'$	$7 + 4 = (7 + 3)'$
$0''''''' + 0'''' = (0'''''''''')'$	$7 + 4 = 10'$

---

* In this example, and henceforth in any context in which recursive functions are being discussed, 0 will be considered a natural number. This contextual redefinition of a term is not essential, but it simplifies the discussion of recursive functions and their application. Mathematically, the question as to whether 0 is a natural number is unimportant. However, historically it is important. Since part of this book is concerned with history, we were prevented from defining *natural number* to include 0 from the outset.

$$(0''''''''')' = 0''''''''''$$

$$10' = 11$$

$$0'''''' + 0'''' = 0''''''''''$$

$$7 + 4 = 11$$

$$0'''''' + 0'''' = (0'''''' + 0''')'$$

$$7 + 5 = (7 + 4)'$$

$$0'''''' + 0'''' = (0''''''''')'$$

$$7 + 5 = 11'$$

$$(0''''''''')' = 0''''''''''$$

$$11' = 12$$

$$0'''''' + 0'''' = 0''''''''''$$

$$7 + 5 = 12$$

One's first reaction to such lists is that they are an unnecessarily complicated way of doing something for which easier ways are taught in grade schools. However, remembering that the theory of recursive functions is very important, let's try to understand what is going on. The two lists of equations are identical in meaning, the difference consisting in the fact that the right-hand list makes use of familiar abbreviations and therefore is easier to read. Abbreviations are, of course, allowed in recursive function theory so long as they are a mere notational device that is theoretically eliminable (by the principle of substituting equals for equals).

The comments below will refer to the right-hand list because of its greater perspicuity. We get the first equation from the addition schema with the numeral '7' taking the place of 'x'. We get the second line from the addition schema by substituting '7' for 'x' and '0' for 'y' in the second line of the schema (of course, '1' is an abbreviation of '0''). We get the third line from the first and second by substituting '7' for '$7 + 0$' in the second line (here applying the rule of substitution). The fourth line introduces a notational abbreviation. The fifth line comes from the third and fourth by substitution of equals for equals. The sixth line comes from the second line of the addition schema. Now the pattern repeats, as can easily be checked.

The definition of addition given above represents a typical recursive definition. Its characteristic feature is that it becomes a completely explicit definition (by a series of equations) whenever numerals are substituted for the symbols representing the free variables. Moreover, this series of equations is always finite and an upper bound on the number of equations could always be specified for any given pair of numerals.

Problem 58. Once we have defined *addition*, we can define *multiplication* recursively from it as follows:

$$x \cdot 0 = 0$$

$$x \cdot y' = x \cdot y + x$$

Assuming this definition, give a recursive definition of *exponentiation* (that is, of 'x^y').

We have given examples of definition by recursion. Now we turn to a typical example of proof by recursion (that is, mathematical induction).

Theorem: $x + 0 = 0 + x$.

Proof: The theorem is true for $x = 0$, since $0 + 0 = 0 + 0$ by definition of equality. Thus we must prove $x' + 0 = 0 + x'$ from the induction assumption $x + 0 = 0 + x$. $x' + 0 = x' + 0$ by definition, and since $x + 0 = x$ by the addition schema, we have $x' + 0 = (x + 0)' + 0$ by substitution. $(x + 0)' + 0 = (x + 0)'$ by the addition schema and hence $x' + 0 = (x + 0)'$ by substitution. By the induction assumption, $x + 0 = 0 + x$, and hence $x' + 0 = (0 + x)'$ by substitution. $(0 + x)' = 0 + x'$ by the addition schema and hence $x' + 0 = 0 + x'$ by substitution.

We may begin to appreciate the philosophical importance of recursive function theory if we compare the usual definition of divisibility with the one in recursive arithmetic. The usual one is ("$x \mid y$" means "x divides y"):

$$\frac{y}{x} = (\exists z)(x = y \cdot z).$$

As Skolem points out, "Such a definition involves an infinite task—that means one that cannot be completed—since the criterion of divisibility ... [involves] *successive trials through the entire number sequence*" [1923 310]. To see this, consider a specific example, say dividing 8 by 3. We then have

$$3 \mid 8 = (\exists z)(8 = 3 \cdot z),$$

which becomes

$$3 \mid 8 \equiv (8 = 3 \cdot 0) \lor (8 = 3 \cdot 1) \lor (8 = 3 \cdot 2) \lor (8 = 3 \cdot 3) \lor (8 = 3 \cdot 4) \lor \cdots$$
$$\lor (8 = 3 \cdot n) \lor \cdots.$$

For each of the infinite number of disjuncts we would have to compute the multiplication and see whether the equality holds. Of course after a certain point it becomes pointless to continue the computation, for we can be sure that the computation would not reveal an equality. At what point in general? It is clear that the number of disjuncts will never have to be more than one greater than the dividend. This is just what is embodied in the recursive definition of divisibility:

$$\frac{y}{x} = (\exists z)((z \leq x) \cdot (x = y \cdot z)).$$

The use here of the bound (apparent) variable is completely inessential, since it can be eliminated for any specific pair of numbers. In our example the definition reduces to:

$$3 \mid 8 \equiv (8 = 3 \cdot 0) \lor (8 = 3 \cdot 1) \lor (8 = 3 \cdot 2) \lor (8 = 3 \cdot 3) \lor (8 = 3 \cdot 4) \lor$$
$$(8 = 3 \cdot 5) \lor (8 = 3 \cdot 6) \lor (8 = 3 \cdot 7) \lor (8 = 3 \cdot 8).$$

This in turn can be checked in a finite number of steps by writing out in full the corresponding equations for each disjunct. For example, consider the corresponding equations for the second disjunct:

$$3 \cdot 0 = 0$$
$$3 \cdot 1 = (3 \cdot 0) + 3$$
$$3 \cdot 1 = 0 + 3$$
$$0 + 3 = 3 + 0$$
$$3 + 0 = 3$$
$$3 \cdot 1 = 3$$

Since it follows by definition that $8 \neq 3$ (that is, $0''''''''' \neq 0'''$), the falsehood of the second disjunct is established. After the truth or falsehood of each of the disjuncts is determined, the truth-table technique will suffice to decide the formula.

In the theory of recursive functions, only free variables are used. The meaning of formulas with free variables is understood to be the same as that of the universal closure of the formulas. For example,

$$x + y = y + x$$

is equivalent in meaning to

$$(\forall x)(\forall y)(x + y = y + x).$$

[Note that in $\mathscr{L}$ we can derive one from the other (by use of (3(e)), (4(a)), and (4(b)) of §17) and thus they are logically equivalent.] On the other hand, existential quantifiers may occur only if there is some bound on the variable in the quantifier; in other words, only if the existential quantification is inessential.

We may summarize the situation by saying that while the usual definition of a function defines it explicitly by giving an abbreviation of that expression, the recursive definition defines the function explicitly only for the first natural number, and then provides a rule whereby it can be defined for the second natural number, and then the third, and so on.

The philosophical importance of a recursive function derives from its relation to what we mean by an *effective finite procedure*, and hence to what we mean by *algorithm* or *decision procedure*. Algorithms had been *used* for thousands of years (Euclid gives one for finding the greatest common divisor [cf. 1926 II 298–299]), but it wasn't until the twentieth century that an attempt was made to explicate exactly what an algorithm is in general. Skolem's paper marked the first development of an area of mathematics which was to provide an adequate analysis of the concept of an effective finite procedure. Skolem's initial motivation, however, was not directly concerned with this; rather his interest was to avoid the paradoxes without epistemological complexity. This issue is important, but we shall postpone discussion of it (cf. §26). It is necessary first to see some application of the theory in order to best appreciate why, after millenniums of disinterest, logicians turned to the problem of explicating the concept of an effective finite procedure.

We are now ready to present some definitions which will be used later on. Given that $f(x) = x'$, then f will be called the *successor function*. Given that $f(x_1, \ldots, x_n) = a$, where a is a natural number, f will then be called a *constant function*. Given that

$$f(x_1, \ldots, x_n) = x_i \qquad (1 \leq i \leq n),$$

then f is called an *identity function*. Given that

$$f(x_1, \ldots, x_n) = g(h_1(x_1, \ldots, x_n), \ldots, h_m(x_1, \ldots, x_n)),$$

where g and h_i ($1 \leq i \leq m$) are given functions, then f will be said to be *defined by substitution* from $g, h_1, \ldots, h_m$. Given that

$$\left\{ \begin{array}{l} f(0, x_2, \ldots, x_n) = g(x_2, \ldots, x_n) \\ f(y', x_2, \ldots, x_n) = h(y, f(y, x_2, \ldots, x_n), x_2, \ldots, x_n) \end{array} \right\},$$

where g and h are given functions, then f will be said to be *defined by recursion (or induction) from g and h*.

We now come to the important definition which depends on the above definitions: A function f is called *primitive recursive* if there is a finite sequence of functions ending with f such that each function is a successor, constant, or identity function, or is defined from preceding functions in the sequence by substitution or recursion. Observe that this definition stays quite close to Skolem's intention to assume only four notions: natural number (which corresponds with constant function), successor of x (which corresponds with successor function), substitution of equals for equals (which corresponds with definition by substitution), and the recursive mode of thought (which corresponds with definition by recursion). Only the identity function has been added, because this function has been found convenient in subsequent work. But the addition is well within the spirit of Skolem's pioneering paper.

Problem 59. Consider the following sequence of functions:

$f_1(x) = x'$ f_1 is the successor function.

$f_2(x) = x$ f_2 is the identity function with $i = 1$.

$f_3(y, z, x) = z$ f_3 is the identity function with $i = 2$.

$f_4(y, z, x) = f_1(f_3(y, z, x))$ f_4 is defined by substitution from f_1 [$= g$] and f_3 [$= h_1$].

$$\left.\begin{array}{l} f_5(0, x) = f_2(x) \\ f_5(y', x) = f_4(y, f_5(y, x), x) \end{array}\right\} \qquad \left.\begin{array}{l} f_5 \text{ is defined by recursion from } f_2 \text{ [}= g\text{]} \\ \text{and } f_4 \text{ [}= h\text{].} \end{array}\right\}$$

This sequence of equations proves that f_5 is primitive recursive and hence that addition is primitive recursive, that is, $f_5 = +$. (Note in particular that if we substitute equals for equals, the last two equations become identical with the addition schema as we had earlier defined it.) Now the problem is to continue this sequence of functions in such a way as to prove that exponentiation is primitive recursive.

We shall call the theory we are developing *recursive arithmetic*. The theory of recursive functions is a more encompassing theory. For example, it includes recursive analysis, in which the domain of definition of the free variables is the real numbers, whereas in recursive arithmetic the domain is the natural numbers. Recursive arithmetic is a powerful theory in spite of its rather lean assumptions. However, it must not be thought that it is equivalent to an arithmetic with full use of bound variables. Its expressive power is limited.

To show this, we need a definition: A property $P(x_1)$ or relation $P(x_1, \ldots, x_n)$ of natural numbers is called *primitive recursive* if there exists a primitive recursive function $f(x_1, \ldots, x_n)$ such that for all $x_1, \ldots, x_n$:

$$P(x_1, \ldots, x_n) \supset (f(x_1, \ldots, x_n) = 0),$$

$$\overline{P(x_1, \ldots, x_n)} \supset (f(x_1, \ldots, x_n) = 1).$$

For an example of a primitive recursive relation, consider the following list of equations:

$$\left.\begin{array}{l} Pd(0) = 0 \\ Pd(x') = x \end{array}\right\} \qquad \left.\begin{array}{l} Pd(x) \text{ is the predecessor of } x \text{ if } x > 0, \\ 0 \text{ if } x = 0. \end{array}\right\}$$

$$\left.\begin{array}{l} x \div 0 = x \\ x \div y' = Pd(x \div y) \end{array}\right\} \qquad \left.\begin{array}{l} x \div y \text{ is } x - y \text{ if } x \geq y, 0 \\ \text{if } y \geq x. \end{array}\right\}$$

$$|x - y| = (x \div y) + (y \div x) \qquad |x - y| \text{ is the absolute value of } x - y.$$

$$\left.\begin{array}{l} Sg(0) = 0 \\ Sg(x') = 1 \end{array}\right\} \qquad Sg(x) \text{ is the signum of } x.$$

It is easy to see that each of the functions on the left is primitive recursive, and that they correspond to the functions named on the right. Now consider the following function:

$$Sg(|x - y|).$$

It is primitive recursive because it is defined by the substitution of one primitive recursive function into another. Finally we have, for all x and y,

$$(x = y) \supset (Sg(|x - y|) = 0),$$

$$(x \neq y) \supset (Sg(|x - y|) = 1).$$

Hence equality is a primitive recursive relation.

Problem 60. Show that $x < y$ is a primitive recursive relation. [*Hint*: First define another simple function.]

We are now ready to show that the expressive power of primitive recursive arithmetic is not equal to that of an arithmetic with full use of bound variables. For example, Euclid proved a theorem that there exist infinitely many prime numbers [cf. 1926 II 413]. In modern notation the argument would proceed as follows: Suppose that there are only a finite number of primes: $a_1, a_2, \ldots, a_n$. Consider a new number b defined by:

$$b = (a_1 \cdot a_2 \cdot \ldots \cdot a_n) + 1.$$

The number b is greater than any of the a's and thus is composite. However, if it is divided by any of the primes $a_1, \ldots, a_n$, the remainder will always be 1. Therefore b is prime. This contradiction establishes the infinitude of primes.

Now it is quite possible to express (and prove) Euclid's theorem in recursive arithmetic. Let '$Pr(x)$' stand for 'x is a prime'. It can be shown that Pr is a recursive property and $\supset$ is a recursive relation. Hence with a little further effort we can show that Euclid's theorem can be expressed in primitive recursive arithmetic as:

$$Pr(x) \supset (\exists y)(Pr(y) \cdot (x < y \leq (1 \cdot 2 \cdot 3 \cdot \ldots \cdot x) + 1)).$$

The use of the existential quantifier is again inessential.

However, suppose that we try to express the statement that there are infinitely many twin primes [cf. (2) of §25]. Since we do not know whether this is true or false, there is no way of putting a bound on the existential quantifier. Its expression in primitive recursive arithmetic would be something like:

$$(Pr(x) \cdot Pr(x + 2)) \supset (\exists y)(Pr(y) \cdot Pr(y + 2) \cdot ((x + 2) < y \leq —)),$$

where the '—' must be replaced by a primitive recursive function of x. In contrast to this, suppose that we allow unrestricted use of quantifiers. We would then have ordinary *arithmetic*, and we could express the infinitude of twin primes by:

$$(\forall x)((Pr(x) \cdot Pr(x + 2)) \supset (\exists y)(Pr(y) \cdot Pr(y + 2) \cdot ((x + 2) \leq y))).$$

The question as to whether all achieved results in arithmetic can be translated into recursive arithmetic is problematical, and depends on the meaning of "achieved results." For example, Vinogradoff in working on Goldbach's conjecture [cf. (4) of §25] proved that there exists a natural number such that every natural number greater than it can be represented as the sum of four primes. However, we are not given any way to establish the size of that natural number, because the proof was of the *reductio ad absurdum* type under the assumption that there are infinitely many integers which could not be expressed as the sum of four primes. Since this proof is accepted as *bona fide* by most mathematicians and since it can't be reproduced within the confines of recursive arithmetic, there are "achieved results" for which recursive arithmetic is inadequate. However, it must be emphasized that there is a small group of mathematicians who do not recognize this proof (and others like it) as valid.

§23 THE METATHEORY OF ARITHMETIC

A. Preliminaries

We are now ready to return to metatheory, in particular to the metatheory of arithmetic. It turns out that arithmetic plays a central role in logic and in a way that is quite unexpected. We can begin to appreciate its crucial significance by considering the justifiably most famous results of contemporary logic: Gödel's first and second incompleteness theorems. We may start by supposing that we have a formal system which includes the first-order predicate calculus but is enriched by further symbols, formation and transformation rules, and initial formulas. We can think of this system as a square in which by some miracle all the sentences of the system are written out. If this enrichment were of such an extent that it allowed us to express the Liar paradox, the system would become inconsistent. That is, we would have:

$$\boxed{\begin{array}{c} \vdots \\ \text{This sentence is not true.} \\ \vdots \end{array}}$$

Thus, however we are going to enrich the predicate calculus, we must not do it to such an extent that the Liar proposition becomes expressible.

In any case, it seems unlikely that we would ever be able to define the notion of truth in a *formal* system. Truth has to do with the relationship between an expression and some reality, whereas an uninterpreted formal system deals only with relationships among expressions. Yet there is a relationship between an interpreted formal system and truth inasmuch as we try to set up the system in such a way that any interpretation which makes all the axioms true makes all the theorems true (correctness), and that any truth expressible in the system is provable (completeness). In other words, we intend that the set of true expressions of the system under any interpretation that makes all the axioms true will be identical with the set of provable expressions or theorems (which includes the axioms). Our attempt is to make provability be a syntactical counterpart of the semantical notion of truth.

So let us forget about truth and concentrate on the notion of provability. Suppose that we enrich the predicate calculus enough so that we could express—not the Liar proposition—but rather a modified version of the Liar proposition with 'true' replaced by 'provable'. That is, we would have:

$$\vdots$$

This sentence is not provable.

$$\vdots$$

Let 'G_p' be the name of the explicit sentence in the square. Does our new system become inconsistent just because it contains G_p, as the previous system became inconsistent just because it contained the Liar sentence? The answer is no, but something else occurs which is disconcerting. We can see by considering the following possibilities:

a)   Suppose that G_p is true. Then what it states is the case, so it is not provable. Hence, if G_p is true, then it is not provable.

b)   Suppose that G_p is not true. Then what it states to be the case is not so, and thus is provable. Hence, if G_p is not true, it is provable.

a')   Suppose that G_p is provable. Then, since it states that it is unprovable, it is not true. Hence, if G_p is provable, it is not true.

b')   Suppose that G_p is not provable. Then, since it states that it is not provable, it is true. Hence, if G_p is not provable, it is true.

Problem 61.   Determine the relation of (a') to (a), and (b') to (b).

What we can conclude from these possibilities is that if we have a system rich enough to express G_p, then it will contain a sentence (G_p itself) which is true if and only if it is not provable. Thus the cozy relationship between truth and provability which we attempt to achieve in a formal system—namely, that the set of true sentences* and the set of provable sentences be

---

* That is, true under any interpretation which makes all the axioms true.

identical—is destroyed. The Liar has disappeared but, like the Cheshire cat, his grin remains behind. Now what Gödel showed is that any system rich enough to express arithmetic—or better, primitive recursive arithmetic—contains the grin; that is, sentences analogous to G_p. Hence any system which contains primitive recursive arithmetic either proves sentences which are false (under some interpretation) or it leaves unproved sentences which are true (under some interpretation). Such—in very rough outline—is the reasoning and statement of Gödel's first incompleteness theorem.

At this level of analysis we can still be somewhat more exact. There are two terms of G_p which can be made more definite. The first is 'this'. It seems doubtful that we could ever reproduce an equivalent of 'this' in the system. Indeed, because of the ambiguity of demonstratives such as 'this', the ability to define an equivalent in the system would lead to inconsistency. For example, let 'A' stand for 'This sentence contains exactly eight words'. Then '$\bar{A}$' would stand for 'This sentence does not contain exactly eight words'. '$\bar{A}$' is true because it is the negation of a false sentence; but is also false because what it states is false. As we shall see, the 'this' can be eliminated from our sentence (without destroying its spirit) by creating a description which only the sentence fulfills.

The second term which can be made more definite (without being technical) is *provable*. What it means is *provable-in-the-system*. The notion of *provable* is a relative one: What is provable in one system may not be provable in another one. The Pythagorean theorem is provable in Euclidean geometry but not provable in Lobachevskian geometry. Hence, whenever we use the word *provable*, we must ask the question: Provable from what assumptions, by what rules of inference?

In light of our observations on these two terms, we may revise our diagram to become:

System $\mathscr{S}$

$$
\boxed{\begin{array}{l}
\vdots \\[4pt]
\text{The sentence in the square on page 162 of } A \\
\textit{Profile of Mathematical Logic is not provable-in-}\mathscr{S}. \\[4pt]
\vdots
\end{array}}
$$

Problem 62. Let '$G_\mathscr{S}$' be the name of the explicit sentence in the System $\mathscr{S}$ square. $G_\mathscr{S}$ could be added to the system as an initial formula. Then the troublesome $G_\mathscr{S}$ would be taken care of. We could raise two objections to this:

1) "If $G_\mathscr{S}$ were added to the system as an initial formula, it would be provable (since it is an initial formula). But because it states that it is

unprovable, it is not true. Thus we cannot add $G_\mathscr{S}$ to the system on pain of proving sentences which are not true."

2) "If $G_\mathscr{S}$ were added to the system, we could find a new sentence which would function in the same way as $G_\mathscr{S}$."

See if you can detect a flaw in the reasoning in (1), and then try to guess how the "new sentence" would be constructed, using the analogy of $G_\mathscr{S}$ as a guide.

We are now ready to consider a more exact analysis. Practically all misunderstandings of Gödel's first incompleteness theorem have been due to the drawing of conclusions from an analysis like the one that has so far been presented. This kind of explanation is extremely helpful in presenting the overall idea which motivates the proof. However, to properly understand the *proof itself*, it is very important to make distinctions which were not drawn in the heuristic explanation.

The most important of these distinctions is the distinction between the object language, the syntactical metalanguage, and (informal) arithmetic. The object language will be $\mathscr{A}$ [cf. §18]. Its intended interpretation is arithmetic. However, let us temporarily forget about this fact and only consider $\mathscr{A}$ to be what it is—a formal uninterpreted system—from which it follows that its formulas are neither true nor false. Since its terms do not denote anything, it is not about anything.

In contrast, the syntactical metalanguage is informal and in English. It is about the object language; that is, the names in the language denote the symbols of the object language and the relations among them. Its sentences are true or false.

Finally, arithmetic is informal and is in the symbolism of recursive arithmetic. It includes recursive arithmetic, but allows unrestricted use of quantifiers [cf. end of §22]. It is about natural numbers and their arithmetic relations; hence its sentences are either true or false. (On pages 178–179 there is a table—Table 15—whose purpose is to summarize the concepts, distinctions and correspondences necessary for understanding Gödel's first incompleteness theorem. Make frequent reference to this table. By the time you reach it, it is hoped that you will understand not only the table, but also Gödel's first incompleteness theorem.)

The crucial idea which Gödel was the first to notice (following a hint of Hilbert's) is that syntactical metamathematical statements can be interpreted arithmetically. This is done by associating some natural number with each symbol. This numbering of each symbol (now called *Gödel numbering*) is analogous to Descartes' numbering of the x- and y-axis in the Cartesian plane; it is the basis for showing similarity of structure between two apparently different realms. And just as the geometric statements correspond to algebraic statements, so syntactical metamathematical statements correspond to arithmetic statements. The analogy might be made clear by the following diagram:

Plane geometry	*Algebra*
point	pair of numbers
line	linear equation
circle	quadratic equation

Syntactical metamathematics	*Arithmetic*
symbol	number
formula	sequence of numbers
proof	sequence of sequences of numbers

What this correspondence enables us to do is to interpret syntactical meta-mathematical statements about relations between signs as arithmetical statements about relations between numbers. There is, in other words, an isomorphism between at least part of syntactical metamathematics and part of arithmetic. Thus, just as earlier we could not tell from a formal point of view whether we were playing ACE or HOT, so there is no formal way of telling whether we are playing the game of syntactical metamathematics or arithmetic. It is important to keep in mind that the arithmetic involved is not about the object language any more than a quadratic equation is about a circle. Or, since *about* is a vague word, perhaps it would be more appropriate to say that arithmetic is "about" the object language only in the very extended sense that a quadratic equation can be "about" a circle.

We shall discuss the exact nature of the correspondence between arith-metic and metamathematics in the next section. However, let us first explore the question of how it happened that two such different realms turned out to be isomorphic. The answer, I think, arises from similarity of intention. The notion of proof, as we have seen, was invented by such Pre-Socratics as Pythagoras and Zeno, and it was partly codified by Aristotle in the theory of the syllogism. The idea was to make knowledge no respecter of persons, to take it out of the theological realm and into the secular. Thus a proof, in order to be a proof, had to be recognizable *as* a proof by anyone. This attitude received its classical expression in Euclid's *Elements.*

Then, after a long hiatus, the notion of proof was further clarified in the development of the formal axiomatic method in the late nineteenth and early twentieth centuries. However, during this time the set-theoretical paradoxes were discovered and a number of artificial devices and dubious assumptions were needed to get around them. These devices and assump-tions had the effect of destroying the universality of proof, and thus of introducing something analogous to the theological element of old. *If* one believed in the axiom of reducibility [one of these doubtful assumptions (found in *Principia Mathematica*)], then such-and-such was proved. *If* one believed in the law of the excluded middle applied to infinite sets (as in Vinogradoff's proof), then such-and-such was proved, etc. This development raised in the minds of a number of mathematicians and logicians the question as to whether there is any justification for calling such so-called proofs "proofs." Instead, they reasoned, one should stick to the immediately evident, to that which can be completely exhibited. As we have seen, it was with this kind of intention that Skolem inquired into recursive arithmetic. In a concluding remark in his paper, Skolem states:

> This paper was written in the autumn of 1919, after I had studied the work of Russell and Whitehead. It occurred to me that already the use of the logical variables that they call "real" would surely suffice to provide a foundation for large parts of mathematics. .... The justification for introducing apparent variables ranging over infinite domains therefore seems very problematic; that is, one can doubt that there is any justification for the actual infinite or the transfinite [1923 332].

A typical reaction to such a statement is that since we can't derive a significant portion of classical mathematics without appealing to the actual infinite, the actual infinite thereby becomes justified and no serious attention need be given to those who propose keeping this concept out of mathematics. It must be pointed out, however, that the finitists in mathematics see themselves as trying to rid mathematics of the doubtful and unknowable and, just as the Pre-Socratics battled theological superstition, so the finitists see themselves as battling the mathematical superstition which depends on such dubious principles as the axiom of reducibility or the axiom of choice.

Hilbert, however, wanted to take a middle course between the finitists and transfinitists: He wished to prove the consistency of transfinite mathematics by finitary metamathematics. Thus, in retrospect, it seems a natural step for Gödel to consider the possibility that a structural similarity might exist between finitary syntactical metamathematics and a new area of mathematics—primitive recursive arithmetic—which was created with the purpose of avoiding any assumption of the actual infinite. According to Aristotle, the Pythagoreans thought that "reason" was "numerically expressible" [cf. §2]. In the work of Gödel we have this Pythagorean intuition embodied in an exact and beautiful way.

B. The Incompleteness of Arithmetic

In order to carry out the proof of the incompleteness of arithmetic, we need to show four things:

First, that the metamathematical names and sentences which were used in the heuristic explanation of Gödel's proof have corresponding names and sentences in primitive recursive arithmetic.

Second, that all names and sentences of primitive recursive arithmetic can be associated with terms and formulas of $\mathscr{A}$ such that, when a sentence of primitive recursive arithmetic is true, its corresponding formula in $\mathscr{A}$ is provable; and when a sentence of primitive recursive arithmetic is false, the negation of its corresponding formula in $\mathscr{A}$ is provable.

Third, that if the system $\mathscr{A}$ is consistent, it contains at least one formula such that neither it nor its negation is provable. Hence $\mathscr{A}$ is incomplete.

Fourth, that the incompleteness of $\mathscr{A}$ is directly related to the incompleteness of arithmetic.

The initial step in solving our first problem is to assign numbers to our primitive signs. This may be done in a number of ways. One such way is as follows:

1	3	5	7	9	11	13	17	19	23
'0'	','	'—'	'⊃'	'∀'	'('	')'	'x'	'y'	'z'

29	31	37	41	43	47	53	.	.	.
'='	'+'	'·'	'x_1'	'y_1'	'z_1'	'x_2'	.	.	.

The sequence is somewhat arbitrary. The '=', '+', '·' are out of what might be the expected order only to keep our numbering closer to Gödel's original numbering. His system—being more powerful than $\mathscr{A}$—could introduce '=', '+', and '·' as defined terms, and thus they were not primitive symbols in it. It is intended that in the sequence of numbers greater than 9, all primes and only primes greater than 9 be included. The fact that 9 is the only composite number is again accidental. It must be emphasized that there are many systems of Gödel numbering, some simpler than the one we are explaining. Our system is close to Gödel's for the sole purpose of easing the task of those who would like to read his original 1931 paper.

Each symbol corresponds to a natural number. Each formula corresponds to a finite sequence of numbers. For example:

Formula

Corresponding sequence of numbers

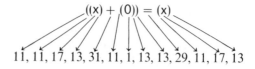

$((x) + (0)) = (x)$

11, 11, 17, 13, 31, 11, 1, 13, 13, 29, 11, 17, 13

A proof will then correspond to a finite sequence of finite sequences of numbers. For example:

Proof	Corresponding sequence of sequences of numbers
$((x) + (0)) = (x)$	11, 11, 17, 13, 31, 11, 1, 13, 13, 29, 11, 17, 13
$(\forall x)(((x) + (0)) = (x))$	11, 9, 17, 13, 11, 11, 11, 17, 13, 31, 11, 1, 13, 13, 29, 11, 17, 13, 13
$((\forall x)(((x) + (0)) = (x)) \supset ((0) + (0)) = (0))$	11, 11, 9, 17, 13, 11, 11, 11, 17, 13, 31, 11, 1, 13, 13, 29, 11, 17, 13, 13, 7, 11, 11, 1, 13, 31, 11, 1, 13, 13, 29, 11, 1, 13, 13
$((0) + (0)) = (0)$	11, 11, 1, 13, 31, 11, 1, 13, 13, 29, 11, 1, 13

Observe that each formula is either an initial formula or follows from earlier lines by a transformation rule of $\mathscr{A}$. Note further that the relationships between signs of $\mathscr{A}$ are preserved in relationships between numbers. For example, the last line of the proof is gotten from the second and the third by an application of *modus ponens*. The rule could be translated into a corresponding rule about a sequence of sequences of numbers.

For our purposes it will be convenient to have a single number correspond to a given formula and a single number correspond to a given proof. To this end we must remember what is called the *fundamental theorem of arithmetic*, according to which, if we ignore order, a natural number can be factored into a product of primes in exactly one way. This was probably known to Euclid, but it was first clearly stated and proved by Gauss. We shall assume it here.

Suppose, for example, that we take the number 1000. It is equal to $2^3 \cdot 5^3$ or $5^3 \cdot 2^3$. Let us call the order in which primes appear in the number sequence their *natural order*. Then there is just one way a composite number can be factored into a product of primes in their natural order. For example,

$$1000 = 2^3 \cdot 5^3, \qquad 99 = 3^2 \cdot 11, \qquad 142 = 2 \cdot 71.$$

Any sequence of natural numbers $a_1, a_2, \ldots, a_n$ may be made to correspond to a unique natural number:

$$b_1^{a_1} \cdot b_2^{a_2} \cdot \ldots \cdot b_n^{a_n},$$

where $b_1 = 2, b_2 = 3, \ldots, b_n = (n\text{th prime number})$. Thus for the sequence of numbers which corresponds to '$((x) + (0)) = (x)$' we have:

$$2^{11} \cdot 3^{11} \cdot 5^{17} \cdot 7^{13} \cdot 11^{31} \cdot 13^{11} \cdot 17^1 \cdot 19^{13} \cdot 23^{13} \cdot 29^{29} \cdot 31^{11} \cdot 37^{17} \cdot 41^{13}.$$

We shall use 'c_1' to designate this number. It is very large, but theoretically, if given c_1, we could break it down into its prime divisors, put the primes in their natural order, and then recapture the sequence of numbers. The number c_1 will be called the *Gödel number* of the formula '$((x) + (0)) = (x)$'. Similarly, we may get the Gödel number of our proof. Suppose that c_1, c_2, c_3, c_4 are, respectively, the Gödel numbers of the formulas (in order) in the above proof. Then the Gödel number of the proof is $2^{c_1} \cdot 3^{c_2} \cdot 5^{c_3} \cdot 7^{c_4}$, which we shall denote by 'd'.

Problem 63. How can we be sure that the Gödel number of a formula will not ever equal a Gödel number of a proof?

Hence there is a one-to-one correspondence between Gödel numbers and symbols, sequences of symbols (including formulas), and sequences of sequences of symbols (including proofs). But we must show that there is a structural identity. For some metamathematical statements this is easy. For example, the true metamathematical sentence, ' '(' is not the same as ')' ' corresponds to the true arithmetic sentence, '$11 \neq 13$'. But for the meta-mathematical statements which involve concepts such as initial formula or proof, the process is more difficult. What Gödel did was to start with the sequence of recursive functions and predicates that we have defined so far— that is, $x + y$, $x \cdot y$, x^y, $x = y$, $x < y$ [cf. §22]—and add 45 more definitions of primitive recursive functions or primitive recursive relations, which enabled him to create an arithmetical correlate of the metamathematical concept of proof. For our system the metamathematical statement is '**u** is a proof-in-$\mathscr{A}$ of **v**', where '**u**' stands for a variable whose domain of definition is sequences of sequences of symbols, and '**v**' stands for a variable whose domain of definition is sequences of symbols. The correlate in primitive recursive arithmetic is symbolized by: 'xPy'. This relation is true if and only if x is a Gödel number of a proof-in-$\mathscr{A}$ of the formula with Gödel number y. For example, 'dPc_4' is a true arithmetic sentence. The relation xPy is of the same type as $x + y = 4$; of course, it is more complicated. If we substitute any two numerals for 'x' and 'y', we can determine in a finite number of steps whether or not the relation holds. In fact, the primitive recursiveness of P gives an absolute guarantee that a proof-in-$\mathscr{A}$ is recognizable as a proof. From P we can get the *proof predicate* for $\mathscr{A}$ as follows:

$$Prov(x) \equiv (\exists y)(yPx).$$

'$Prov(x)$' is true if and only if x is a Gödel number of a formula provable-in-$\mathscr{A}$. Since there is no limitation on the bound variable, this definition is not recursive. Note that we have not yet proved that there is no primitive recursive proof predicate for $\mathscr{A}$; we have just not given one. The decision problem for $\mathscr{A}$ would be solved if a primitive recursive definition of the proof predicate could be found.

Problem 64. Determine the conditions under which the following are true:
a) $53P8$
b) $(\forall x)(xPc_4)$
c) $(\exists x)(xPc_4)$
d) $Prov(c_4)$

Besides the primitive recursive relation P, we shall need to define a number of primitive recursive functions. The first function is denoted by 'Sb' and is meant to serve as the arithmetic correlate of the metamathematical concept of substitution. It is somewhat harder to understand than P, but it may be made clear by taking a special case. Consider

$$Sb\left(x\,\begin{matrix}y\\z\end{matrix}\right).$$

It names the Gödel number of the formula which results from the formula with Gödel number x when the free variable with Gödel number y is replaced by the term with Gödel number z. If x is not the Gödel number of a formula, or y is not the Gödel number of a free variable in that formula, or z is not the Gödel number of a term, then

$$Sb\left(x\,\begin{matrix}y\\z\end{matrix}\right) = x.$$

The obscurity of the last two sentences may be overcome by thinking of a specific illustration:

Formula in $\mathscr{A}$	*Gödel number*
$((x) + (0)) = (x)$	c_1
$((0) + (0)) = (0)$	$Sb\left(c_1\,\begin{matrix}17\\1\end{matrix}\right)$

Hence

$$Sb\left(c_1\,\begin{matrix}17\\1\end{matrix}\right)$$

names the same number as c_4, that is,

$$c_4 = Sb\left(c_1\,\begin{matrix}17\\1\end{matrix}\right).$$

It should be clearly understood that

$$Sb\left(c_1\,\begin{matrix}17\\1\end{matrix}\right)$$

names a particular number just as $(5 \cdot 10) + 2$ names a particular number.

Similarly,

$$Sb\left(x \, {y \atop z}\right)$$

is a primitive recursive function just as $(x \cdot y) + z$ is a primitive recursive function. The function Sb may be substituted in itself [cf. the definition of primitive recursive function in §22]. For example, we might have

$$Sb\left(Sb\left(x \, {y_1 \atop z_1}\right) \, {y_2 \atop z_2}\right).$$

In such a case we shall write

$$Sb\left(x \, {y_1 \ y_2 \atop z_1 \ z_2}\right),$$

and similarly for more variables. If $y_1 \neq y_2$, then

$$Sb\left(x \, {y_1 \ y_2 \atop z_1 \ z_2}\right) = Sb\left(x \, {y_2 \ y_1 \atop z_2 \ z_1}\right).$$

The other three functions that we shall need are easy to understand. The first will be denoted by 'Nml', to remind us of *numeral*. '$Nml(x)$' names the Gödel number of the numeral (in $\mathscr{A}$) for x. For example, $Nml(0) = 1$, $Nml(1) = (2^{11} \cdot 3^1 \cdot 5^{13} \cdot 7^3)$, etc. The second function is denoted by 'Neg' and corresponds to the operation of *negation*. '$Neg(x)$' names the Gödel number of the negation of the formula with Gödel number x. Finally, the third function is denoted by 'Gen' and corresponds with the operation of *universal generalization*. '$xGeny$' names the Gödel number of the formula which results from the operation of universally generalizing the formula with Gödel number y with respect to the variable with Gödel number x. For example, $17Genc_1$ names the Gödel number of '$(\forall x)(((x) + (0)) = (x))$'. That is, $c_2 = 17Genc_1$.

Problem 65. Describe the numbers named by, or the conditions for the truth of, the following:

a) $Nml(2)$

b) $Neg(c_4)$

c) $Neg\left(Sb\left(c_4 \, {19 \atop Nml(1)}\right)\right)$

d) $Sb\left(c_1 \, {17 \ 19 \atop 1 \ \ 1}\right)$

e) $(\exists x)\left(xPSb\left(c_1 \quad \begin{matrix} 17 \\ 1 \end{matrix}\right)\right)$

f) $(\forall x)\left(xPSb\left((2^{11} \cdot 3^1 \cdot 5^{13} \cdot 7^{29} \cdot 11^{11} \cdot 13^{11} \cdot 17^1 \cdot 19^{13} \cdot 23^3 \cdot 29^{13}) \quad \begin{matrix} 17 \\ 1 \end{matrix}\right)\right)$

g) $Sb\left(c_1 \quad \begin{matrix} 17 \\ Nml(1) \end{matrix}\right)$

h) $Sb\left((17Genc_1) \quad \begin{matrix} 17 \\ Nml(1) \end{matrix}\right)$

i) $17Gen\left(Sb\left(c_1 \quad \begin{matrix} 19 \\ Nml(1) \end{matrix}\right)\right)$

j) $Prov\left(Neg\left(Sb\left((17Genc_1) \quad \begin{matrix} 19 \\ Nml(1) \end{matrix}\right)\right)\right)$

We have shown how to find expressions in primitive recursive arithmetic which are true if and only if their corresponding syntactical meta-mathematical statements are true. Hence, we are ready for our second problem, namely, to describe the kind of correspondence which holds between primitive recursive arithmetic and the system $\mathscr{A}$. To do this we shall state a *correspondence lemma* for $\mathscr{A}$:

Let $g_1, \ldots, g_n$ be the Gödel numbers of the first n variables in alphabetic order. (For example, $g_1 = 17$, $g_2 = 19$, $g_3 = 23$, $g_4 = 41$, etc.) Then for every primitive recursive relation $P(x_1, \ldots, x_n)$, there exists (in $\mathscr{A}$) a formula (with Gödel number r) which contains free variables with Gödel numbers $g_1, \ldots, g_n$ such that for all $x_1, \ldots, x_n$:

$$P(x_1, \ldots, x_n) \supset Prov\left(Sb\left(r \quad \begin{matrix} 17 & 19 & \cdots & g_n \\ Nml(x_1) & Nml(x_2) & \cdots & Nml(x_n) \end{matrix}\right)\right)$$

$$\overline{P(x_1, \ldots, x_n)} \supset Prov\left(Neg\left(Sb\left(r \quad \begin{matrix} 17 & 19 & \cdots & g_n \\ Nml(x_1) & Nml(x_2) & \cdots & Nml(x_n) \end{matrix}\right)\right)\right).$$

Consider what this means in a particular case, say, $1 + 2 = 3$. Since $x + y = z$ is a primitive recursive relation, there exists a formula in $\mathscr{A}$ with free variables 'x', 'y' and 'z'—namely, '((x) + (y)) = (z)'—such that when the numerals for 1, 2, and 3 are respectively substituted for 'x', 'y', and 'z', the result is a provable formula; that is, '(((0)') + (((0)')')) = ((((0)')')')' is provable-in-$\mathscr{A}$. It must be emphasized that the conditionals in the lemma are statements in primitive recursive arithmetic, not in $\mathscr{A}$.

Problem 66. Write out the first conditional for the special case $1 + 0 = 1$, using $x + 0 = x$ as the recursive predicate.

We shall not carry out the proof of the correspondence lemma here. But it might be made plausible by comparing the definition of *recursive* with the initial formulas and transformation rules of $\mathscr{A}$. Note in particular the similarity of the definition of *proof* and the definition of *recursive function*. Of course, $\mathscr{A}$ was set up with the intention that to every sentence of primitive recursive arithmetic there would correspond a formula in $\mathscr{A}$ which is provable if and only if the sentence is true. The fulfillment of this requirement does not guarantee that models for every interpretation of $\mathscr{A}$ will be isomorphic to the natural numbers. But if it were impossible to fulfil at least this requirement, the system could hardly be adequate for arithmetic. What the correspondence lemma shows is that provability-in-$\mathscr{A}$ is a syntactical counterpart of the semantical notion of primitive recursive truth.

We are now ready for our third problem, namely, to show that if $\mathscr{A}$ is consistent it contains an undecidable formula, that is, one which is neither provable nor refutable in $\mathscr{A}$. (A formula is *refutable* if its negation is provable.) Our aim is to construct in $\mathscr{A}$ a formula G which is analogous to G_p and $G_{\mathscr{G}}$.

Consider the following special case of the correspondence lemma:

1)
$$xP\left(Sb\left(y \quad \genfrac{}{}{0pt}{}{19}{Nml(y)}\right)\right) \supset Prov\left(Sb\left(q \quad \genfrac{}{}{0pt}{}{17 \quad 19}{Nml(x)\,Nml(y)}\right)\right)$$

2)
$$xP\left(Sb\left(y \quad \genfrac{}{}{0pt}{}{19}{Nml(y)}\right)\right) \supset Prov\left(Neg\left(Sb\left(q \quad \genfrac{}{}{0pt}{}{17 \quad 19}{Nml(x)\,Nml(y)}\right)\right)\right)$$

The formula with Gödel number q has two free variables, 'x' and 'y'. Consider the universal generalization of that formula with respect to 'x', that is, the formula with Gödel number $17Genq$. Let p be the Gödel number of that formula (hence $p = 17Genq$). Consider the following infinite sequence of substitution instances of (1):

3)
$$\overline{0P\left(Sb\left(p \quad \genfrac{}{}{0pt}{}{19}{Nml(p)}\right)\right)} \supset Prov\left(Sb\left(q \quad \genfrac{}{}{0pt}{}{17 \quad 19}{Nml(0)\,Nml(p)}\right)\right)$$

$$\overline{1P\left(Sb\left(p \quad \genfrac{}{}{0pt}{}{19}{Nml(p)}\right)\right)} \supset Prov\left(Sb\left(q \quad \genfrac{}{}{0pt}{}{17 \quad 19}{Nml(1)\,Nml(p)}\right)\right)$$

$$\overline{2P\left(Sb\left(p \quad \genfrac{}{}{0pt}{}{19}{Nml(p)}\right)\right)} \supset Prov\left(Sb\left(q \quad \genfrac{}{}{0pt}{}{17 \quad 19}{Nml(2)\,Nml(p)}\right)\right)$$

$$\vdots$$

Suppose that each of the antecedents is true. Then it follows that

$$(\forall x)\left(\overline{xP\left(Sb\left(p \begin{array}{c} 19 \\ Nml(p) \end{array}\right)\right)}\right).$$

Let 'G_A' denote this arithmetical expression. If G_A is true, it follows that the formula in $\mathscr{A}$ which has Gödel number

$$Sb\left(p \begin{array}{c} 19 \\ Nml(p) \end{array}\right)$$

is not provable-in-$\mathscr{A}$. Let 'G' denote that formula. Finally, consider the syntactical metamathematical sentence: 'G is not provable-in-$\mathscr{A}$'. Let 'G_M' denote that sentence.

We may summarize the relationships as follows: (G_M is true) if and only if (G_A is true) if and only if (G is not provable-in-$\mathscr{A}$). At this point the argument breaks down into two cases:

Case 1. Suppose that G is provable-in-$\mathscr{A}$. Then G_A is false and there exists an n which is the Gödel number of the proof-in-$\mathscr{A}$ of G. We then know

$$nP\left(Sb\left(p \begin{array}{c} 19 \\ Nml(p) \end{array}\right)\right).$$

But then by (2) and *modus ponens*, we get

4)
$$Prov\left(Neg\left(Sb\left(q \begin{array}{cc} 17 & 19 \\ Nml(n) & Nml(p) \end{array}\right)\right)\right).$$

On the other hand, if G is provable-in-$\mathscr{A}$, this means that the formula with Gödel number

$$Sb\left(p \begin{array}{c} 19 \\ Nml(p) \end{array}\right)$$

is provable-in-$\mathscr{A}$. But

$$Sb\left(p \begin{array}{c} 19 \\ Nml(p) \end{array}\right) = Sb\left((17Genq) \begin{array}{c} 19 \\ Nml(p) \end{array}\right) = 17Gen\left(Sb\left(q \begin{array}{c} 19 \\ Nml(p) \end{array}\right)\right).$$

But if the formula with Gödel number

$$17Gen\left(Sb\left(q \begin{array}{c} 19 \\ Nml(p) \end{array}\right)\right)$$

is provable-in-$\mathscr{A}$, by (3(e)) and (4(a)) of §17 the formula with Gödel number

$$Sb\left(q \begin{array}{cc} 17 & 19 \\ Nml(n) & Nml(p) \end{array}\right)$$

is also provable-in-$\mathscr{A}$. Hence

5) $$Prov\left(Sb\left(q\ \frac{17}{Nml(n)}\ \frac{19}{Nml(p)}\right)\right).$$

By (4) and (5) it follows that if $\mathscr{A}$ is consistent, G is not provable-in-$\mathscr{A}$.

Case 2. Suppose that $\overline{G}$ is provable-in-$\mathscr{A}$. By Case 1, if $\mathscr{A}$ is consistent, G is not provable-in-$\mathscr{A}$. Hence G_A is true and so each of the consequents in (3) is true; that is, we have

$$(\forall x)\left(Prov\left(Sb\left(q\ \frac{17}{Nml(x)}\ \frac{19}{Nml(p)}\right)\right)\right).$$

On the other hand, if $\overline{G}$ is provable-in-$\mathscr{A}$, the formula with Gödel number

$$Neg\left(Sb\left(p\ \frac{19}{Nml(p)}\right)\right)$$

is provable-in-$\mathscr{A}$. But

$$Neg\left(Sb\left(p\ \frac{19}{Nml(p)}\right)\right) = Neg\left(Sb\left((17Genq)\ \frac{19}{Nml(p)}\right)\right).$$

This result is curious, for we now have a formula by which it is possible to prove both every instance of it and also the negation of its universal generalization. A formal system is called *ω-inconsistent* if there is some formula **P(x)** such that the following are all true:

$$\vdash \overline{(\forall x)P(x)}, \quad \vdash P(0), \quad \vdash P(1), \quad \vdash P(2), \ldots$$

Otherwise it is called *ω-consistent*. It is possible for a system to be consistent without being *ω-consistent*. For example, suppose that $\mathscr{A}$ is consistent and we add $\overline{G}$ as an initial formula. This system would be consistent (since G would not be provable) but *ω-inconsistent* (by the above argument). A formal system in which, for some **P(x)**, it is possible to prove **P(0)**, **P(1)**, **P(2)**, ... but not **(∀x)P(x)** is called *ω-incomplete*. If in such an *ω*-incomplete system **(∀x)P(x)** could not be proved, we would not have *ω*-inconsistency, but rather a system in which a general formula exists such that every instance of the formula is provable without the formula itself being provable. We may thus conclude: If $\mathscr{A}$ is *ω*-consistent, $\overline{G}$ is not provable-in-$\mathscr{A}$. Our results may be summed up in a chart (Table 14).

So far we have shown that some formal system $\mathscr{A}$ has an undecidable formula G under the assumption of *ω*-consistency. We shall postpone showing that an undecidable formula exists under the weaker assumption of consistency and turn to our fourth problem, namely, to show how the

Table 14

Assume $\mathscr{A}$ is consistent. Then:	*Assume $\mathscr{A}$ is ω-consistent. Then:*
G is not provable-in-$\mathscr{A}$.	G is not provable-in-$\mathscr{A}$.
Hence G_M and G_A are both true.	Hence G_M and G_A are both true.
Each and every instance of G is provable-in-$\mathscr{A}$ (since all are primitive recursive).	Each and every instance of G is provable-in-$\mathscr{A}$ (since all are primitive recursive).
Thus $\mathscr{A}$ is ω-incomplete.	Thus $\mathscr{A}$ is ω-incomplete.
$\overline{G}$ may or may not be provable-in-$\mathscr{A}$.	$\overline{G}$ is not provable-in-$\mathscr{A}$.
If $\overline{G}$ is provable-in-$\mathscr{A}$, $\mathscr{A}$ is ω-inconsistent, since every instance of G is provable-in-$\mathscr{A}$.	
If $\overline{G}$ is not provable-in-$\mathscr{A}$, then $\mathscr{A}$ is incomplete.	Hence $\mathscr{A}$ is incomplete.

incompleteness of $\mathscr{A}$ is related to the incompleteness of arithmetic. Recall (the temporarily forgotten fact) that the intended interpretation of $\mathscr{A}$ is arithmetic. We know that if $\mathscr{A}$ is ω-consistent, every instance of G is provable, but G itself is not. Hence G_A is true. But under the intended interpretation, G means the same as G_A. Hence G under its intended interpretation is mathematically undecidable, but it is metamathematically decidable, and, in fact, we have proved it true.

"But," it might be objected, "this only shows that $\mathscr{A}$ under interpretation contains an undecidable sentence. Perhaps there is some other system for which the correspondence lemma is provable. And in that system there might be no undecidable formulas." The answer to this objection is that the Gödelian argument will work for all the well-known formal systems for which the correspondence lemma holds. As we have seen, if the correspondence lemma does not hold, the system can hardly be called adequate for arithmetic.

"But," it might again be objected, "the important point is not whether the Gödelian argument can be applied to this system or that system or even all known systems. The important question is: Can the Gödelian argument be applied to every *possible* formal system for which the correspondence lemma holds?" This is an important question, but we shall postpone discussion of it [cf. §26].

The situation may be summed up as follows: There are two versions of Gödel's first incompleteness theorem. The first is *syntactical*, in which the formulas of the formal system are left uninterpreted. As we have seen, there is no self-reference involved: G_M is about G, G_A about natural numbers and the way they can or cannot be factored, and G is not about anything.

The second is *semantical*, in which G is interpreted as G_A. Hence G, under the intended interpretation, is about natural numbers and the way they can or cannot be factored. There is a self-reference involved but it is not—like the Liar—of the destructive kind.

An analogy may make it clear. Suppose that there was a complicated equation in algebra. It would, of course, be about real numbers and relations among them. However, suppose that under the isomorphism of analytic geometry, the equation corresponded to a certain geometrical entity. Finally, suppose that that geometrical entity was the equation itself. This is the Liar's grin. In short, the understanding of Gödel's incompleteness theorem requires that we keep in mind a distinction between self-reference with a semantical predicate (such as is involved in the Liar) and the extended sense of self-reference which operates via an isomorphism (such as is involved in the semantical version of Gödel's first incompleteness theorem).*

In a footnote to his 1931 paper Gödel states that "any epistemological antinomy could be used for a similar proof of the existence of undecidable propositions" [598]. We might here note how some others could be used. Consider the Dualist form of the Liar paradox [cf. §12]. This argument could be expressed in arithmetical symbolism as follows: Let r be the Gödel number of the formula which corresponds to

$$(\forall x)\left(xP\left(Sb\left(\overline{y\ \genfrac{}{}{0pt}{}{19}{Nml(z)}\ \genfrac{}{}{0pt}{}{23}{Nml(y)}}\right)\right)\right)$$

and s be the Gödel number of the formula which corresponds to

$$(\exists x)\left(xP\left(Sb\left(y\ \genfrac{}{}{0pt}{}{19}{Nml(z)}\ \genfrac{}{}{0pt}{}{23}{Nml(y)}\right)\right)\right).$$

Consider the two arithmetic propositions:

$$(\forall x)\left(xP\left(Sb\left(\overline{s\ \genfrac{}{}{0pt}{}{19}{Nml(r)}\ \genfrac{}{}{0pt}{}{23}{Nml(s)}}\right)\right)\right)$$

and

$$(\exists x)\left(xP\left(Sb\left(r\ \genfrac{}{}{0pt}{}{19}{Nml(s)}\ \genfrac{}{}{0pt}{}{23}{Nml(r)}\right)\right)\right).$$

We shall let 'S_A' and 'T_A' respectively denote them and 'S' and 'T' respectively denote their correlates in $\mathscr{A}$.

---

* This distinction is not the only one necessary to understand self-reference. It would not, for example, account for the relevant, distinct differences among the examples given in the last two paragraphs of §12. Nevertheless, it is the most important one as far as mathematical logic is concerned. For further discussion see Parsons and Kohl 1960. Gödel's own views on self-reference may be found in Davis 1965 63–64.

The argument would proceed roughly as follows: Suppose that S is provable. Then T_A is true and T is not provable. But under the intended interpretation T is true if and only if S is provable. Hence there would exist under the intended interpretation a true sentence which is not provable. On the other hand, suppose that S is not provable. Then T_A is false and T (hopefully) cannot be proved. But then S under the intended interpretation is true and we again have a true formula which is not provable. Other paradoxes could be used. Indeed, it could be argued that the *God exists* version of the Liar paradox [cf. §12] bears the closest resemblance to Gödel's argument, because the "God exists" in the argument could be replaced by any necessarily true proposition as, for example, the conjunction of the axioms for arithmetic.

The relation of the Richard paradox to the argument for undecidable sentences may be seen by comparing them point by point.

Richard argument	*Gödel–Rosser argument*
1) Consider a natural language in which the properties of natural numbers can be defined.	1) Consider a formal system in which the properties of natural numbers can be defined.
2) Order these definitions by length and, within definitions of the same length, alphabetically: $$1 \leftrightarrow D_1(x)$$ $$2 \leftrightarrow D_2(x)$$ $$\vdots$$	2) Order these definitions by length and, within definitions of the same length, alphabetically: $$1 \leftrightarrow A_1(x)$$ $$2 \leftrightarrow A_2(x)$$ $$\vdots$$
3) For any natural number m, call m *non-Richardian* if and only if '$D_m(m)$' is true. In the contrary case, m is *Richardian*.	3) For any natural number n, call n *non-Gödelian* if and only if '$A_n(n)$' is provable (where 'n' is the numeral for n). In the contrary case, n is *Gödelian*.
4) The definition of 'Richardian'—symbolize it by '$D_r(x)$'—is itself somewhere in the list. Therefore '$D_r(x)$' is true if 'x is Richardian' is true. The definition of 'non-Richardian'—symbolize it by '$\overline{D_r(x)}$'—is also somewhere in the list. Therefore '$\overline{D_r(x)}$' is true if and only if 'x is non-Richardian' is true.	4) The definition of 'Gödelian' is not itself in the list, but there does exist a formula—symbolize it by '$A_g(x)$'—in the list such that '$A_g(x)$' is provable if and only if 'x is Gödelian' is true. The definition of 'non-Gödelian' is also not itself in the list, but there does exist a formula—symbolize it by '$\overline{A_g(x)}$'—in the list such that '$\overline{A_g(x)}$' is provable if and only if 'x is non-Gödelian' is true.

Table 15

	Syntactical metalanguage		Arithmetic		Object language	
Characteristics	Informal and in English. About symbols and their syntactical properties and relations. Names or name forms denote symbols, properties, and relations of the object language. Sentences are true or false. When variables are replaced by names, name forms become names and predicates become sentences.		Informal and in the symbolism of recursive arithmetic, which it includes. About natural numbers and their arithmetic properties and relations. Constants or functions denote natural numbers, properties or relations among natural numbers. Sentences are true or false. When variables are replaced by constants, functions become names and predicates become sentences.		Formal and in the symbolism of $\mathscr{A}$. Not about anything, since it is uninterpreted. Terms do not denote anything because they are uninterpreted. Formulas are neither true nor false, but are provable or not provable. (Under the intended interpretation of $\mathscr{A}$ as arithmetic, terms become either constants or functions of natural numbers, and formulas become either sentences about natural numbers or predicates with variables whose domain of definition is the natural numbers.)	
	Name or name forms	Sentences or predicates	Constants or functions	Sentences or predicates	Terms	Formulas
Examples	'′' '((((0)′)′)′' the proof formed by adjoining the formula **v** to the proof **u**.	'′' is a symbol. '(0)′' is a numeral for 1. '((0) + (0)) = (0)′' is provable. **u** is a proof-in-$\mathscr{A}$ of **v**. G is provable.	4 $(5 \cdot 10) + 2$ $(x \cdot y) + z$ $Nml(x)$ $Neg(c_1)$ $17 Genq$ $Sb\left(x^{y}_{z}\right)$ $Neg\left(Sb\left(p^{19}_{Nml(p)}\right)\right)$	$1 + 2 = 3$ $(x + y) = 4$ xPy $Prov(c_4)$ $\overline{\left(xP\left(Sb\left(y^{19}_{Nml(y)}\right)\right)\right)} \supset$ $Prov\left(Sb\left(q^{17}_{Nml(x)} {}^{19}_{Nml(y)}\right)\right)$ $17Genq = p$ $\overline{(\forall x)\left(xP\left(Sb\left(p^{19}_{Nml(p)}\right)\right)\right)(yPx)(y\mathbf{A})(x\mathbf{E})}$	$((((0)′)′)′$ x $((x) \cdot (y)) + (z)$ $(y)′$	$(((0)′) + ((0)′)′) =$ $(((((0)′)′))$ $((x) + (y)) =$ $(((((0)′)′))$ $((y) \cdot (x)\mathbf{A})(\mathbf{A})(x\mathbf{A})(((x) + (y)) =$ $((y) + (x))))$

Correspondences	To each sentence of syntactical metamathematics there corresponds a sentence of arithmetic which is true if and only if the syntactical metamathematical sentence is true.	To each sentence of recursive arithmetic there corresponds a formula of the object language which is provable if and only if the sentence of recursive arithmetic is true.	To each sentence of syntactical metamathematics there corresponds a formula of the object language which under the intended interpretation is true if and only if the syntactical metamathematical sentence is true.
Examples	('‛' differs from '¬' is true) if and only if ('$3 \neq 5$' is true) if and only if ('$\overline{((((0)')')') = (((((0)')')')')')}$' is provable-in-$\mathscr{A}$). (G$_M$ is true) if and only if (G$_A$ is true) if and only if (G is not provable-in-$\mathscr{A}$).		

Gödel's first incompleteness theorem (syntactical version)	Gödel's first incompleteness theorem (semantical version)
The system $\mathscr{A}$ is understood as being uninterpreted. Hence G, as a formula of $\mathscr{A}$, is not about anything and is neither true nor false. G_A is about natural numbers and the way they can or cannot be factored. G_M is about G and states that G is not provable-in-$\mathscr{A}$. If $\mathscr{A}$ is consistent, then G_A and G_M are both true, and G is not provable-in-$\mathscr{A}$. No self-reference is involved.	The system $\mathscr{A}$ is understood under its intended interpretation as arithmetic. Hence G is interpreted as G_A. Both are about natural numbers and the way they can or cannot be factored. If $\mathscr{A}$ is consistent, then G, G_A, and G_M are all true. G involves self-reference in the following sense: G states that a certain relation between natural numbers holds. But this relation can only hold provided that G_M is true. G_M is about G and says that G is not provable-in-$\mathscr{A}$. Hence an extended sense of self-reference is involved analogous to that which operates in the Dualist version of the Liar paradox.

The relation of Gödel's first incompleteness theorem to the Liar paradox

The formula with Gödel number p is the formal correlate in $\mathscr{A}$ of $(\forall x)\overline{\left(xP\left(Sb\left(y\,{}^{19}_{Nml(y)}\right)\right)\right)}$. $Sb\left(p\,{}^{19}_{Nml(p)}\right)$ is the Gödel number of the formula which results from the formula with Gödel number p when the free variable with Gödel number 19 is replaced by the numeral for p. Thus $Sb\left(p\,{}^{19}_{Nml(p)}\right)$ is the Gödel number of the formal correlate of $(\forall x)\overline{\left(xP\left(Sb\left(p\,{}^{19}_{Nml(p)}\right)\right)\right)}$, which is to say, the formal correlate of G_A. But if G_A is true, then G, the formal correlate of G_A, is not provable-in-$\mathscr{A}$. Under interpretation of $\mathscr{A}$ as arithmetic, G expresses the same relationship between natural numbers as G_A, a relationship which holds only if G is not provable-in-$\mathscr{A}$. If we assume $\mathscr{A}$ is consistent, this relationship holds, that is, G is true of the natural numbers and therefore under interpretation G expresses a true fact which is inconsistent with its provability-in-$\mathscr{A}$. Hence G is analogous to G_p and $G_{\mathscr{S}}$ [cf. §23A].

5) Suppose that '$D_r(r)$' is true. Then r is non-Richardian (by 3). But then '$\overline{D_r(r)}$' is true (by 4).

5) Suppose that '$A_g(g)$' is provable. Then g (which is represented by the numeral 'g') is non-Gödelian (by 3). But then '$\overline{A_g(g)}$' is provable (by 4).

6) Suppose that '$\overline{D_r(r)}$' is true. Then r is non-Richardian (by 4). But then '$D_r(r)$' is true (by 3).

6) Suppose that '$\overline{A_g(g)}$' is provable. Then g is non-Gödelian (by 4). But then '$A_g(g)$' is provable (by 3).

7) Conclusion: '$D_r(r)$' is true if and only if '$\overline{D_r(r)}$' is true.

7) Conclusion: if the formal system is consistent, neither '$A_g(g)$' nor '$\overline{A_g(g)}$' is provable.

Problem 67. There is a fallacy in (4) of the Richard argument. Explain as carefully as you can in what this fallacy consists.

Problem 68. The right-hand column is called the *Gödel–Rosser* argument, because five years after Gödel's 1931 paper J. B. Rosser devised a proof in which the condition of ω-consistency is replaced by simple consistency. That is, we can assert that if $\mathscr{A}$ is consistent, there exists an undecidable formula in $\mathscr{A}$. Metamathematically, the Rosser formula asserts the following: If there is a proof of this formula in $\mathscr{A}$, then there is an equal or shorter proof of its negation. Let 'R_A' stand for the corresponding formula in arithmetic. Making use of such symbols as '$\forall$', '$\exists$', 'P', 'Sb', '$\leq$', etc., express R_A in a way analogous to the expression of G_A as

$$(\forall x)\left(\overline{xP\left(Sb\left(p \begin{array}{c} 19 \\ Nml(p) \end{array}\right)\right)}\right).$$

[*Hint*: First devise a number analogous to p (denote it by 't').]

C. Consistency and Categoricalness

But is $\mathscr{A}$ consistent? This is the subject of Gödel's second incompleteness theorem. By the argument of Case 1 of his first theorem, we have: If $\mathscr{A}$ is consistent, G is not provable-in-$\mathscr{A}$. Denote the antecedent by 'C_M'. Hence $C_M \supset G_M$. What arithmetic sentence corresponds to C_M? One candidate is $(\forall x)(\overline{xPNeg(c_4)})$. Denote it by '$C_A$'. C_A is true if and only if $\mathscr{A}$ is consistent. For if $\mathscr{A}$ is consistent, C_A is true, since the formula with Gödel number c_4 is provable-in-$\mathscr{A}$ [cf. §23B]. Conversely, if C_A is true, there is a formula not provable-in-$\mathscr{A}$, and so $\mathscr{A}$ is consistent. Let the formula in $\mathscr{A}$ which corresponds with C_A be denoted by 'C'. The argument by which $C_M \supset G_M$ was proved can be arithmetized to give us $C_A \supset G_A$, that is,

$$(\forall x)(\overline{xPNeg(c_4)}) \supset (\forall x)\left(\overline{xP\left(Sb\left(p \begin{array}{c} 19 \\ Nml(p) \end{array}\right)\right)}\right).$$

The formula in $\mathscr{A}$ corresponding to this arithmetic sentence is $C \supset G$. It is possible to prove $C \supset G$ in $\mathscr{A}$, although we shall not carry out the proof. If it could not be carried out, $\mathscr{A}$ would be an inadequate formalization of arithmetic. Hence we have $\vdash C \supset G$. Suppose that we could prove C in $\mathscr{A}$, that is, $\vdash C$. By *modus ponens* [4(a) of §17] we have $\vdash G$, and the argument of Case 1 of the first incompleteness theorem gives us the inconsistency of $\mathscr{A}$. The upshot of all this is Gödel's second incompleteness theorem: If $\mathscr{A}$ is consistent, C is not provable-in-$\mathscr{A}$; if $\mathscr{A}$ is inconsistent, C is provable-in-$\mathscr{A}$ (since every formula is provable).

The theorem gets its significance from the fact that C is true (under the intended interpretation) if and only if $\mathscr{A}$ is consistent. Further, all the arguments about the application of Gödel's first incompleteness theorem to other systems also apply to the second.

Problem 69. A reader who compares the proofs of the two theorems might make the following observation: "According to the argument of the second theorem, the informal argument that the consistency of $\mathscr{A}$ implies that G is not provable can be formalized to $\vdash C \supset G$. But we can do the same thing with respect to Case 1 in the proof of the first theorem. There it was argued that if G is provable, then $\overline{G}$ is provable. Hence we have $\vdash \overline{G} \supset G$. By the truth-table technique, we can see that $(\overline{G} \supset G) \supset G$ is a tautology, and therefore is a theorem. Thus, by *modus ponens*, we have $\vdash G$, from which it follows that $\mathscr{A}$ is inconsistent." Explain the fallacy in this argument.

Perhaps by using methods not formalizable in $\mathscr{A}$ we could prove the consistency of $\mathscr{A}$. As we shall see, this can be done [cf. §29]. However, as far as proving the consistency of arithmetic goes, such a proof, even if formally valid, has this defect: It commits the fallacy of begging the question. If we are in doubt about the consistency of arithmetic, then we shall not have our doubts relieved by a metamathematical proof which uses stronger (and therefore more doubtful) principles. Of course, in a trivial sense, even if we could prove the consistency of arithmetic by using methods no stronger than arithmetic ones, there is a *petitio principii*, since we are using the principles of arithmetic to prove the consistency of arithmetic. This brings us back to the Aristotelian question of whether everything can be proved. The answer, as we have seen, is no. If we are going to inquire at all, we must assume some basic principles of inquiry. Here mathematical logic is in the great Greek tradition of assuming the possibility of inquiry, in contrast to, say, the Buddhist tradition, which suggests that inquiry is pointless. Relying on a tradition that begins with the Pythagoreans, as well as observation of numerous diverse cultures, mathematical logicians have taken the most elementary parts of arithmetic (that is, primitive recursive arithmetic) as their basis for inquiry. The apparent inference that we may draw from Gödel's second incompleteness theorem is that this basis does not appear to have the virtue of being self-validating.

Problem 70.   Determine the effect on consistency caused by:
a) Adding C to the initial formulas of $\mathscr{A}$.
b) Adding $\overline{\text{C}}$ to the initial formulas of $\mathscr{A}$.

We have so far shown that if arithmetic is consistent it is incomplete, and that there does not appear to be a proof for its consistency which does not commit—in a nontrivial way—the fallacy of begging the question. If we assume consistency, we can show the independence of the initial formulas of $\mathscr{A}$, although we shall not prove that here. Instead, we turn to the more interesting question of categoricalness.

By Gödel's completeness theorem, we know that any formula expressible in the predicate calculus is valid if and only if it is provable. We know that the inferences of arithmetic are expressible in the language of the predicate calculus. Consider this question: Is the inference from the initial formulas of $\mathscr{A}$ to the formula G a valid one? If it is, then G is provable and $\mathscr{A}$ is inconsistent. Thus *if* $\mathscr{A}$ is consistent the inference from the initial formulas of $\mathscr{A}$ to the formula G is not valid. By the definition of *validity*, we then have the existence of a model in which the initial formulas of $\mathscr{A}$ are true but G is false. But, as we have seen, if $\mathscr{A}$ is consistent, G is true under the intended interpretation. So there exists a nonintended model of $\mathscr{A}$ which is not isomorphic to the intended model. Hence $\mathscr{A}$ is not categorical.

Actually, any formal system which has a model whose cardinality is $\geq \aleph_0$ has a model of any transfinite cardinality. So if $\mathscr{A}$ is consistent, it has a model of any transfinite cardinality. The comments which follow are intended to apply to denumerable models.

A subset of any nonintended model is isomorphic to the natural numbers, as can easily be proved (N.B. the recursive definitions of '+' and '·' are part of the initial formulas). Because there are nonintended models, we know that some models of $\mathscr{A}$ contain objects other than those in the subset isomorphic to the natural numbers. These objects we shall call *unnatural numbers*. It can be shown that they are all larger than any natural number. Furthermore, it can be shown that all nonintended models have the same order type, namely,

$$\omega + (^*\omega + \omega)\eta.$$

What '$\omega + (^*\omega + \omega)\eta$' means is difficult to clarify in a brief fashion. 'ω' refers to the ordering structure of the natural numbers, that is, a set with a denumerable number of elements containing a first element, no last element, and every element having a unique successor. '$^*\omega$' refers to the ordering structure of the negative integers, that is, a set with a denumerable number of elements containing a last element, no first element, and every element having a unique predecessor. 'η' (eta) refers to the ordering structure of the rational numbers, that is, a set with a denumerable number of elements containing no first element and no last element, and having another element between

every two elements. The following diagram may be of help in understanding
'$\omega + (*\omega + \omega)\eta$'.

	Order type	Example of ordered set
1)	ω	$\{1, 2, 3, \ldots\}$
2)	$*\omega$	$\{\ldots, -3, -2, -1\}$
3)	$*\omega + \omega$	$\{\ldots, -3, -2, -1, 0, 1, 2, 3, \ldots\}$
4)	η	$\{\ldots, -1\frac{1}{4}, \ldots, -\frac{1}{3}, \ldots, \frac{1}{5}, \ldots, 37\frac{6}{13}, \ldots\}$
5)	$\omega + (*\omega + \omega)\eta$	$\{1, 2, 3, \ldots \quad \ldots -1_{-1\text{-}1/4}, 0_{-1\text{-}1/4}, 1_{-1\text{-}1/4} \cdots \quad \cdots$
		$\{-1_{-1/3}, 0_{-1/3}, 1_{-1/3}, \cdots \quad \cdots -1_{1/5}, 0_{1/5}, 1_{1/5}, \ldots$
		$\{\ldots -1_{37\text{-}6/13}, 0_{37\text{-}6/13}, 1_{37\text{-}6/13}, \cdots \quad \cdots\}$

In (4) one should think of all the rationals as being ordered by size; in (5)
one should think that after the natural numbers comes an infinite set such
that for each element in (4) there corresponds a sequence like (3). Thus
$15_{1/4}$ comes before $3_{1/2}$ but after $16_{-1/3}$.

If we assume that $\mathscr{A}$ is consistent, G is true in the domain of natural
numbers, but is false in some nonintended model with order type $\omega + (*\omega +$
$\omega)\eta$; hence $\overline{G}$ is false in the domain of natural numbers, but is true in some
nonintended model with order type $\omega + (*\omega + \omega)\eta$. On one hand, if we
add G to the initial formulas of $\mathscr{A}$, G is true in the domain of the natural
numbers and provable in the new system; however, in this new system there
will be another formula which will be false in the domain of natural numbers,
but true in some nonintended model with order type $\omega + (*\omega + \omega)\eta$. On
the other hand, if we add $\overline{G}$ to the initial formulas of $\mathscr{A}$, the resulting system
is ω-inconsistent, that is, there is a predicate $\mathbf{P(x)}$ such that the following
propositions are all true:

$$\vdash(\exists x)\mathbf{P(x)}, \vdash\overline{\mathbf{P(0)}}, \vdash\overline{\mathbf{P(1)}}, \vdash\overline{\mathbf{P(2)}}, \ldots.$$

One of the formulas named by the first proposition is just $\overline{G}$ itself, and hence
the others could name each individual instance of G. However, under
interpretation, the object whose existence is asserted by $\overline{G}$ is not a natural
number (otherwise we would have an inconsistent system) but rather an
unnatural number. Hence the phenomenon of ω-inconsistency represents a
defect only if our intention is to formalize the natural numbers (or some
other ω sequence). If we intend to have a formal system adequate for a
model with order type $\omega + (*\omega + \omega)\eta$ (or, of course, many other models),
ω-inconsistency is not a defect, it is an essential virtue.

Now assume that $\mathscr{A}$ can be interpreted as arithmetic and that we add
G to the initial formulas of $\mathscr{A}$ to form system $\mathscr{A}_2^1$. Then, in a similar way,
we add $\overline{G}$ to the initial formulas of $\mathscr{A}$ to form system $\mathscr{A}_2^2$. If $\mathscr{A}$ is consistent,
so are $\mathscr{A}_2^1$ and $\mathscr{A}_2^2$ by Gödel's first incompleteness theorem. By Gödel's
completeness theorem, both will possess a denumerable model. We apply
Gödel's incompleteness argument to each system to get G_2^1 and G_2^2,

respectively. Now the process can be repeated by successively adding G_2^1 and $\overline{G}_2^1$ to $\mathscr{A}_2^1$, and G_2^2 and $\overline{G}_2^2$ to $\mathscr{A}_2^2$. This process can be repeated indefinitely, as suggested by the following tree pattern:

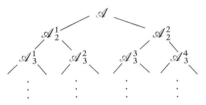

No model of any system of a given row can be a model of any other system in that row, but models for any of the systems are also models for $\mathscr{A}$.

Consider the sequence of systems $\mathscr{A}$, $\mathscr{A}_2^1$, $\mathscr{A}_3^1, \ldots, \mathscr{A}_n^1, \ldots$. We shall assume each $\mathscr{A}_i^1$ is formed by adding the formula which comes out true under the Gödelian argument applied to system $\mathscr{A}_{i-1}^1$ ($i \geq 3$). All the systems in the tree pattern have denumerable models with order type $\omega + (*\omega + \omega)\eta$; except for the sequence of systems $\mathscr{A}$, $\mathscr{A}_2^1$, $\mathscr{A}_3^1$, etc., this is the only order type represented. Each of the systems in the latter sequence —in addition to models of order type $\omega + (*\omega + \omega)\eta$—has models of order type ω—the only order type we intended when we set up $\mathscr{A}$. The upshot is that all the nonintended denumerable models are isomorphic relative to the order predicate $<$, but they are not in general isomorphic with regard to their other predicates. In other words, from a formal point of view, we can get sets which contain only natural numbers or their structural equivalent (namely, finite sets), and we can get sets which contain all the natural numbers or their structural equivalent [namely, sets with order type $\omega + (*\omega + \omega)\eta$], but we cannot get sets which contain all and only the natural numbers or their structural equivalent (namely, sets with order type ω). Formally speaking, we can never get the natural numbers without getting unnatural numbers.

Yet the sequence of systems $\mathscr{A}$, $\mathscr{A}_2^1$, $\mathscr{A}_3^1$, etc., represents an increasingly accurate approach to arithmetic in this sense: For any system (except the first) in this sequence, there are sentences which are true of the natural numbers and which are provable in that system, but are not provable in any preceding system, while every sentence which is both true of the natural numbers and provable in some system is provable in every succeeding system. The situation is in some respects analogous to the problem of incommensurate magnitudes which the Pythagoreans faced: they could achieve only very rough approximations of the real number $\sqrt{2}$. By using other methods, Eudoxus was able to solve the problem [cf. §10A]. This suggests that the problem of unnatural numbers might be amenable to new methods, which if they were exploited would enable us to show that some formal system can categorically characterize the natural numbers. No known formal system can. The question of whether any possible system can will be postponed [cf. §26].

The fact that no known system will categorically define the natural numbers was discovered by Skolem in the thirties without reference to Gödel's 1930 and 1931 papers. The upshot of this part of the work of these two men comes to this: Gödel's completeness theorem shows that formal arithmetic is as complete as possible (any formula not provable is false under some interpretation); Gödel's incompleteness theorem and the work of Skolem show that there exist an infinite number of nonintended models; Skolem and others [cf. Kemeny 1958] have shown that all nonintended denumerable models are of order type $\omega + (*\omega + \omega)\eta$. In short, if we adopt only the point of view of a formal axiomatic system, we can never be sure we are talking about the natural numbers or their structural equivalent.

We are ready to consider the question of a decision procedure. Note carefully that the results presented so far leave open the question of whether there is a decision procedure for arithmetic. On the basis of what has been presented, it is still conceivable that there is a finitary method which would tell us whether an arbitrary formula is a theorem, its negation is a theorem, or it is undecidable. Before we can answer the question of a decision procedure for $\mathscr{A}$, we shall need to characterize the notion of a general recursive function. This concept is due to Gödel, who built on a suggestion in a letter from Herbrand. Every primitive recursive function is general recursive, but not vice versa. Here is an example given by Gödel in 1934 (others were known as early as 1928). Suppose that $f(x)$ and $g(y)$ are given primitive recursive functions. Let $h(x, y)$ be defined recursively as follows:

$$h(0, y) = g(y)$$

$$h(x + 1, 0) = f(x)$$

$$h(x + 1, y + 1) = h(x, h(x + 1, y))$$

It can be shown that for some primitive recursive functions f and g, h is not primitive recursive. And yet it can readily be seen that for every ordered pair of numbers, the function h determines exactly one number and does so in a finitary way (for fixed primitive recursive functions f and g). What makes h fail to be primitive recursive is the simultaneous induction on two variables. A function f is called *general recursive* if there is a finite sequence of functions ending with f such that each function is a primitive recursive function, or is defined from preceding functions in the sequence by substitution, or is defined by recursion (with any finite number of variables) from preceding functions. A general recursive predicate is defined in a way that is exactly analogous to the way that a primitive recursive predicate is defined. From this point on, the use of the word *recursive* without qualification will mean *general recursive*.

Let us return to the question of a decision procedure. In order to prove the existence of a decision procedure for a system, one has to find an effective finite method for determining whether or not a formula is a theorem. For such a proof it is not necessary to define the general concept of an effective finite method. It suffices to merely exhibit a method which everyone accepts as effective (that is, no inventiveness is needed beyond what is learned in the procedure) and finite (that is, that anyone could finish (barring death, etc.) in a finite amount of time). For a proof that no decision procedure exists, however, there must be an accurate definition of the concept.

An analogy will make this clear. To bisect an angle in plane Euclidean geometry is easy; one does not have to know everything that can be done with a straightedge and a compass. All one has to do is to construct a bisector. In contrast, to show that it is impossible in general to trisect an angle requires that one have a precise definition of *all possible constructions with a straightedge and a compass.* Because the ancient Greeks lacked such a precise definition, they were never in a position to prove the impossibility of trisection. Hence, in general, impossibility proofs are more fundamental and more difficult than proofs of construction.

If we are going to consider the possibility that arithmetic might not have a decision procedure, we must precisely define decision procedure and, consequently, effective finite method. The purpose of an effective finite method is to determine the truth value of some given instance of a proof predicate for some system, which would include calculating functions. Our problem, then, is to know what an effectively calculable function is, and what an effectively decidable predicate is. In 1936 Alonzo Church proposed that we identify the intuitive notion of an effectively calculable function with the mathematically exact notion of a general recursive function and hence also the intuitive notion of an effectively decidable predicate with the mathematically exact notion of general recursive predicate. This identification is called *Church's thesis* and we shall discuss it in detail in Chapter 5 [cf. §26].

Assuming the thesis, Church was able to show both that there is no decision procedure for arithmetic and that there is no decision procedure for the predicate calculus. The proof for these conclusions depends in an essential way on the device of Gödel numbering and the diagonal argument. We shall not give a sketch of a proof of these conclusions,* but their plausibility is suggested by the fact that the set of predicates is nondenumerable, while the set of recursive predicates is denumerable [cf. Kleene 1967 246–247, Webb 1968 164].

If we assume Church's thesis, the results of metatheory can be translated into the theory of recursive functions. To make this translation, we shall need several definitions.   A set of numbers for which there is a

---

* For a proof, see Patton 1965 and the references therein.

recursive predicate which is true of just the numbers in the set (and no others) is called a *recursive set*. For example, since $\overline{Sg}$ is a recursive function [cf. answer to Problem 60], it follows that the following function is recursive:

$$Ev(0) = 0$$

$$Ev(x') = \overline{Sg}(Ev(x)).$$

Now let '$E(x)$' be true if and only if $Ev(x) = 0$. Thus '$E(0)$' is true, '$E(1)$' is false, '$E(2)$' is true, '$E(3)$' is false, etc. Finally, let $(x \in S)$ if and only if '$E(x)$' is true. Then S is a recursive set, and is, of course, the set of even numbers. We shall call a set *recursively enumerable* if there exists a recursive function f such that the sequence $f(0), f(1), f(2), \ldots$ contains exactly the same elements as the set.

Assume that we have applied the technique of Gödel numbering to the propositional calculus and the predicate calculus. Then we have the following results: For the propositional calculus the set of Gödel numbers for valid formulas (tautologies) is recursive, as is the set of Gödel numbers for invalid formulas (Post's completeness theorem). Hence the proof predicate for the system is recursive and we have a decision procedure.

For the predicate calculus, the set of Gödel numbers for valid (or provable) formulas is recursively enumerable (Gödel's completeness theorem) and the set of Gödel numbers for invalid (nonprovable) formulas is not recursive (Church's theorem). From a result of Post to the effect that if a set and its complement are recursively enumerable then the set is recursive, it follows that the set of Gödel numbers for invalid (nonprovable) formulas is not recursively enumerable. Hence the proof predicate for the system is not recursive and we have no decision procedure. For arithmetic, the set of Gödel numbers for true sentences is neither recursive nor recursively enumerable, a result which also holds for false sentences (Gödel's incompleteness theorem). Hence the proof predicate for the system is not recursive and we have no decision procedure.

§24 THE METATHEORY OF SET THEORY

Once the metatheory of the predicate calculus and arithmetic has been worked out, the metatheory of set theory can be rather briefly described. Since the incompleteness results of Gödel hold for any system containing primitive recursive arithmetic, and since set theory contains primitive recursive arithmetic, it is incomplete and its consistency can apparently not be established. Further, there can be no decision procedure for set theory because this would give us a decision procedure for arithmetic which is inconsistent with Church's theorem. As for independence, it is believed that the system is independent, but the proofs which established this point are not all finitary.

The most interesting question as far as set theory goes concerns categoricalness. Set theory can be presented in the language of the predicate calculus, which contains no propositional or predicate variables and only one predicate constant: '$\in$'. For example, the axiom of extensionality could be rendered as

$$(\forall x)(\forall y)((\forall z)((z \in x) \equiv (z \in y)) \supset (\forall x_1)((x \in x_1) \equiv (y \in x_1))),$$

or the axiom of power set as

$$(\forall x)(\exists y)(\forall z)((z \in y) \equiv (\forall x_1)((x_1 \in z) \supset (x_1 \in x)))$$

[cf. §18 and Fraenkel and Bar-Hillel 1958 274–275]. Of course, if we wished to add these axioms to the initial formulas of $\mathscr{L}$, we would need to eliminate the defined symbol '$\equiv$' as well as write, for example, '$\in(x, y)$' instead of the more familiar '$x \in y$'. The axioms could then be easily translated into the symbolism of $\mathscr{L}$.

Once we know that set theory can be expressed in general logic, we know that Gödel's completeness theorem applies. In particular, the last part of the proof applies, that is, that part which states that if a formula is invalid, it is invalid in the domain of natural numbers [cf. §21]. This is called the *Löwenheim–Skolem theorem*, after Leopold Löwenheim, who first proved a version of it in 1915, and Skolem, who generalized and provided a more rigorous proof of it in the twenties. What makes it interesting is the following circumstance. It is possible to prove within the system that there is a nondenumerable set of objects; we can prove this essentially by using the axiom of infinity and the axiom of power set and giving a formal equivalent to the informal diagonal argument as applied to real numbers [cf. §11].

Yet now we have a curious situation: From within the formalism we can prove that there exists a nondenumerable number of elements; yet from a metamathematical point of view we know that there exists a denumerable model of the formalism. This curious situation is known as *Skolem's paradox*. It is not a paradox in the sense of a contradiction, but rather a paradox in the sense of an unexpected irregularity. It represents the reverse of the problem that the Pythagoreans faced with the incommensurate. They wanted to deal only with whole numbers and ratios between them. Yet their assumptions forced them to recognize an irrational magnitude. On the other hand, as heirs to the work of Eudoxus, Dedekind, and Cantor, mathematical logicians want to deal with real numbers and sets of higher cardinality. But the assumptions they make in formal systems always allow an interpretation in the domain of natural numbers. Had the Pythagoreans known about Gödel numbering and the Skolem paradox, perhaps Pythagoreanism would still be a vigorous sect [cf. end of §23A].

The correct explanation of Skolem's paradox is controversial, but we may gain some insight into the problem by examining the following considerations. If we interpret all the symbols of our formalized set theory (such as '∀', '⊃', '⁻', etc.) in their standard ways and have '∈' uninterpreted, the Löwenheim–Skolem theorem *still* holds. Now suppose that we interpret '∈' as *is a member of*. Since this is the only predicate letter, there is essentially just this one interpretation. In this system it is possible to define a set of objects in such a way that they become isomorphic with the natural numbers. In particular, one can prove as theorems the equivalent of the formulas in (D) of §18.

But now let us suppose that we leave '∈' uninterpreted. Then we know that there are nonintended interpretations of the symbols in the system which are intended to represent the natural numbers. The denumerable model whose existence is asserted by the Löwenheim–Skolem theorem produces a nonintended interpretation of '∈' which yields a nonintended interpretation of natural number. There appears to be a contradiction. For within the system there is a proof of the nondenumerability of the set of all subsets of the "natural numbers." From an informal, external point of view there is a denumerable model, that is, a model that can be put into one-to-one correspondence with the natural numbers. But the contradiction is resolved when it is realized that *denumerable* does not mean the same thing in each case (since *natural number* does not).

Nevertheless, set theory is obviously noncategorical, as we should have expected from the discussion of the metatheory of arithmetic. The source of the noncategoricalness appears to be the formula which is equivalent to (D(1)). As we noted [cf. §18], this formula does not represent mathematical induction. The reasoning is that because our vocabulary is denumerable,

$$(\mathbf{P(0)} \cdot (\forall \mathbf{x})(\mathbf{P(x)} \supset \mathbf{P(x')})) \supset (\forall \mathbf{x})\mathbf{P(x)}$$

allows the interpretation that there are only a denumerable number of predicates of the objects represented by

$$0 \qquad (0)' \qquad ((0)')' \qquad (((0)')')'$$

etc. Thus $\mathbf{P(x)}$ may be interpreted in other ways than as representing a predicate. Hence we have noncategoricalness. If we try to get away from this expressive limitation and express the full meaning of mathematical induction, we can, but only in a second (or higher) order predicate calculus. There we could have

$$(\forall \mathbf{P})((\mathbf{P(0)} \cdot (\forall \mathbf{x})(\mathbf{P(x)} \supset \mathbf{P(x')})) \supset (\forall \mathbf{x})\mathbf{P(x)}).$$

However, then the "logic" or "deductive apparatus" of the system would have to be enriched to deal with quantification over predicates. Such an enrichment would allow the correspondence lemma to hold for the "logic." Hence the "logic" would be incomplete, and so we again have noncategoricalness. This is why we classify only the first-order predicate calculus by the name *logic* [cf. §18]. We may sum up the situation by saying that not only is there no formal categorical characterization of the notion of natural number, but there is also none of set (and consequently, none of the related notions of subset, or one-to-one correspondence).

Miguel de Unamuno (1864–1936), the Spanish existentialist, once asserted that "the supreme triumph of reason is to cast doubt on its own validity" [1954 109]. Unamuno was ignorant of the work of Skolem, Gödel, Church, etc., but we might well ask whether the remark was prophetic of the direction mathematical logic has taken. As we have seen, mathematicians in the nineteenth century developed a bewildering variety of methods and systems, and questions naturally arose as to their cogency. Then came the paradoxes, which raised further severe doubts. So mathematical logicians created new methods which set the most rigorous standards for cogency, explicitness, and perspicuity. Indeed, it is hard to conceive going much further in this direction than has already been achieved.

The result? The two core concepts of contemporary mathematics— that of natural number and set—were found to be unapproachable by the rigorous standards which were laid down. By using these formal methods with respect to the concept of set or natural number, we find that we literally don't know what we are talking about. On the other hand, if we abandon the formal methods, there appears to be only makeshift protection against the paradoxes.

The doubts described in the last paragraph were vaguely expressed, but the anxiety which gave rise to them is real enough. It is the purpose of the next chapter to inquire carefully into the larger, more philosophical questions which surround mathematical logic.

CHAPTER 5

PHILOSOPHICAL IMPLICATIONS OF
MATHEMATICAL LOGIC

§25 INTRODUCTION

There are many philosophical problems having to do with mathematical logic. Some of these problems antedate the birth of the subject by thousands of years, as, for instance, the philosophical problem of determining the ontological status of the natural numbers. There are also numerous philosophical positions concerning the nature of mathematics; for example, logicism, formalism, and intuitionism. But these philosophical positions were all created when the important results of metatheory were unknown. What we shall do in this chapter is to approach the problems of logic and mathematics afresh, by taking as our starting point the *metatheorems* of these subjects. Great achievements in science have often revolutionized philosophic thought, as can be seen in the work of Newton, Darwin, or Einstein. The philosophic impact of metatheory is likewise profound, and we must not expect that we may easily assimilate its impact with categories from the past. Instead, taking our cue from the history of science, we shall expect not only new points of view, but intellectual excitement and puzzlement from even the first faltering steps at exploring those points of view.

Hence the topics of this chapter are controversial. But controversy is part of the very core of our philosophical heritage and there is no need to avoid it. In any case, at this date it would be impossible, for there is no "received view" on the philosophical interpretation of the results of metatheory. Nevertheless, there is a widespread pedagogical theory that the beginner should be steered clear of philosophical issues, presumably because it is assumed that these issues cloud his thinking. So let the beginning reader be twice warned: on the pedagogical side and on the philosophical side. He can protect himself pedagogically by resolving to form only tentative opinions until he knows more logic. There is no need to give the impression, as many introductory texts do, that logic is what Kant thought it to be: complete and uncontroversial. The reader can protect himself philosophically by resolving to explore the views of others. The purpose of this chapter, then, is to stimulate interest, inquiry, and—better philosophical thoughts.

For purposes of illustration, it will be helpful later in the chapter to be able to refer to some particular large finite number, as well as to some unsolved problems of elementary arithmetic. The purpose of the rest of this introductory section is to describe this number and the problems.

First let us take the number. Consider the following function:

$$f_1(x) = x^{x^{\cdot^{\cdot^{\cdot^{x}}}}}$$

where the number of exponents is equal to x. For example,

$$f_1(1) = 1^1 = 1, \qquad f_1(2) = 2^{2^2} = 16, \qquad f_1(3) = 3^{3^{3^3}},$$

$$f_1(4) = 4^{4^{4^{4^4}}},$$

etc. Note that this function gets very large very quickly. $f_1(10)$, for example, is a number larger than is needed in any (applied) scientific theory. If we were to try to write out the numeral corresponding to the number of ordinary arabic notation, and if each numeral took up only a billionth of an inch with as little space between numerals, there would not be enough space to do so between here and the farthest known star. Now consider the following function:

$$f_2(x) = f_1(x)^{f_1(x)^{\cdot^{\cdot^{\cdot^{f_1(x)}}}}}$$

where the number of exponents is equal to $f_1(x)$. Of course, $f_2(x) = f_1(f_1(x))$. Continuing this process, let us define

$$f_n(x) = f_{n-1}(x)^{f_{n-1}(x)^{\cdot^{\cdot^{\cdot^{f_{n-1}(x)}}}}}$$

Finally let 'c_1' be defined as in §23B. Now, in this book, let us refer to the number $f_{c_1}(c_1)$ by the word *zillion*.

Turning to the issue of describing some unsolved problems of elementary arithmetic, let us first note that part of the continued fascination of this subject is the ease with which extremely difficult problems are expressed. The following are four famous ones:

1.   *The problem of perfect numbers.* A *perfect number* is understood to be a natural number which is equal to the sum of its divisors (excluding itself). For example, 6, the smallest perfect number, equals $1 + 2 + 3$. The next six perfect numbers in order are 28, 496, 8128, 33550336, 8589869056, 137438691328. The problem, at least as old as Euclid and probably dating back to the time of Pythagoras, is to determine whether there are a finite or an infinite number of them.

2.   *The twin-prime conjecture.* This problem, also known in ancient Greek times, is to decide whether there is a finite or infinite number of pairs of primes

which differ by 2. For example, the following are examples of twin primes (as these pairs are called): 3, 5; 17, 19; 101, 103; 7027, 7029; 21491, 21493; 55049, 55051; 1000000009649, 1000000009651.

3. *Fermat's last theorem.* Fermat claimed to have a proof for the statement

$$x^{n+2} + y^{n+2} \neq z^{n+2},$$

where x, y, z, n are variables whose domain of definition is the natural numbers. All of Fermat's other claims about mathematics have since been confirmed or refuted. Hence the name *last theorem*.

4. *Goldbach's conjecture.* In 1742 Goldbach (1690–1764) conjectured that every even number is the sum of two primes. For example, $20 = 13 + 7$, $88 = 5 + 83$, $7000 = 3 + 6997$. The problem is to prove this or find an even number which can't be so represented.

To each of these problems an enormous amount of time has been devoted by both professional and amateur mathematicians. These attempts have produced many partial solutions and have even opened up new areas of mathematics. Yet no full solutions have been found.

§26 CHURCH'S THESIS

At this point in our account one can imagine someone interrupting with the following argument: "I object to drawing any philosophical conclusions at all from the work of Skolem, Gödel, or Church, which has been described above. In particular: Gödel has shown that, provided a certain calculus is ω-consistent, there are undecidable formulas; Church has shown that, for that calculus, there is no decision procedure; and Skolem has shown that the calculus does not categorically represent the natural numbers. But this is hardly any more philosophically significant than the fact that we cannot devise a winning strategy in chess for white with a king and a knight to checkmate black with only a king. Or—to take a mathematical example— the impossibility of squaring a circle or trisecting an angle is not philo- sophically interesting. It is perfectly possible to do it, but not with just a straightedge and a compass. Similarly, when Gödel proves that there are undecidable formulas, or Church that there is no decision procedure, or Skolem that a calculus doesn't categorically represent the natural numbers, all this comes to is that *using the means which they have selected* (just as Euclid selected a straightedge and compass), their conclusions follow. There is no philosophical import to this; it just suggests that mathematicians and logicians must look for other means."

This objection would have a great deal of sting were it not for one circumstance: *There do not appear to be any other means.* To see this, let's consider briefly the requirements we make in order to claim deductive knowledge. We first require that all the "symbols" which are used be publicly and effectively recognizable. The word 'symbols' is in double quotes because it is not required that they be symbols in the usual sense of the word, that is, marks on paper or a blackboard, etc. They might be sounds or colors or electric charges. The requirement includes not only being able to tell the difference between a symbol and something else (for example, we should be able to tell the difference between a parenthesis and a broken fishhook), but being able to effectively recognize repeated occurrences of the same symbol type [for example, '(' is of the same symbol type as '(']. The 'all' in our requirement means that the number of symbols be finite (for example, a zillion symbols) or, if infinite, that there be an ordering (called the alphabetic order above) such that only a finite number of symbols precede any given symbol in that ordering.

Second, we require that the length of any given sentence be finite and that a sentence be effectively recognizable as such. We make the same requirement of rules of inference. As for the axioms, we require that they be either finite or, if infinite, that there be an ordering such that only a finite number of axioms precede any given axiom in that order. For the theorems, we do not require that there be an effective method to recognize them, but only that there be a constructive program which, if carried out, will produce any given theorem. Finally, we require that for any given sentence we be able to effectively recognize a proof of it if one should be presented. The above are all syntactical requirements. There is only one semantical requirement: that the theorems all be true under every interpretation which makes the axioms true.

Consider the relation of these requirements for deductive knowledge to Church's thesis. We speak of "effectively recognizable" or "constructive program." These are psychological terms. The first term means that human beings can be taught an effective finite method for recognition of certain objects. Our formation rules for $\mathscr{P}$ or $\mathscr{L}$ are examples. The second term means that human beings can be taught a program which produces objects of a certain kind, and that given any object of that kind the human following that program will eventually produce it. The theorems of $\mathscr{P}$ or $\mathscr{L}$ are examples. A simple mathematical example is working out the decimal expansion of π, where, if we wanted to know whether there are 7 consecutive 7's in its expansion, we have a procedure which will produce them—*if* there is such a sequence. If not, the program would have no termination. So, if we view the concept of a formal system from this perspective, we can see that it is a psychological concept, having to do with what humans are psychologically capable of doing.

An analogy may again be helpful. The Greeks understood geometry as in part involving what a human could ideally construct with a straightedge and a compass. This involves psychological and mechanical concepts.

For example, one is not allowed to construct an infinite number of lines in a geometrical demonstration. To talk about all possible constructions requires a mathematical analysis of these psychological and mechanical concepts, an analysis which involves an analog of Church's thesis. Now what Church's thesis implies is that we can identify the psychological concept of effective with the mathematically defined concept of recursive, and, hence, the psychological concept of constructive with the mathematically defined concept of recursively enumerable.

It should be clearly understood that Church's thesis is an empirical thesis, not a mathematical theorem. It is a claim which is subject to confirmation or refutation by empirical methods. However, the evidence in its favor is so great that it should really be called *Church's law* since, if true, it is a natural law [cf. Post's remark in Davis 1965 291]. Nevertheless, since almost all writers refer to it as *Church's thesis*, we shall follow this practice. We cannot avoid the question of truth and falsehood by considering the identification proposed by Church to be a definition. For the identification must be in the form of a lexical definition, in order to avoid triviality. Lexical definitions require evidence that they are appropriate. In the case of important core concepts of science, the evidence required for such definitions is substantial (for example, in the definition of *atom*). So evidence is necessary, and we shall examine some of it.

First, however, let us suppose that we accept Church's thesis. Then the requirements for deductive knowledge can be stated in the language of recursive function theory. For example, we can say that the set of Gödel numbers for symbols, the set of Gödel numbers for formulas, and the set of Gödel numbers for axioms must each be recursive sets, while the set of Gödel numbers for theorems must be recursively enumerable. Any system which meets these requirements and which is adequate for primitive recursive arithmetic is subject to the limitations imposed by Gödel's first incompleteness theorem, Church's theorem, and Skolem's theorem. (The situation with respect to Gödel's second incompleteness theorem is more complicated, as will be described in §29.) These theorems may be stated in generalized form, although we shall not describe the proof of this form:

1. *Gödel's first incompleteness theorem* (generalized version): There exists a predicate such that there is no correct and complete formal system for it.

2. *Church's theorem* (generalized version): There exists a predicate such that there is no correct formal system which contains a decision procedure for both the predicate and its negation.

3. *Skolem's theorem* (generalized version): There is no consistent, categorical formal system having the natural numbers as its intended interpretation.

Since each of these theorems suggests limitations on man's abilities, they have become known as *limitative* theorems. The question immediately arises:

Are there any ways that we can avoid these limitations? The answer is yes, but none of the known ways take the sting out of these results.

For example, we could avoid the incompleteness required by Gödel's theorem by taking the set of axioms to be the set of true sentences of arithmetic. This system is consistent and complete. The only trouble is that it is impossible in general to recognize an axiom as an axiom. In fact, the generalized Gödel theorem as applied to this system would tell us that there exists a sentence for which it would be impossible to decide whether or not it is an axiom. All other ways that have been tried to avoid the limitations—and many are much more subtle than the one just suggested—have had similar outcomes. This in itself is evidence that we are dealing with an important epistemological notion: A great amount of effort has been directed at trying to formulate a system which is not subject to Gödel's, Church's, and Skolem's theorems, and all these efforts have failed. Of course, "successes" analogous to that described in this paragraph are deemed failures.

There is substantial evidence for Church's thesis. First, every recursive function so far produced appears to be effectively calculable in the intuitive sense. Conversely, every function so far examined that would be considered effectively calculable—including many constructed with the specific purpose of creating an effectively calculable nonrecursive function—have all turned out to be recursive. Finally, practically all the suggested mathematical interpretations of *effectively calculable* have been shown to be equivalent. These include *λ-definability* (Church), *reckonability* (Gödel), and *binormality* (Post). We shall examine two other proposals, the first made by Alan M. Turing (1912–1954) and the second by Post. Both these proposals were published in 1936. Although they are identical in import, they nevertheless were formulated independently of one another.

Problem 71. Suppose at this point that there was the following interruption: "Of course there are easy ways to construct an effectively calculable nonrecursive function. All that is necessary is to use Cantor's diagonal method. For example, consider an ordering of all general recursive functions of one variable: $f_1(x)$, $f_2(x)$, etc. Then by Cantor's diagonal procedure there exists a function which is defined by the following equality:

$$g(x) = f_x(x) + 1.$$

This function is effectively calculable for every value of x, if $f_x(x)$ is. Now it is agreed that every general recursive function is effectively calculable. Thus $f_x(x)$ is effectively calculable for every value of x. Since addition is an effectively calculable function, $g(x)$ is also. But $g(x)$ cannot be recursive, since it differs by 1 from every recursive function of one variable. This argument can easily be generalized." Find the difficulty in this ploy of showing Church's thesis to be incorrect.

Both Post and Turing decided to approach the problem of the meaning of *effectively calculable* by first considering a paradigm case of computing, then, second, by ignoring inessential features of that case, and third, by analyzing the essential features into combinations of very simple operations. The following account is in the spirit of Turing and Post with some details changed.

We begin by imagining some human who is faced with a specific computational problem. It might be some such problem as computing the sum of 101 and 102 and 103 and ... and 198 and 199, where the three dots indicate one occurrence each of all the natural numbers between 103 and 198. We assume that he is working according to a finite set of rules which have been fixed before the problem was given and that he is using pencil and paper. We also assume that after a finite amount of time he stops with the correct answer.

If we examine this paradigm case of computing with a view toward eliminating inessential features, a number of such features come to mind, such as the use of pencil and paper or the particular computational problem chosen. However, the most striking one appears to be the *human*: the computer might as well be a machine. Our very language suggests this possibility. A beginning student in some area of mathematics might say that he mechanically went through the procedure of computing such and such but that he didn't understand the procedure. Understanding *how* or *why* a procedure works is not necessary for computation. An analogy might be helpful here. A person ignorant of the products of the industrial revolution—for example, an ancient Egyptian—might think that in order to lift a heavy object every morning and lower it every night it is necessary to have blood, bones, muscles, brains, an eye, etc. However, we know that a certain kind of machine with a photoelectric cell will suffice. Similarly, it might seem that brains or understanding why or how a procedure works are necessary for computation. Yet this is not the case, as anyone familiar with the computer revolution knows. We shall now turn our attention to an ideal computing machine which is known as a *Turing machine*.

We imagine a machine which can scan one square at a time on a tape which is infinite in both directions, that is, it would have the same order type as the positive and negative integers and might look something like this:

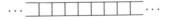

The reason the tape is infinite is to allow for the computation of any problem. For example, if we limited the tape to a zillion squares, we couldn't compute the first 100 zillion primes. The tape represents the symbol space and we assume that only one symbol can be written in each square. Further, we assume that there are only two possibilities: Either a stroke ('|') occurs in a square or the square is blank. Of course, the person with a pencil uses a dozen or more symbols, but this is inessential, since combinations of a single symbol can represent the other symbols.

We assume that the machine is capable of performing the following primitive operations:

1) If the scanned square is blank, to mark it with a stroke.

2) If the scanned square is marked, to erase the stroke.

3) Moving the tape so that the new scanned square is the one immediately to the right of the old scanned square.

4) Moving the tape so that the new scanned square is the one immediately to the left of the old scanned square.

5) Determining whether the scanned square is marked or blank.

Finally, we assume that the machine may be programmed in such a way as to perform operations (1) through (5) according to a finite list of directions and that each of these operations takes a finite amount of time which we shall call a *moment*. This assumption includes the machine performing one way if the outcome of (5) is marked and another way if it is blank.

A Turing machine is different from any actual machine in at least four respects:

1) We assume that it makes no errors.

2) We assume an infinite memory in which can be stored (at any given moment) the description of the function and the answer (if already computed). The memory, however, contains only a finite amount of information at any given time.

3) We assume an infinite amount of time (that is, number of moments) in which the computation takes place. Again, at any given moment, only a finite number of moments in the computation have preceded it.

4) We assume that the machine does not break down, that there is an infinite supply of ink for the strokes, that there is an infinite supply of energy to run the machine, and so forth.

It would not be hard to give a description of a computation as carried out on a Turing machine. We shall not do so here because the description— even for relatively simple functions—is rather long. The reason for this is the extremely simple operations with which we start. A function is called *Turing computable* if it can be computed by a Turing machine; *Turing computably enumerable* if it can be enumerated by a Turing machine. In 1936 Turing proposed the thesis—now called *Turing's thesis*—that the set of effectively computable functions is identical with the set of Turing computable functions.

Since it seems to be part of the meaning of *effectively computable* to be computable by a machine, Turing's thesis would imply that a function which is computable by any computing machine whatsoever is also Turing computable. In 1937 it was shown (by Turing himself) that the class of recursive functions is identical with the class of Turing computable functions. Hence Church's thesis is equivalent to Turing's thesis. We shall refer to both as *Church's thesis.*

At this point let us take stock of where we stand. We have seen that a body of deductive knowledge, in order to claim that name, must be effective or constructive in certain ways. By Church's thesis, the theory of formal systems becomes part of the theory of recursive functions. The theory of recursive functions is equivalent to the theory of Turing machines. It must be remembered that a Turing machine is a mathematical, not a physical object. An actual physical computing machine may be considered a kind of disabled Turing machine; that is, it is subject to limitations in addition to the limitations of Turing machines. Some of these limitations come from the particular physical elements it is made of (for example, aluminum); others might come from the nature of physical things in general (for example, being subject to breakdowns). Of course, Turing machines have some limitations which actual physical machines lack. For example, a particular physical machine might accidentally get the right answer because of the breakdown of a circuit. A Turing machine, however, has no short cuts through short circuits. Nevertheless, an actual computing machine, *qua* computing machine, is subject to all the limitations of a Turing machine. Similarly a human computer, *qua* computer, is subject to the same limitations. Naturally a human is capable of things which a Turing machine is not (for example, feeling a vague sense of anxiety), but not *qua* computer. Thus Church's thesis, if true, becomes a law of psychology (as applied to human computers) and a law of physics (as applied to computing machines). We now have the translations shown in Table 16.

Table 16

Mathematically defined and exact		Empirically defined and inexact	
Language of recursive functions	Language of Turing machines	Language of psychology	Language of physics
Recursive	Turing computable	Effective	Computable
Recursively enumerable	Turing computably enumerable	Constructive	Computably enumerable

From this point of view, both a human computer and a computing machine become approximate embodiments of either particular formal systems or particular Turing machines. Of course, exactly what formal systems or what Turing machines depends on the program and capacities (human or mechanical) of the computer.

Thus the acceptance of Church's thesis forces us to accept the limitative theorems of metalogic (for example, those of Gödel, Church, and Skolem) as laws of psychology on the one hand, and laws of physics on the other. The far-reaching significance of this fact can better be realized by actually translating the language of those theorems into the language of psychology or physics (Table 17).

Table 17

Theorem	Language of psychology	Language of physics
Gödel's first incompleteness theorem	There is no consistent human computer capable of formulating a program which, if carried out, would produce all the true and only the true sentences of arithmetic.	There is no consistent computing machine which can be programmed to produce all the true and only the true sentences of arithmetic.
Gödel's second incompleteness theorem	No consistent human computer can prove his own consistency (as represented by an analog of C).	No consistent computing machine can be programmed to prove its own consistency (as represented by an analog of C).
Church's theorem	There exists a set of problems which no consistent human computer can solve.	There exists a set of problems which no consistent computing machine can be programmed to solve.
Skolem's theorem	No consistent human computer can (categorically) formalize the notion of natural number.	No consistent computing machine can be programmed to produce a (categorical) formalization of the notion of natural number.

Problem 72. Consider the following argument: Gödel's first incompleteness theorem shows that man is different from a computing machine. For any consistent computing machine adequate for recursive arithmetic contains a formula (say, M) which we can prove true (under the intended interpretation),

but which the machine cannot prove. Hence man cannot be a machine, however complicated.

Discover the fallacy in this argument.

At this point many philosophical questions arise as to the import of the limitative theorems for the problems of machines, mathematics, knowledge, etc. These questions arise because of the identification of effectively calculable functions and general recursive functions. Thus this concept becomes central to any theory of knowledge which takes into account the limitative theorems. Gödel himself has stated that

...the great importance of the concept of general recursiveness (or Turing's computability)...is largely due to the fact that with this concept one has for the first time succeeded in giving an absolute definition of an interesting epistemological notion, i.e., one not depending on the formalism chosen. In all other cases treated previously, such as demonstrability or definability, one has been able to define them only relative to a given language, and for each individual language it is clear that the one thus obtained is not the one looked for. For the concept of computability however, although it is merely a special kind of demonstrability or decidability the situation is different. By a kind of miracle it is not necessary to distinguish orders, and the diagonal procedure does not lead outside the defined notion [Davis 1965 84].

In the rest of this chapter we shall consider some of the consequences of the limitative theorems for mathematics, logic, and philosophy. It would be surprising if everyone wholly accepted the account given here. Nevertheless, it is hoped that the arguments given will convince everyone that the limitative theorems must be taken into account by both mathematicians and philosophers.

§27 THE NATURE OF INDETERMINATE STATEMENTS

The central change which the limitative theorems require of all previous theories of the nature of mathematics is the recognition of unanswerable questions in the subject. Previously it was thought that if a question could be made precise, that question had an answer. An instance of this attitude may be found in Hilbert's address, "On the Infinite," delivered in 1925:

As an example of the way in which fundamental questions can be treated I would like to choose the thesis that every mathematical problem can be solved. We are all convinced of that. After all, one of the things that attract us most when we apply ourselves to a mathematical problem is precisely that within us we always hear the call: here is the problem, search for the solution; you can find it by pure thought, for in mathematics there is no *ignorabimus* [384].

The new attitude that seems required by the limitative theorems may best be exemplified by considering the following thought experiment: Imagine a middle-aged man and woman walking into a restaurant and ordering two steak dinners. When this image is clearly in mind, take note of the time (denote this time by '*t*'). Now answer the following questions about your image (as it was at time *t*):

Was the man a giant and the woman a midget? What was the race of the man? When did the woman last visit the dentist? What did the restaurant have to pay for the steaks? Did the man and woman sit or stand to eat their dinner? Are they married? What is the woman's mother's first name?

At this point it is surely clear that many people could answer some of these questions about their image but that no one could answer them all (or others that could be easily formulated). Why? Well, the reality toward which the questions are directed is indeterminate in certain ways. It is not indeterminate in all ways. For example, to the question "Did the man enter the restaurant?" the answer is yes. On the other hand, a statement such as "The restaurant paid $3.87 plus $.12 tax for the steaks" doesn't seem either true or false. Rather, we say that it is undecidable on the basis of the image at time *t*. Of course, at a later time it might be decidable.

The indeterminateness which is found in our experiment is also found in all products of the imagination, including artistic creations (How often did Juliet sneeze in the year before she met Romeo?). In these areas it is pointless to ask questions about things which are not determined by the evidence. Relative to these imaginative creations, physical reality is determinate. For example, all the questions which were raised make good sense when applied to an actual man and woman entering an actual restaurant. We might not know or be able to find out the answer, but we assume that there is one. However, the results of quantum mechanics suggest that physical reality is also indeterminate in certain ways, or, better, that our relation to the reality which we are investigating does not allow us to raise certain kinds of questions.

Just as some of the indeterminateness which was previously thought typical of (and peculiar to) imaginative creations was found in the physical world with the discovery of the quantum theory, so some of this indeterminateness was also found in mathematics with the discovery of the limitative theorems. The impact of this discovery is comparable to the discovery by the Pythagoreans of the incommensurability of the side and diagonal of a square, in that it upsets fundamental conceptions as to the nature of mathematics. To explore the impact, we must be more precise about the indeterminateness which has been introduced.

Perhaps the easiest way to approach the subject is by considering Euclidean geometry. The question of truth or falsehood doesn't arise as long as a system is uninterpreted. So consider a system containing general logic and the formal equivalents of Euclid's first four postulates. Denote it by '$\mathscr{E}_4$'. Now consider it being applied to some model $M_{\mathscr{E}}$ which happens to be a model of Euclidean geometry. Then, in this system, statements equivalent to the fifth postulate are true but unprovable. Or, to put it another way, statements equivalent to the negation of the fifth postulate are false but not refutable. It would seem that the situation with respect to interpreting the formal system $\mathscr{A}$ in a model of arithmetic is analogous. For here if $\mathscr{A}$ is consistent, G is true in the model but unprovable in $\mathscr{A}$, while $\overline{G}$ is false but not refutable. The situation is indeed analogous, but there are two important differences to be noted.

First, if $\overline{G}$ is added to $\mathscr{A}$ as an initial formula, this system appears to be mathematically sterile. It would be as if we took a formal system containing general logic and the formal equivalent of Euclid's *last* four postulates and added the formal equivalent of the negation of the first postulate. That is, we would add the formal equivalent of 'It is not the case that a straight line may be drawn from any point to any point'. This system contains very little in the way of theorems which Euclidean geometry doesn't contain. Of course, you have the new first postulate and its logical equivalents, but these statements do not combine with the other four postulates to produce new theorems in any but a trivial way.

For example, a new theorem might be the conjunction of the new first postulate and the second postulate. Thus, even though G is independent (provided that $\mathscr{A}$ is ω-consistent), it does not generate two kinds of arithmetic, as the fifth postulate generates two kinds of geometry. It generates two kinds of arithmetic only in the sense that the first postulate generates two kinds of geometry. Since it is misleading to express the situation in this way, we shall say that the first postulate of Euclid does not produce any new geometry, and the $\overline{G}$ of Gödel does not produce any new arithmetic. Of course, there may be formulas in $\mathscr{A}$ which do produce new arithmetics, but we are not yet aware of them. In the case of set theory, however, there has probably been discovered a genuine non-Cantorian version, in the sense that the new version will not prove to be mathematically sterile. This new set theory involves denying the Axiom of Choice (cf. §18), but we shall not discuss it here [see Cohen and Hersh 1967 for a popular account of it].

Second, if the formal equivalent of the fifth postulate is added as an initial formula to the system $\mathscr{E}_4$, the system can be made complete. Yet, as we have seen, arithmetic is not made complete by adding G or even an infinite number of new formulas. Every consistent extension of $\mathscr{A}$ is incomplete, a situation we shall refer to as *essential incompleteness*. $\mathscr{E}_4$ is incomplete but not essentially so.

To return now to the question of indeterminacy. In the system $\mathscr{E}_4$ as applied to $M_\mathscr{E}$, the statements equivalent to the fifth postulate are true, *but nothing we assume would enable us to prove them to be true*. We assume that $M_\mathscr{E}$ is a reality whose existence is defined independently of $\mathscr{E}_4$. The indeterminacy then is a result of our *refusal* to make enough assumptions to prove all the truths about a particular reality. From the point of view of someone making only the assumptions of $\mathscr{E}_4$, the statements equivalent to the fifth postulate are just accidentally true of $M_{\mathscr{E}}$. By the phrase *accidentally true statement* we shall mean *a statement which is neither necessary nor impossible*. These concepts are relative to a language. It is necessarily true that similar figures exist in Euclidean geometry and it is impossible that they do in Lobachevskian geometry.

Turning to arithmetic: The indeterminacy is due to our systematic *inability* to make enough or the right kind of assumptions to prove all the truths of arithmetic. We assume the independent existence of the model. Thus, from the point of view which any particular human might take, there must exist accidental truths of arithmetic, that is, statements which are true *but are such that nothing that he assumes would allow him to prove them to be true*. From his point of view they would be neither necessary nor impossible. What the limitative theorems represent then is the discovery of an abstract structure which is of such a sort that it is impossible for any human to make systematically complete and correct assumptions about it. No matter how hard we try, we cannot talk about just the natural numbers, but must always talk about the unnatural numbers, too. In other words, just as our sensory perceptions have limits which can be extended but not eliminated by certain techniques of science (for example, the microscope), so our abstract conceptions are limited and the methods of mathematics which are intended to extend them (for example, mathematical induction) are at best partial in their effect. Our powers of conceptual discrimination are no less limited than our powers of perceptual discrimination.

There doesn't seem to be any way around these conclusions. But it should be noted that the above argument depends on what is known as *realism* (or *platonism*) in mathematics. That is, it is assumed that the abstract structure of arithmetic exists independently of human conceptions about it. The assumption is embedded in classical mathematics, but we might question it. After all, how does arithmetic in particular or mathematics in general differ from the game of ticktacktoe (or ACE or HOT)? There seem to be just two relevant differences: First, mathematics is more useful than ticktacktoe: and second, mathematics (or at least recursive arithmetic) impresses itself upon us as necessary if we are going to think at all. But consider the theoretical possibility that there should be creatures whose environment is such that ticktacktoe (or some such game) is very useful, and whose mental structure is such that the game impresses itself upon them as necessary. We might then take a *nominalistic* position according to which the only realities we recognize are those implied by our symbolism. We invent mathematics the same way we invent games, the only difference being that (recursive) arithmetic seems to be a game we need to invent if we wish to inquire.

If we take this point of view and apply it to the example above, it would follow that the only geometric realities that we recognize are those defined by the system $\mathcal{E}_4$. Then our inability to prove or refute the formal equivalent of the fifth postulate is due to *the indeterminacy of the geometric reality itself*. It is similar to mental images or works of art, in the sense that there are sensible questions about them which can't be answered. If we take such an attitude, the law of the excluded middle no longer universally applies. The law applies to any sentence which, if true, is true on the assumption of the truth of the axioms, or if false, is false on the assumption of the truth of the axioms. For such a sentence there is no third alternative. But a sentence not true or false on the assumption of the truth of the axioms is neither true nor false. Even so it seems that if arithmetic is just one of the games people play, we nevertheless want a game in which the law of the excluded middle applies. This desire apparently springs from certain esthetic ideals which we have toward games which are part of inquiry; that is, that these "games" be more than mere games. We shall return to this subject later, but at this point the difficulty appears to be that we (as nominalists) are platonistic with regard to our epistemological aspirations, and nominalistic with respect to judging epistemological claims.

In any case, the existence of indeterminateness in mathematics upsets traditional disputes as to the nature of mathematical statements. Kant in the eighteenth century argued that mathematical statements are synthetic *a priori*, Mill in the nineteenth that they are synthetic *a posteriori*, and Russell in the twentieth that they are analytic *a priori*. It turns out that as a result of the limitative theorems all three views are inadequate, because mathematics contains all three kinds of statements. We shall be able to see this more clearly if we first explore the question of unsolved problems in mathematics and the question of the consistency of arithmetic.

§28 THE PROBLEM OF UNSOLVED PROBLEMS

Before the problem of unsolved problems of arithmetic may be discussed, we must draw a distinction between the undecidable and the unsolvable. A sentence is *undecidable* in a given system if neither it nor its negation is provable in the system. The concept of undecidable, we may emphasize again, is relative to a system. What is undecidable in one system may be decidable in another. In contrast, the concept of unsolvable is absolute. We shall say that a predicate of natural numbers is *recursively solvable* if it is recursive, otherwise *recursively unsolvable*. The predicate 'x is the Gödel number of a valid formula of $\mathscr{L}$' is recursively unsolvable (Church's theorem). This concept is absolute in the sense that if we accept Church's thesis, anything recursively unsolvable is unsolvable in the ordinary sense of the word. This does not mean that for some given value of the predicate we cannot solve the problem. For example, if g_0 is the Gödel number of '(p ⊃ p)', then 'g_0 is the Gödel number of a valid formula of $\mathscr{L}$' is both true and provable. What *recursive unsolvability* means is that no technique will suffice for an arbitrary numeral taking the place of 'x' in the example above.

In order to understand the difficulty of unsolved problems, we must distinguish between theoretical and practical undecidability, as well as between theoretical and practical unsolvability. Without making these distinctions we cannot appreciate the kind of task with which we *may* be faced in attempting to find a solution to hitherto unsolved problems of elementary arithmetic. Furthermore, these distinctions are also helpful in clarifying the nature of mathematical statements.

1.   *Theoretically undecidable.* G is theoretically undecidable in $\mathscr{A}$, provided that $\mathscr{A}$ is ω-consistent. As we have seen, under its intended interpretation each and every instance of G is true and provable, but G itself is not provable (although true). Consider how this situation might arise in an ordinary mathematical problem, say in trying to prove-in-$\mathscr{A}$ the formal expression for Goldbach's conjecture. A standard technique in mathematics is that of dividing a proof into cases. This is necessary wherever two or more fundamentally different methods are needed to establish a theorem. But suppose that an infinite number of different methods is needed. Then the statement in question would be theoretically undecidable in $\mathscr{A}$. This may be illustrated by supposing that Goldbach's conjecture is a theoretically undecidable statement, and the cases break down according as the even numbers are, or are not, powers of 2 or divisible by certain primes. In other words, we shall divide the cases as follows:

Case 1. Every even number which is a power of 2 is the sum of two primes.

Case 2. Every even number divisible by 3 is the sum of two primes.

Case 3. Every even number divisible by 5 is the sum of two primes.

Case 4. Every even number divisible by 7 is the sum of two primes.

$$\vdots$$
$$\vdots$$

Here we assume that for any given case there is a proof-in-$\mathscr{A}$, but that these proofs are all unlike one another. The qualification that they be unlike one another is important, since if all but a finite number had similar structures, they could be reduced by some means (perhaps mathematical induction) to a finite number of cases. If we assume that the proofs are unlike one another, then we have ω-incompleteness. By Gödel's first incompleteness theorem we know that if $\mathscr{A}$ is consistent, it is ω-incomplete. It might be the case that Goldbach's conjecture—like G—is an illustration of that ω-incompleteness. In such a case we would be able to prove-in-$\mathscr{A}$ the formal expression for each of the following:

2 is the sum of two primes.
4 is the sum of two primes.
6 is the sum of two primes.

$$\vdots$$

But we could not prove-in-$\mathscr{A}$ the formal expression for

$(\forall x)(2x + 2$ is the sum of two primes$)$.

We can gain more insight into the concept of ω-incompleteness by remembering the phenomenon of unnatural numbers. If every even natural number is the sum of two primes and there is no proof-in-$\mathscr{A}$ of that fact, then Goldbach's conjecture is an illustration of the ω-incompleteness of $\mathscr{A}$. For we have a guarantee that we can prove-in-$\mathscr{A}$ the formal expression for any specific instance of the conjecture, if the conjecture is true of the natural numbers. But if the conjecture is true of the natural numbers and unprovable in $\mathscr{A}$, then by Gödel's completeness theorem there must be a model which contains at least one even unnatural number which is not the sum of two primes (as defined in that model).

On the other hand, if we could prove from a metamathematical point of view that Goldbach's conjecture is undecidable, then we would have (as in the case of G) a metamathematical proof of its truth for natural numbers. By constructing another stronger system—essentially one embodying the principles used in the metamathematical proof—we could then give a formal proof of the conjecture. It would thus follow that any informal mathematical proof of Goldbach's conjecture must use non-arithmetic techniques (such as those included in the theory of real numbers).

Problem 73. What guarantee do we have that we can prove-in-$\mathscr{A}$ the formal expression for any specific instance of Goldbach's conjecture, if the conjecture is true of natural numbers?

A similar result would be obtained if it were true that Fermat's last theorem held for every natural number as exponent, but not for every unnatural number; that is, Fermat's last theorem would be true but not provable by arithmetic means. Examples such as these have led some authors to suggest that a metamathematical proof of undecidability always decides the issue of truth and falsehood. But this is not always the case. Both the perfect number and twin prime conjectures could be either true or false even if proofs of undecidability in $\mathscr{A}$ were obtained. Be careful to note that in both the case of Goldbach's conjecture and Fermat's last theorem the statements corresponding to them in the system, if false, are refutable in the system.

For example, in the case of Goldbach's conjecture there would be some even natural prime number N such that it is not the sum of two primes. Once this N has been given (say we find it by chance), it is a recursive procedure to show that it is not the sum of two primes. On the other hand, the problem of perfect numbers is not decidable by a proof of its undecidability in $\mathscr{A}$. For suppose that we gave a metamathematical proof of the undecidability of the perfect number problem in $\mathscr{A}$. If '$A(x)$' stands for 'x is a perfect number', then an arithmetic statement that there is no highest perfect number is

$$(\forall x)(A(x) \supset (\exists y)(A(y) \cdot (x < y))).$$

We shall denote the formal expression in $\mathscr{A}$ for this statement by the letter 'K'. Now suppose that there is no highest natural perfect number, but there is a model with a highest unnatural perfect number. K would then be true of the natural numbers but undecidable in $\mathscr{A}$. On the other hand, suppose that there is a highest natural perfect number, but there exists a model with an infinite number of unnatural perfect numbers. K would then be false as applied to natural numbers, but nevertheless undecidable in $\mathscr{A}$.

Problem 74. In §22 the twin prime conjecture was expressed by:

$$(\forall x)((Pr(x) \cdot Pr(x + 2)) \supset (\exists y)((Pr(y) \cdot Pr(y + 2)) \cdot ((x + 2) \le y))).$$

Show that this conjecture could be either true or false under the assumption that its formal expression is undecidable in $\mathscr{A}$.

2. *Practically undecidable.* A formula will be called *practically undecidable* if it is theoretically decidable, but the shortest proof or refutation of it is too long to actually carry out by any known means. This definition is vague, but it is exact enough for our purposes. We may begin to study the difficulties of the practically undecidable by proving that there is no fixed relation between the length of a formula (defined as the number of symbols in it) and the length of a proof (defined as the number of symbols in it). For suppose that there were some fixed relation so that, say, if there are n symbols in a formula, then there are no more than n to the zillionth power of symbols in the shortest proof. It can then be shown that only a finite number of proofs need be checked in order either to prove or to refute the formula in question. A decision procedure would then exist, which we know is impossible. Thus by examining the length of a formula we can say nothing about the length of its proof.

For example, let's suppose that Goldbach's conjecture is false, but that the first even number for which it is false is the zillionth. Suppose further that the shortest proof of this fact would involve substantially the taking of all prime pairs less than two zillion and showing that they do not add up to it. Goldbach's conjecture would then be practically undecidable in $\mathscr{A}$. Of course a statement practically undecidable in one system might be practically decidable in another. Nevertheless, the phenomenon of the practically undecidable is one that can be annoying, to say the least.

To take another example, suppose that Fermat's last theorem is true, but practically undecidable in $\mathscr{A}$. It might break down into a zillion different cases. Suppose further that this fact is unknown to us and we are investigating the theorem in $\mathscr{A}$. We first try to prove the theorem. We fail. We then try to refute it. We fail. Finally we try to prove it undecidable. Again we fail. In each case we have no way of knowing whether the failure is due to lack of ingenuity or lack of time. And even if we correctly guess the truth we might still fail in finding a system (although there might be one) in which both Fermat's last theorem is practically decidable and the axioms in the interpreted system are all statements we are prepared to accept. Of course, we can trivially find such a system by adding the formal equivalent of the theorem to the initial formulas of $\mathscr{A}$. Our only hope is to look for a proof that the shortest proof-in-$\mathscr{A}$ of the theorem is more than a zillion symbols long. But we have no assurance that we can find such a proof.

3. *Theoretically unsolvable.* As we have seen, the predicate 'x is the Gödel number of a valid formula in $\mathscr{L}$' is recursively unsolvable. Again recall that this concept is absolute in that it doesn't depend on the system. We could require that every valid formula in $\mathscr{L}$ be an axiom of $\mathscr{L}$ and then we would have "solved" the unsolvable problem. The only trouble is that we would not in general be able to recognize an axiom when we saw one. In fact, Church's theorem would guarantee this failure. The system would thus violate our requirements for deductive knowledge. Hence a proof of recursive unsolvability yields an insight into a fundamental limitation on our abilities. There has been a long tradition in Western philosophy which has suggested that our intellectual abilities are fundamentally limited. What is new with mathematical logic is the *precision* with which the unsolvable problems are stated, and the *simplicity* of the problems themselves.

For example, the following "word problem"* is unsolvable. Consider finite concatenations of the letters 'a', 'b', 'c', 'd', and 'e', which will be called *words*. For example, '$aaabd$' is a word and '$baadebcbc$' is a word. Two words will be called equivalent if one can be transformed into the other by the following rules of substitutivity of equivalent expressions: 'ac' is equivalent to 'ca', 'ad' to 'da', 'bc' to 'cb', 'bd' to 'db', '$adac$' to '$abac$', 'eca' to 'ae', 'edb' to 'be'. For instance, '$abbe$' is equivalent to '$abedb$' since 'be' is equivalent to 'edb'. It is an unsolvable problem to decide for an arbitrary pair of words whether they are equivalent.

The difference between an undecidable problem and an unsolvable problem may be illustrated in the following way: An undecidable problem breaks down into an infinite number of subproblems such that each of these problems *may* be decided within the system, but there is no method within the system to collect these into a finite number of proofs; an unsolvable

---

* Due to G. S. Tsentin and D. Scott [cf. Wang 1965].

problem breaks down into an infinite number of subproblems, such that any formal system (or finite series of formal systems) may have among its powers the ability to solve a finite number of these subproblems (or even an infinite number of them, provided that it leaves an infinite number unsolved), but no formal system whatsoever has techniques powerful enough to decide the general problem. In other words, no formal system whatsoever has techniques powerful enough to solve the word problem described above.

Strictly speaking, we should speak of undecidable closed* formulas (or sentences or propositions), and unsolvable predicates (or propositional functions). Any given sentence, or finite sequence of such, is always trivially solvable. For example, consider three sentences of arithmetic (whether solvable or not). Let their symbolic form be represented by 'A', 'B', and 'C'. Consider the following eight combinations of these three formulas:

(1)	(2)	(3)	(4)	(5)	(6)	(7)	(8)
A	A	A	A	$\bar{A}$	$\bar{A}$	$\bar{A}$	$\bar{A}$
B	B	$\bar{B}$	$\bar{B}$	B	B	$\bar{B}$	$\bar{B}$
C	$\bar{C}$	C	$\bar{C}$	C	$\bar{C}$	C	$\bar{C}$

Let system $\mathscr{S}_1$ be arithmetic with the three formulas in (1) added as initial formulas. $\mathscr{S}_2, \ldots, \mathscr{S}_8$ are defined similarly. Then at least one of these systems is consistent and in that system (or systems) the three formulas are trivially decidable. Of course, we may not know which one, but this knowing is not part of the definition. The point is that there exists a consistent formal system in which the formulas are decidable.

On the other hand, what *unsolvable* means is that there is no formal system which is adequate for a predicate. Consider the predicate 'u is equivalent to v', where 'u' and 'v' are variables whose domain of definition consists of the "words" in the above problem. The "words" may first be ordered in some manner into a linear sequence representing all possible combinations of pairs of "words." Now let the formal equivalent of this sequence be represented by 'A_1', 'B_1', 'C_1', 'A_2', 'B_2', etc. For example, 'A_1' might be the formal equivalent of "a' is equivalent to 'a" or C_{13} of "bca' is equivalent to 'ab". As applied to this problem, *any* formal system must be such that (a) it contains as a theorem a formula which under the intended interpretation is false (incorrect) or (b) it contains as a nontheorem a formula which under the intended interpretation is true (incomplete).

Nor could we make the general statement that some system contains as axioms all the true statements in the sequence of statements of equivalence of word pairs. For the problem is recursively unsolvable and we would not (in general) know an axiom when we saw one. Nor can we follow the method

---

* A formula is *closed* if it contains no free variable.

used above in the finite case of taking all possible combinations of formulas in different systems. Suppose that we tried. Just as there are 2^3 systems in the finite case, so there would be $2^{\aleph_0}$ systems in an infinite case. $2^{\aleph_0}$ is the cardinal number of the real numbers, so that we could use real numbers for subscripts for our "formal systems." For instance, we might have $\mathscr{S}_\pi$, $\mathscr{S}_{.325}$, or $\mathscr{S}_{\pi-.78}$. There would be a nondenumerable number of these "formal systems." And here is the central difficulty, because we would have no way of identifying the axioms of a given system. What, for example, are the initial formulas of $\mathscr{S}_\pi$?

To sum up, if a predicate (or propositional function or infinite set of problems) is unsolvable, every formal system for it is either incorrect or incomplete. By Church's thesis *recursively unsolvable* becomes co-extensive with *unsolvable* in the intuitive sense, that is, unsolvable by a human or unsolvable by a machine. Never before in the history of thought has a human intellectual limitation been stated so exactly, or been exhibited at such an elementary level. The question naturally arises whether some easily understandable problem of elementary arithmetic is unsolvable. Although we don't yet know, there seems to be no reason why it couldn't be true. For example, the predicate 'x is a divisor of a perfect number' might very well be unsolvable.

4. *Practically unsolvable.* A predicate will be called *practically unsolvable* if it is solvable but the formal system (or techniques of solution) is so impractically complicated that it cannot be actually used by any known means. Again the definition is vague, but it is exact enough for our purposes. What makes this phenomenon so frustrating is that we would have no sure way of identifying it in some given arbitrary case.

Consider a predicate which we successively try to show to be solvable and unsolvable, and fail both times. This leaves open three possibilities: (a) that it is practically solvable (but we didn't use enough ingenuity); (b) that it is practically unsolvable (and we lacked the time); or (c) that it is theoretically unsolvable (but we didn't use enough ingenuity). Thus, for example, the predicate 'x is a divisor of a perfect number' might be solvable, but the simplest formal system which would solve it would have a zillion separately stated axioms, each of which has at least a zillion symbols, and for good measure requires a zillion rules of inference. This might be put another way. The ability of both men and machines to deal with complications is limited, whereas arithmetic presents us with problems whose solutions are arbitrarily complicated. We have as yet no way of being sure whether or not some of the simply formulated problems of arithmetic have solutions of only the most extreme complexity. In general, there is no fixed way, by examining a predicate, to determine the complexity of its solution. Our only hope would be to look for a proof that, if there were a solution of a certain set of problems, the simplest solution would be complicated beyond any possible, practical means that we might use. But again we have no assurance that we can find such a proof.

To sum up: The phenomena of theoretical and practical undecidability of sentences and of theoretical and practical unsolvability of predicates (problems) occur in arithmetic in such a way that it appears that there is no way of avoiding them. It also seems possible that some quite simple and short sentences of arithmetic are undecidable, and some quite simple and short predicates are unsolvable. The impression we might draw from results such as these is that the prospects for the mathematician and logician are not promising. The following considerations should be taken into account in determining whether such a conclusion is a hasty one.

In the first place, logicians have had some successes in their attempts to find decision procedures for some logical and mathematical theories. We have shown that the propositional calculus possesses one. So does the first-order predicate calculus, where all predicates represent only properties (which would include the whole theory of the syllogism). So does the theory of addition of natural numbers or the theory of multiplication of natural numbers. Altogether there are perhaps a dozen theories of some logical or mathematical consequence which have decision procedures. The most remarkable result in this area is due to Alfred Tarski: the discovery of a decision method for elementary algebra and, consequently, elementary geometry. He discovered it in 1930, but the results were not published until 1948. We might at first think that this result is in conflict with Church's theorem, since we would have a decision procedure which would decide Fermat's last theorem. Unfortunately Fermat's theorem is a statement of arithmetic and not algebra. The predicate 'x is a natural number' cannot be expressed in Tarski's system (as we know from Skolem's result). In fact, neither can the notions of integer, rational number, or algebraic number. Any particular natural number (or integer, etc.) may be expressed, but not the notion of all natural numbers. Thus

$$(\forall x)(\forall y)(\forall z)(\exists x_1)(xx_1 + yx_1 + zx_1 = 0),$$

$$31 + 73 = 104,$$

and

$$(\forall x)(\forall y)(\forall z)(\forall x_1)(x^{x_1+2} + y^{x_1+2} \neq z^{x_1+2})$$

are all statements of elementary algebra. However,

$$(\forall x)(\forall y)(\forall z)(\forall x_1)((N(x) \cdot N(y) \cdot N(z) \cdot N(x_1)) \supset (x^{x_1+2} + y^{x_1+2} \neq z^{x_1+2})),$$

where '$N(x)$' means 'x is a natural number', is not. A sentence of geometry is elementary if it corresponds by some coordinate system to a sentence of elementary algebra. It turns out that there are interesting undecided problems of elementary algebra, but that the method is so long that they cannot yet be resolved. It is hoped that by improvements in the method and in the technology of digital computers, hitherto unsolved problems may be solved [cf. Fraenkel and Bar-Hillel 1958 311 and Smart 1968 31]. Thus it seems worth while to ask of any mathematical system or subsystem whether it has a decision procedure. This is true even though the general problem of deciding for an arbitrary formal system whether it has a decision procedure is itself unsolvable [cf. Fraenkel and Bar-Hillel 1958 314].

In the second place, the methods that we do have—while not complete— are nevertheless very potent and there is good evidence that this potency has not been fully exploited. An indication of this is provided by Hao Wang, who states that he has "examined the theoretically undecidable domain of the predicate calculus and managed to make an IBM 704 prove all theorems (over 350) of *Principia Mathematica* in this domain in less than 9 minutes; this suggests that we usually do not use the full power of strong mathematical methods and should not be prevented from trying to handle an area on account of pessimistic abstract statements of the more difficult cases in the region" [cf. Braffert and Hirschberg 1963 5–6]. In fact, it can be shown that to each formula in $\mathcal{A}$ there corresponds one in $\mathcal{L}$ such that the formula in $\mathcal{A}$ is provable-in-$\mathcal{A}$ if and only if the formula in $\mathcal{L}$ is unprovable-in-$\mathcal{L}$. Thus to Fermat's last theorem there corresponds a formula in general logic whose unprovability is equivalent to the truth of the last theorem. As Kleene notes,

... in effect, mathematicians have labored unsuccessfully for over 300 years to settle the question whether this one particular formula of the predicate calculus is unprovable or provable. ... The predicate calculus is such a rich system that a host of particular problems that are commonly considered in mathematics, using the predicate calculus as a tool in successive short arguments, can be clothed entirely in the pure predicate calculus. Indeed ... this is the case for all problems whether a given statement holds in a formal axiomatic theory whose axioms are finite in number and expressible in the symbolism of the predicate calculus with predicate, (individual) and functional symbols [1967 283].

From these considerations we can conclude that the question still remains open whether there is some one formal system (or Turing machine) which best represents us in the sense that our mathematical and logical limitations are the same as the mathematical and logical limitations of the system (or Turing machine). A common argument against this possibility is fallacious, as we have seen (see Problem 72).

To recapitulate: The argument that we do not have the same disabilities as a Turing machine (because for any given Turing machine we can prove a formula—say, M—which it cannot) is incorrect. We can only prove M on the assumption of the consistency of the machine, and this the machine can do also. In fact, this is how Gödel's second incompleteness theorem is proved. We haven't established that there is a Turing machine (or formal system) which best represents us, only that it still makes sense to look for one. We even have some idea of what it would be like. It must be able to prove the formal equivalents of all the mathematical sentences that we have proved. Further, all those mathematical problems which we haven't been able to solve must be unprovable in it (although we might not be able to prove that they are unprovable in it). The machine could be progressively defined, according as we are able to solve new hitherto-unsolved problems. Such a progressive definition would merely indicate that our mathematical limitations are not constant.

In the third place, we cannot overlook the possibility of "probability" proofs, where the possibility of "deterministic" proofs is lacking. For example, it is well known that some specific questions about primes have, up to the present time, yielded only probabilistic answers. A typical problem of this sort is to find some order in the distribution of primes; or more specifically, to find a polynomial function of two variables—say, $f(x, y)$—such that for any given pair of natural numbers it yields the number of primes between them (inclusively). For example,

$$f(3, 8) = 3, \qquad f(1, 17) = 8, \qquad \text{etc.}$$

Despite centuries of attempts, no such function has yet been described. It might be the case that this problem is recursively unsolvable, that is, that the function exists but is "infinitely complicated" (if we take the platonistic point of view) or that such a function does not exist (if we take the nominalist point of view). No one has yet proved that it is recursively unsolvable.

The fact that it may be is in no way inconsistent with the result that a probabilistic or approximation formula does exist. There is an important difference between empirical and mathematical probability. Empirical probability results when we inspect part of a class and make a conjecture about the whole class. Such an assertion is subject to revision in light of later experience. Consider, for example, the well-known function $x^2 - 79x + 1601$. One can imagine calculating the value of the function for $0, 1, \ldots, 79$. In each case it would yield a prime number. We might then conjecture that it yields a prime number for every value of x. The conjecture is shown to be false if we calculate the value of the function for $x = 80$. Empirical conjectures are not valueless, however, because they are often important heuristically in helping a mathematician to decide what statements he should try to prove. And they often turn out to be true. Empirical probability statements may be exact in their content ($x^2 - 79x + 1601$ yields only primes), but then the "probability" aspect concerns the degree to which we are justified in asserting them.

On the other hand, there is no doubt in general about the assertion of mathematical probability statements; it is the content of the statement which brings in the "probability" aspect. For example, near the end of the eighteenth century Gauss made a conjecture about the distribution of primes. Let p_n be the nth prime number. Gauss conjectured that

$$\frac{n \times \left(\dfrac{1}{1} + \dfrac{1}{2} + \cdots + \dfrac{1}{n} \right)}{p_n}$$

becomes arbitrarily close to 1 as n increases. This does not mean that this function starts fairly close to 1 and continually gets closer as n increases. It does not mean that it is very close at n = zillion. All it means is that at *some* point in the natural number series the ratio from there on becomes arbitrarily close to 1. This conjecture of Gauss was an *empirical* one based on the prime number tables with which he was familiar.

Finally, near the end of the nineteenth century, Hadamard and de la Vallée Poussin, working independently, succeeded in mathematically proving it by making use of very advanced methods. The *empirical* then changed into a *mathematical* probability statement. Thus, although we are still ignorant about the *exact* distribution of primes, we are not ignorant about their approximate distribution. It may be that some of the unsolvable predicates or undecidable statements of arithmetic would lend themselves to a probabilistic approach. Perhaps we could even obtain a sequence of successively more powerful systems—analogous to $\mathscr{A}$, $\mathscr{A}_2^1$, $\mathscr{A}_3^1$, etc., of §23C—such that we could find a probabilistic formula which would tell us where in the sequence of systems an arbitrary formula could be proved. Our probabilistic formula might tell us, for example, that it is highly improbable that Goldbach's conjecture could be proved in system $\mathscr{S}_1$, but that it is very likely it could be proved in system $\mathscr{S}_{\text{zillion}}$.

In the fourth place, the category of the unsolvable turns out to be interesting in its own right. The analogy to the infinite cardinals is a good one (used by Kleene 1967 267). The work of Cantor shows that the category of the infinite is structured, and that there are different sizes of infinite sets. Similarly the work of Turing, Kleene, Post, and Mostowski has shown that the category of the unsolvable is structured; there are, to use Post's expression, degrees of recursive unsolvability. It turns out that this structure is very complex and is mathematically very interesting. The working out of a structure so that any abstract problem may be classified in a systematic way is certainly not yet complete. Many interesting and unanswered questions still exist.

Let us now sum up at least four reasons why we should not take a pessimistic attitude as a result of the limitative theorems: (1) Some positive results are available, and perhaps others might be found for areas of mathematics in which the question is still open. (2) The mathematical means at our disposal have apparently not been anywhere near fully exploited. (3) Probabilistic techniques are also available. (4) The systematic investigation of the structure of the unsolvable is itself very interesting and significant.

Nevertheless, these considerations do not answer all doubts about the value of further logical and mathematical research; in particular, the problem of consistency must still be taken up.

§29   THE QUESTION OF CONSISTENCY

On perhaps no question has the pessimism generated by the limitative theorems been felt more acutely than on the question of the consistency of arithmetic. The usual account goes something as follows. By Gödel's second incompleteness theorem we know that any system which contains primitive recursive arithmetic cannot prove its own consistency. Any system (purporting to be a formalization of arithmetic) which didn't contain primitive recursive arithmetic would be so inadequate as not to deserve the name *arithmetic*. Thus the best we can do is to have faith that arithmetic is consistent. Such is the usual account. For our purposes here we shall distinguish four different types of consistency proofs and show that the usual account is at best misleading.

1.   *Empirical proofs of consistency.* We have good empirical evidence that arithmetic is consistent. In spite of many attempts by many different inquirers over a long period of time, no inconsistency has been discovered. Of course, this is not conclusive evidence that there is no inconsistency, but it does raise the issue of the consistency of arithmetic out of the realm of mere conjecture. Furthermore, arithmetic has been used and investigated (without looking for an inconsistency) on an enormous variety of problems and no inconsistency has turned up accidentally. Finally, even if an inconsistency should turn up, it might be unimportant. This can best be explained by an analogy. The differential and integral calculus as originally proposed (independently by Newton and Leibniz) was inconsistent. However, this inconsistency was unimportant, since the calculus could be recast with the loss of nothing but the inconsistency and its undesirable consequences. If someone today were to say that the calculus is inconsistent, his remark would be unjustified, because there are consistent formulations of the theory which Newton and Leibniz intended to formulate. Similarly, even though our present versions of arithmetic might prove to be inconsistent, this would not necessarily imply that a consistent version of the intended theory could not be devised. Thus, from an empirical point of view it seems very unlikely that someone will prove arithmetic inconsistent.

2.   *Probabilistic mathematical proofs of consistency.* We have seen that probabilistic statements may be made about the distribution of primes;

for example, we can say that if we pick a large number n at random, the chances are

$$\frac{1}{2} + \frac{1}{3} + \cdots + \frac{1}{n}$$

to 1 that it is not a prime. Could some variant of this technique give some probabilistic answer to the question of consistency? So far as I know, the problem has never been approached from this point of view. For instance, all possible formal systems (consistent and inconsistent) might be divided into an infinite number of classes each containing an infinite number of systems. It might then be possible to show that, for an arbitrary choice of a formal system in one of the sets, the chances of it being consistent are π to 1. It might be possible to show that arithmetic belongs to that set. Furthermore, by more refinements of these techniques, it might be possible to increase the odds to any desired level. There are trivial answers which might be obtained in this way. For example, if the set containing arithmetic also contained only variants of the propositional calculus, then the probability that arithmetic is consistent would be 1. Nevertheless *some* variant of this general approach might yield significant results.

3. *Deterministic proofs of consistency.* The reason that no probabilistic approach was tried was no doubt due to the fact that deterministic proofs exist. Within five years after the publication of Gödel's second incompleteness theorem, Gerhard Gentzen (1909–1945) published a proof of the consistency of elementary arithmetic.

The reason this proof hasn't been universally accepted is due to the following circumstance. Gentzen makes use of a generalized version of mathematical induction called *transfinite induction*. Transfinite induction was devised by mathematicians to take care of inductions on sets other than those with order ω. To see that it is necessary, consider the following set of order type $\omega + \omega$:

$$\{1, 2, 3, \ldots, -1, -2, -3, \ldots\}.$$

By using only ordinary induction, we can prove that every element is positive (that is, greater than 0). For the first element—1—is greater than 0. Next, it is true that for any n, if the nth element is greater than 0, then the $(n + 1)$st element is greater than 0. (Recall that in a conditional statement the falsehood of the antecedent guarantees the truth of the conditional.) Hence, by ordinary mathematical induction, every element is greater than 0.

We can avoid this invalid argument by revising the induction statement to read this way: For any n, if every element less than n has a property, then n has that property too. The revision would simultaneously get rid of the invalid induction proving that all elements are positive and yet allow cogent inductions. This revision gives us transfinite induction. It can be used for both definition and proof. Different uses of transfinite induction can be classified according to how far in the sequence of ordinals the induction goes. Thus we can speak of 'induction up to ω^2' or 'induction up to ε_0'.

Just what kinds of transfinite inductions are to be considered finitary is debatable. Transfinite induction up to an arbitrary ordinal is certainly not finitary. However, it can be shown that certain transfinite inductions are reducible to ordinary mathematical inductions. For example, induction up to ω^ω is reducible to ordinary induction. Gentzen in his proof used transfinite induction up to ε_0. Now transfinite induction up to ε_0 *is* reducible to ordinary mathematical induction, but this reduction is very complicated and *cannot* be carried out in elementary arithmetic. Since all transfinite inductions up to some particular ordinal less than ε_0 are reducible to ordinary induction and can be carried out in elementary arithmetic, it seemed that Gentzen went as little as possible beyond the arithmetical system in order to prove consistency. For all the other methods used by Gentzen are formalizable in the arithmetical system. If induction up to ε_0 were also formalizable, then the proof of C (or its analog) could be carried out and the system would be inconsistent, by the reasoning of Gödel's second incompleteness theorem. Thus it appears that Gentzen has devised a finitary method which is adequate for proving the consistency of number theory. (It seems to me that those who would deny the name *finitary* to Gentzen's methods should be obliged to show where the nonfinitary enters in the proof of the reduction of transfinite induction—up to ε_0—to ordinary induction. To my knowledge, this has not been done.)

The true source of dissatisfaction with Gentzen's proof should not be over the issue of "finitary" (which, after all, was only rather vaguely defined in the first place), but on the issue of begging the question. In the proofs of consistency of the propositional calculus and the predicate calculus, methods were used which transcended the system, but these methods were of such simplicity as hardly to be open to doubt, in fact, these methods were recursive. On the other hand, the finitary methods used by Gentzen are very complicated, and it is a serious question whether this complication is of such a degree that the proof is no longer of value. For the value of a proof is to convince a person of something about which he is not initially convinced. For example, it is perfectly possible for someone to simultaneously accept the common notions and postulates of Euclidean geometry and yet doubt the Pythagorean theorem. The purpose of the proof is to show that the doubt is unjustified.

This is what Gentzen's proof does not succeed in doing. Although formally cogent, it is informally not effective because of *petitio principii*. It represents, it seems to me, a halfway house between not knowing about the consistency of arithmetic and knowing in a way which removes all legitimate doubt. In this sense, it is a phenomenon similar to that of the result on the distribution of primes: It is a remarkable result, which represents a high mathematical achievement, but we should be able to do better. And just as "better" would represent a deterministic answer to the question of the distribution of primes, so here it would mean a proof of consistency making use of methods no more complicated than those of arithmetic itself.

In judging the value of Gentzen's proof, we must distinguish between uninterpreted formal systems and systems interpreted as arithmetic. Here the analogy with other systems is a good one. In proving the consistency of $\mathscr{P}$ or $\mathscr{L}$ we used methods which transcended the system. However, these methods are of such simplicity that everyone accepts them. To anyone who accepts transfinite induction up to ε_0—and this means almost all mathematicians—Gentzen's proof is perfectly valid for showing the consistency of a system like $\mathscr{A}$. For example, suppose that a logician sets up a quite complicated system $\mathscr{B}$ for which arithmetic is the intended interpretation. He might, however, have doubts that his particular system $\mathscr{B}$ is consistent. If he accepted transfinite induction up to ε_0, these doubts would be allayed by a Gentzen-type proof. But he could not conclude from this that $\mathscr{B}$ interpreted as arithmetic was consistent. To do so would be begging the question.

The value of the proof is like the nineteenth-century relative proofs of consistency of geometry (cf. §9), in which one system is proved consistent *if* another is. On the basis of Gentzen's proof, we can only conclude that an arithmetic system is consistent if the method of induction up to ε_0 is consistent. This is far from trivial; in particular, Gentzen's proof reveals exactly how far we have to go beyond arithmetic techniques to get a proof of the consistency of an arithmetic system. But it does not tell us everything we want to know.

Problem 75. How might it be possible for $\mathscr{A}$ to be consistent and yet have arithmetic be inconsistent?

4. *Proofs of consistency capable of being expressed in the system.* As we have seen, a common paraphrase of Gödel's second incompleteness theorem is that it states that if a system is consistent, then the consistency of that system is not provable in the system. According to this paraphrase, the present paragraph, if not nonexistent, should at least be very short. It should merely state that there is no such possibility. The common paraphrase is an incorrect one, however. The actual situation is more complicated.

To make this definite, suppose that $\mathscr{A}$ is consistent. Since

$$((0) + (0)) = (0)$$

is provable-in-$\mathscr{A}$, the negation of this formula is not provable-in-$\mathscr{A}$. The Gödel number of that negation is $Neg(c_4)$ (cf. §23B). We are then able to prove-in-$\mathscr{A}$ the formal expressions for the following arithmetic sentences (since each is true and primitive recursive):

$$(\overline{0PNeg(c_4)}), \quad (\overline{1PNeg(c_4)}), \quad (\overline{2PNeg(c_4)}), \quad \dots$$

However, we could not prove the formal expression for $(\forall x)(\overline{xPNeg(c_4)})$, for then we would prove C and the system would be inconsistent. So we see again that if $\mathscr{A}$ is consistent, it is ω-incomplete.

By Gödel's completeness theorem, we know that there must be some model which contains an unnatural number α such that the formula in $\mathscr{A}$ corresponding to $\alpha PNeg(c_4)$ holds. For otherwise, the formal expression for $(\forall x)(\overline{xPNeg(c_4)})$ would be true under all interpretations and thus provable-in-$\mathscr{A}$. So there is some interpretation under which the formal expression for $(\exists x)(xPNeg(c_4))$ is true and some interpretation under which it is not true. But if this is the case, the formulas corresponding to P do not under every interpretation represent what we intuitively mean by *provable*. This follows because we are assuming that $\mathscr{A}$ is consistent and hence the formal expression for $(\exists x)(xPNeg(c_4))$ is false under the intended interpretation. So under some nonintended interpretation it is true, and consequently the formulas corresponding to P cannot express the intuitive concept of *provable*. This is the reason C is not provable-in-$\mathscr{A}$, provided that $\mathscr{A}$ is consistent. For there must, by Gödel's completeness theorem, be an interpretation of C which makes it false.

We thus have the following result: Either $\mathscr{A}$ is inconsistent or C does not under every interpretation express the consistency of $\mathscr{A}$. The reason C cannot be proved-in-$\mathscr{A}$ is the same as the reason consistency cannot be proved within $\mathscr{P}$ or $\mathscr{L}$, namely, consistency cannot be expressed within $\mathscr{P}$ or $\mathscr{L}$ and C does not express consistency in $\mathscr{A}$. What makes $\mathscr{A}$ different is that a formula in $\mathscr{A}$—C—expresses the consistency of the system under *one interpretation of that formula*. If there were some other formula which expressed the consistency of $\mathscr{A}$ under *every* interpretation, it would be true under every interpretation and, by Gödel's completeness theorem, provable. Hence the trick is to find an expression of consistency in $\mathscr{A}$ which is not fouled up by the phenomenon of unnatural numbers. Can this be done?

To make a judgment, we must understand that we are faced with a two-fold problem: first, to find a formula in $\mathscr{A}$ which we can show expresses the consistency of $\mathscr{A}$, and second, to demonstrate that that formula is provable-in-$\mathscr{A}$. The second problem is a mathematically precise one. The first problem is not mathematically precise; it depends on what we mean by *express the consistency of* $\mathscr{A}$. So our problem becomes: Can we find a formula which is provable-in-$\mathscr{A}$ and for which it is "reasonable" to say that it expresses the consistency of $\mathscr{A}$? The answer is yes, but it does not follow that we thereby increase our confidence in the consistency of $\mathscr{A}$. To see this, let '$xP'y$' be defined as

$$(xPy) \cdot (\overline{xPNeg(c_4)}).$$

Since there is a proof-in-$\mathscr{A}$ of the formula with Gödel number c_4 (cf. §23B), the set of natural numbers which makes 'xPy' true is the same as the set which makes '$xP'y$' true, provided that $\mathscr{A}$ is consistent. [Observe that if $\mathscr{A}$ is inconsistent,

$$(\exists x)(xPNeg(c_4))$$

is true, whereas

$$(\exists x)(xP'Neg(c_4))$$

is false.] Now let 'C'_A' be defined as

$$(\forall x)(\overline{xP'Neg(c_4)}),$$

and let 'C'' denote its formal expression in $\mathscr{A}$. C' is provable-in-$\mathscr{A}$, as may be made plausible by noting that

$$(\overline{xP'Neg(c_4)})$$

is equivalent to

$$xP(Neg(c_4)) \supset xP(Neg(c_4))$$

(cf. §16). Hence all we need to prove is $(P \supset P)$ and then use universal generalization.

The situation is somewhat analogous to Rosser's version of Gödel's first incompleteness theorem. By changing the definition of the undecidable formula, Rosser was able to weaken the requirement of ω-consistency to simple consistency. Similarly, by changing the definition of the formula which we take to express consistency, we reduce the requirement of transfinite induction to the methods available in the system. But the analogy is not exact, for whereas Rosser gained an improved result, the present proof of consistency is less convincing than Gentzen's. The reason is that the proof works whether or not $\mathscr{A}$ is consistent. Yet what is doubtful is not the proof. Rather it is whether C'_A expresses consistency, and we know it does only on the hypothesis that $\mathscr{A}$ is consistent! (Recall that if $\mathscr{A}$ is inconsistent, C_A is false but C'_A is true.)

So the problem becomes: Can we find an arithmetic sentence which we can identify as expressing the consistency of $\mathscr{A}$ in an *effective* or *constructive* way and which is such that its formal expression is provable-in-$\mathscr{A}$? The answer apparently is no, for it follows from a result of Feferman that for any formula which effectively or constructively expresses consistency, Gödel's second incompleteness theorem holds. So we can now add a generalized form of that theorem to our earlier list (cf. §26). *Gödel's second incompleteness theorem* (generalized version):

No consistent formal system contains a theorem which is an effective or constructive expression of the consistency of that system.

Hence any formula in $\mathscr{A}$ which is an effective or constructive expression of consistency allows for another interpretation in which it does not express consistency.

We may now state an improved translation of the theorem into the language of psychology or physics (cf. §26):

Language of Psychology	*Language of Physics*
No consistent human computer can prove an effective or constructive expression of his own consistency.	No consistent computing machine can be programmed to prove a computable or computably enumerable expression of its own consistency.

These results, as well as the generalized form of the theorem, depend on Church's thesis. So we see yet another place in which the philosophical interest of metatheory derives from the import of that thesis. Our conclusion is that the question of consistency is open to the extent that Church's thesis is open.

We are now in a position to see that there is no single quality which characterizes mathematical statements which would allow us to decide who is right: Kant, Mill, or Russell. Mathematical statements such as '0 = 0' are certainly analytic *a priori*, their truth being wholly dependent on general logic and definition. On the other hand, the phenomenon of empirical probability in mathematics, as well as the practically undecidable and unsolvable, give us examples of synthetic *a posteriori*. Finally, C under its intended interpretation could reasonably be classified as *synthetic a priori*, synthetic because it does not follow by definition and general logic, *a priori* because if arithmetic is consistent, it must be necessarily consistent.

Yet more than anything else, the limitative theorems impress upon us that the whole taxonomy of the dogmatic idealist (Kant) and empiricist (Mill, Russell) past is inadequate to deal with the complications of present knowledge. A new classificatory scheme is suggested by mathematical logic involving such notions as recursive, recursively enumerable, decidable, degrees of recursive unsolvability, etc. Although far from complete, this classificatory scheme has been extensively investigated and it makes the analytic-synthetic distinction outmoded. Is G, for example, analytic or synthetic? This is somewhat like asking whether a virus is living or dead.

The point is that the old system is no longer helpful or even interesting. For example, suppose that there were two technical papers, the first claiming to show that the problem that Einstein set himself in the unified field theory is analytic, the second claiming to show that in some clear sense the problems involved in obtaining a unified field theory are recursively unsolvable. My guess is that physicists might find the first paper at best mildly interesting, but would be very excited by the second.

§30 LOGIC AND PHILOSOPHY

The impact of mathematical logic on contemporary philosophy is comparable to the impact that traditional logic and Newtonian physics had on a philosophy such as Kant's. The limitative theorems have in fact provided a kind of critique of pure reason far beyond Kant's in both certainty and precision. We now know with much greater assurance the way in which we must proceed if we are to have knowledge.

The central notion for knowledge is that of a recursive function, and it is with this concept that would-be epistemologists must deal. At the end of his great 1944 paper, Emil Post boldly declares that "if general recursive function is the formal equivalent of effective calculability, its formulation may play a role in the history of combinatory mathematics second only to that of the formulation of the concept of natural number" [336]. I think it is also fair to say that if Church's thesis proves to be true, the notion of a general recursive function will have an impact on philosophy comparable to Plato's notion of a Form. We might briefly explore how a philosophy which takes the concept seriously might proceed.

The decision to inquire in principle is not rational, but, as it were, pre-rational. Aristotle's argument that we ought to philosophize is not cogent because the third premiss is false. For his argument is: "Either we ought to philosophize or we ought not. If we ought, then we ought. If we ought not, then also we ought [i.e. in order to justify this view]. Hence in any case we ought to philosophize" [cf. Kneale 1962 97]. There is a widespread attitude which rejects inquiry not by the self-defeating method of inquiring into whether or not one should inquire, but rather by the perfectly legitimate way of ignoring inquiry. The Buddha, for example, counselled that the pursuit of knowledge is a waste of time. It was in opposition to views of this kind that the Pre-Socratics laid the foundations for logic by their *decision* to inquire. Once it has been decided that knowledge is to be pursued, the question then arises as to how to go about it.

The results of mathematical logic suggest that our starting assumption should be whatever is necessary for the theory of recursive functions. In the past, philosophers have taken elementary arithmetic, elementary geometry, and elementary logic as paradigms of certain knowledge. By taking the theory of recursive functions as basic, we actually gain the equivalents of all three. For example, we might then treat mathematical logic itself as a branch of the theory of recursive functions. Such an approach would require a different order to topics than the more or less orthodox one we have presented. In particular, the elementary parts of recursive theory would come before any "logic." Or again, the definition of validity in the predicate calculus makes reference to set-theoretical ideas and therefore Gödel's completeness theorem is less elementary than his first incompleteness theorem, which is clearly finitary. The approach to logic via the theory of recursive functions would bring this into greater relief.

But there is a more profound reason for taking this theory as basic, namely, the impossibility of completely formalizing and thus axiomatizing this theory. It constitutes the intuitive theory which, while it resists all formalization, is nevertheless necessary for the very purpose of setting up formal systems. All attempts to avoid this core theory and instead to derive it from something else appear to commit the fallacy of begging the question.

The book in philosophy which describes what presuppositions are involved in the development of the theory of recursive functions has yet to be published. The theory presupposes a number of very simple mathematical operations. But more and different kinds of presuppositions are needed. For example, all but the shortest kinds of proofs involve the reliability of memory. For longer proofs, a certain orderliness in the world is necessary. The marks which we make on paper cannot change in a haphazard fashion when we aren't looking. Furthermore, the existence of non-recursive recursively enumerable sets requires that there be a time sequence in which such sets can be defined. For example, the notion of validity is inherently constructive and thus time-consuming, since there can be no immediate recognition of a valid formula. Again, we must presuppose that we have the ability to discover and correct our mistakes. This presupposes that we have access to our own past, the knowledge of which can influence our future behavior.

Such, then, is a sample of the kind of presuppositions involved in the pursuit of knowledge via the theory of recursive functions. But there is another aspect to the situation which has not been mentioned. When a pursuit is referred to, something is sought and in this case that something is arithmetic truth. What is its nature? From the limitative theorems we know that it does not have a recursive (= effective = computable) nature or even a recursively enumerable (= constructive = computably enumerable) nature. Because it represents an ideal toward which we strive, we may yield to the temptation to give it the names *ideal or prospective* (both terms are used in Myhill 1952). It represents a property whose definition involves both psychology and history.

Insofar as the notion of arithmetic truth corresponds with either effective or constructive properties, it is a purely mathematical property. Yet we know that this correspondence is only approximate. Over and above this, what counts as an arithmetic truth is dependent on what people decide is arithmetically true. At this point there is an analogy between a natural language (say, English) and arithmetic truth: Just as there can be no complete systematic abstract account of why we speak English the way we do, so there can be no complete account as to why we accept one rather than another statement as arithmetically true. Both English and true are in part historically defined properties. Arithmetic truth, like effectively computable, is an inherently informal notion. Yet, unlike effectively computable, it can have no exact formal counterpart. The formal concept of provability can at best be an approximation to it (such is part of the import of Gödel's first incompleteness theorem). In fact, there is a truth theorem due to Tarski which states that "the notion of truth (the set of all true sentences, the truth set) of a consistent formalized system containing recursive number theory is not definable in this system" [Fraenkel 1953 306]. We shall not describe its proof; it is in fact a sophisticated version of the argument about the Liar paradox given in §23A. But the question of truth becomes increasingly pressing: Is the ideal of arithmetic truth perhaps a will-o'-the-wisp?

Ortega, the Spanish existentialist, has stated that "man ... has no nature; what he has is ... history. Expressed differently: what nature is to things, history ... is to man" [Ortega 217]. What I would like to suggest is that in this regard prospective notions (and in particular arithmetic truth) are like man; that is, they have not just a nature but also a history.

This is very rough, so let's consider it in more detail. As we have seen, truth does not have a single nature: The notion of an elementary algebraic or geometric truth has a recursive (= effective = computable) character, the notion of logical truth (= valid in general logic) has a recursively enumerable (= constructive = computably enumerable) character, and the notion of arithmetic truth has a prospective character. Even within arithmetic systems there are decision procedures for large classes of formulas. For example, there is a decision procedure for all formulas which contain only universal quantifiers which apply to the whole unquantified formula. But the notion of arithmetic truth in general—and therefore truth in general—is prospective.

The analogy of a game is useful in explaining the notion of the prospective. It often happens that a game is invented (and the rules are laid down which define that game), but at a later time a circumstance occurs for which the rules give no guidance. At this point a *decision* has to be made as to what will *henceforth* be the rule concerning that circumstance. The decision might be made on the basis of fairness, whether it makes it a better spectator sport, whether it increases the danger, etc. However, it cannot be made on the basis of the rules of the game because they are incompletely defined. Now part of the impact of the limitative theorems is that the rules by which we discover mathematical truth not only are, but must be, incompletely defined. We are thus forced to define the notion of arithmetical truth historically; that is, it cannot be explicated once and for all but must be continually redefined. We have seen how both Gauss and Lobachevsky came to the conclusion that the problems of truth in non-Euclidean geometry required that they go beyond the data of pure geometry. In an analogous way we must go beyond the data of mathematics to define mathematical truth. Man has invented a game of mathematics which is incomplete apparently because of the incommensurability of man's ideals and his abilities.*

Yet the temptation is very strong to say that although mathematics might be a game, it is a game in which we learn about reality, abstract reality. This point of view suggests that there is an incommensurability between reality and our ability to understand it completely. Mathematical reality is always partly elusive and therefore, perhaps, continually fascinating. Had Hilbert's program actually been carried out, mathematics would have become theoretically boring. Mathematics, then, demands an intrinsic creativity in order to progressively define the reality to which it refers.

Nevertheless, if we go beyond our knowledge and postulate a reality, this postulation acquires the character of a myth. There is a long tradition in the history of philosophy which suggests that when we try to define the ultimate nature of reality, a myth is the best that can be obtained. Plato, for example, denied that there could be any nonpoetic discourse about the Form of the Good. The Form is beyond reason, yet it is just our aspiration toward it which makes reason and truth possible. Kant is another example. He suggests that there is no cognitive way in which the noumenal may be apprehended. Kant's Ideals of Reason are "goals to which our reason irresistibly compels us even when that reason itself proclaims to us their unattainability" [Myhill 1952 190]. As Myhill points out, this is surely close to the notion of the prospective.

Or consider Nietzsche's statement that

... the periphery of science has an infinite number of points. Every noble and gifted man has, before reaching the mid-point of his career, come up against some point of the periphery that defied his understanding, quite apart from the fact that we have no way of knowing how the area of the circle is ever to be fully charted. When the inquirer, having pushed to the circumference, realizes how logic in that place curls about itself and bites its own tail, he is struck with a new kind of perception: a tragic perception which requires, to make it tolerable, the remedy of art [Nietzsche 1954 95].

---

*For some views on the relation of history and the empirical sciences to logic and mathematics see Bernays 1965, Church 1966, Finkelstein 1969, Kac and Ulam 1969, Mehlberg 1962, Mostowski 1955, Myhill 1952 and 1960, Nagel and Newman 1958, Putnam 1969, Turquette 1950.

In other words, art is a necessary complement to science and logic. Mathematical logic, it seems to me, brings us to a position which is within the spirit of these views of Plato, Kant, and Nietzsche.

To explore these ideas in more detail, we might recall that it follows from Tarski's truth theorem that no formal system is rich enough to state its own semantics. But what is the difference between an interpreted formal system and an ordinary scientific theory? The only apparent one is that of rigor. Therefore, it seems to me that this result applies to all comprehensive theories whatsoever. Any fixed comprehensive account of reality which states its own truth-conditions could not possibly be true, but only mythical or fictional. No nonpoetic account of the totality of which we are a part can be adequate. What is characteristic of poetic discourse about something is that conflicting and contradicting accounts about it are permissible. If one understands them as literature there is no need to reject the book of Genesis in order to read the Iliad. One of the functions of literature is to provide that vision within which it is possible to have a science at all, that is, to provide a goal and framework in which the morale and energies of man may be maintained. The ideal or prospective notion of mathematical truth is what sustains interest in mathematics as a path to knowledge. Great mathematicians—such as Archimedes, Gauss, and Russell—have often spoken and acted extravagantly with regard to the esthetic delights of mathematics. These reactions are generally understood to be mere epiphenomena of the pursuit of mathematical truth. But perhaps the esthetic visions are part of the heart of the matter.

At this elevated point, metaphysics can merge with literature or art; we don't *necessarily* have to accept one metaphysical system and at the same time reject others. The logical positivists earlier in this century called metaphysics meaningless in order to deprecate it. I call some of it fictional not to deprecate metaphysics but rather to suggest that some literature has a more than hedonistic function. An example of this is Plato's myth of the Forms which has provided the framework in terms of which many mathematicians have interpreted their own activity.

To summarize: There is no entirely nonpoetic and nonfictional account of the world in general. Our understanding and the totality of the universe is in principle incommensurate, but it is this very incommensurability which provides the fascination which generates the never-ending demands on our creativity.

EPILOG

"To poke fun at philosophy," Pascal once said, "is to be a true philosopher." He was suggesting, I believe, the comic gulf between the aspirations and the achievements of philosophers. Mathematical logic has defined—at least partly—the nature and extent of this gulf. It has also shown that this gulf provides an occasion for our mathematical, philosophical and artistic inventiveness. The forbidding symbolism, the pedantic distinctions, and the technical complications of mathematical logic become justified by the rich cultural harvest.

Yet there is another view in contemporary philosophy which suggests that the pursuit of a technical and abstract subject is at best trivial, and usually a tragic waste of human potential. Many who hold such a view would be classified as existentialists. To my knowledge, not one of the leading existentialists—Heidegger, Sartre, Jaspers, Merleau-Ponty—has even mentioned the limitative theorems. The existentialists disagree among themselves about many things, but they would all agree in not assigning a technical subject such as mathematical logic a high place—let alone a central one— in the theory of knowledge. Yet they wish to talk about proof or truth or the limitations of man.

Being as generous as possible, let us suppose that in some sense their low estimation is correct. At the very least we should expect this low estimation to be the result of familiarity with the subject. We are not impressed by the astrologer who rejects astronomy in complete ignorance of Newton and Einstein. So far as I know, the leading existentialists are completely innocent of any knowledge of the limitative theorems. St. Thomas Aquinas argued that semen was caused by excess food. What condemns his viewpoint is not only that he was wrong but that the methods which he adopted to answer the question are inadequate to determine the answer. No biologist today is tempted to turn to St. Thomas for answers to biological questions. It seems unfortunately true that existentialists today are in the position of Aquinas on semen: namely, even if they are right on some issue or other, the way they arrived at their view would probably have to be radically revised in order to be acceptable. Acceptable, that is, to someone not innocent of the intellectual achievements of contemporary science. This is unfortunate because the existentialists have obvious talents and, I believe, something to contribute to philosophy.

As against the existentialists, mathematical logic is often taken as clearly connected with the philosophy known as logical positivism. This is also very unfortunate because there is no particular connection. To demonstrate this, consider what Wittgenstein, often thought to be the fountainhead of logical positivism, has said about logic. In *Tractatus Logico-Philosophicus*, published in 1921, Wittgenstein claims that "there can never be surprises in logic" [165]. The limitative theorems make this claim as false as St. Thomas' remarks on semen. There is even good evidence that Wittgenstein never understood the limitative theorems (on this point see Anderson 1958, Kreisel 1958, Bernays 1959). Or again, the logical positivists have made the claim that there is a sharp distinction between the analytic and synthetic and that all mathematical statements are analytic. As we have seen, this claim is inconsistent with the limitative theorems.

Mathematical logic, then, is neither a waste of time nor dependent on some philosophic school. Its importance stems from the central role it can play in knowledge. According to ancient tradition, Plato required a mastery of geometry as a prerequisite to the study of philosophy. When Plato wrote, geometry represented the only body of deductive knowledge and he presumably made the requirement because of his conviction that anyone who lacked the discipline to master geometry lacked the discipline to master philosophy. Surely mathematical logic (including the theory of recursive functions) occupies such a position today. All disciplines—whether in the humanities, social sciences, or sciences—rest on inferences, and mathematical logic has no rivals in the study of inferences. We should therefore explore mathematical logic for its philosophical presuppositions and implications.

Yet the history of philosophy is replete with philosophers—such as Aristotle or Kant—who took too seriously the results of the science of their day and, as a result, many of their statements now appear foolish. Perhaps future developments will also put mathematical logic in a quite different light. Hence in making our exploration we should be as thorough as possible, but we should also not forget a lesson from the history of philosophy, namely, to always keep a sense of humor and perspective.

A LOGICAL PARADOX*

By Lewis Carroll

"What, *nothing* to do?" said Uncle Jim. "Then come along with me down to Allen's. And you can just take a turn while I get myself shaved."

"All right," said Uncle Joe. "And the Cub had better come too, I suppose?"

The "Cub" was *me*, as the reader will perhaps have guessed for himself. I'm turned *fifteen*—more than three months ago; but there's no sort of use in mentioning *that* to Uncle Joe: he'd only say "Go to your cubbicle, little boy!" or "Then I suppose you can do cubbic equations?" or some equally vile pun. He asked me yesterday to give him an instance of a Proposition in *A*. And I said "All uncles make vile puns". And I don't think he liked it. However, that's neither here nor there. I was glad enough to go. I *do* love hearing those uncles of mine "chop logic," as they call it; and they're desperate hands at it, *I* can tell you!

"That is not a logical inference from my remark," said Uncle Jim.

"Never said it was," said Uncle Joe: "it's a *Reductio ad Absurdum*".

"An *Illicit Process of the Minor!*" chuckled Uncle Jim.

That's the sort of way they always go on, whenever *I'm* with them. As if there was any fun in calling me a Minor!

After a bit, Uncle Jim began again, just as we came in sight of the barber's. "I only hope *Carr* will be at home," he said. "Brown's so clumsy. And Allen's hand has been shaky ever since he had that fever."

"Carr's *certain* to be in," said Uncle Joe.

"I'll bet you sixpence he *isn't!*" said I.

"Keep your bets for your betters," said Uncle Joe. "I mean"—he hurried on, seeing by the grin on my face what a slip he'd made—"I mean that I can *prove* it, logically. It isn't a matter of *chance*."

"Prove it *logically!*" sneered Uncle Jim. "Fire away, then! I defy you to do it!"

"For the sake of argument," Uncle Joe began, "let us assume Carr to be *out*. And let us see what that assumption would lead to. I'm going to do this by *Reductio ad Absurdum*."

---

* Reprinted from Carroll 1894.

"Of course you are!" growled Uncle Jim. "Never knew any argument of *yours* that didn't end in some absurdity or other!"

"Unprovoked by your unmanly taunts," said Uncle Joe in a lofty tone, "I proceed. Carr being out, you will grant that, if Allen is *also* out, *Brown* must be at home?"

"What's the good of *his* being at home?" said Uncle Jim. "I don't want *Brown* to shave me! He's too clumsy."

"Patience is one of those inestimable qualities—" Uncle Joe was beginning; but Uncle Jim cut him off short.

"*Argue!*" he said. "Don't *moralise!*"

"Well, but *do* you grant it?" Uncle Joe persisted. "Do you grant me that, if Carr is out, it follows that if Allen is out Brown *must* be in?"

"Of course he must," said Uncle Jim; "or there'd be nobody to mind the shop."

"We see, then, that the absence of Carr brings into play a certain Hypothetical, whose *protasis* is 'Allen is out,' and whose *apodosis* is 'Brown is in'. And we see that, so long as Carr remains out, this Hypothetical remains in force?"

"Well, suppose it does. What then?" said Uncle Jim.

"You will also grant me that the truth of a Hypothetical—I mean its *validity* as a logical *sequence*—does not in the least depend on its *protasis* being actually *true*, nor even on its being *possible*. The Hypothetical 'If you were to run from here to London in five minutes you would surprise people,' remains true as a *sequence*, whether you can do it or not."

"I *ca'n't* do it," said Uncle Jim.

"We have now to consider *another* Hypothetical. What was that you told me yesterday about Allen?"

"I told you," said Uncle Jim, "that ever since he had that fever he's been so nervous about going out alone, he always takes Brown with him."

"Just so," said Uncle Joe. "Then the Hypothetical 'if Allen is out Brown is out' is *always* in force, isn't it?"

"I suppose so," said Uncle Jim. (He seemed to be getting a little nervous, himself, now.)

"Then, if Carr is out, we have *two* Hypotheticals, 'if Allen is out Brown is *in*' and 'if Allen is out Brown is *out*,' in force at once. And two *incompatible* Hypotheticals, mark you! They ca'n't *possibly* be true together!"

"*Ca'n't* they?" said Uncle Jim.

"How can they?" said Uncle Joe. "How *can* one and the same *protasis* prove two contradictory *apodoses*? You grant that the two *apodoses*, 'Brown is *in*' and 'Brown is *out*,' *are* contradictory, I suppose?"

"Yes, I grant *that*," said Uncle Jim.

"Then I may sum up," said Uncle Joe. "If Carr is out, these two Hypotheticals are true together. And we know that they *cannot* be true together. Which is absurd. Therefore Carr *cannot* be out. There's a nice *Reductio ad Absurdum* for you!"

Uncle Jim looked thoroughly puzzled: but after a bit he plucked up courage, and began again. "I don't feel at all clear about that *incompatibility*. Why shouldn't those two Hypotheticals be true together? It seems to me that would simply prove '*Allen* is in'. Of course it's clear that the *apodoses* of those two Hypotheticals are incompatible—'Brown is in' and 'Brown is out'. But why shouldn't we put it like this? If Allen is out Brown is *out*. If Carr and Allen are *both* out, Brown is *in*. Which is absurd. Therefore Carr and Allen ca'n't be *both* of them out. But, so long as Allen is *in*, I don't see what's to hinder Carr from going *out*."

"My dear, but most illogical, brother!" said Uncle Joe. (Whenever Uncle Joe begins to "dear" you, you may make pretty sure he's got you in a cleft stick!) "Don't you see that you are wrongly dividing the *protasis* and the *apodosis* of that Hypothetical? Its *protasis* is simply 'Carr is out'; and its *apodosis* is a sort of sub-Hypothetical, 'If Allen is out, Brown is *in*'. And a most absurd apodosis it is, being hopelessly incompatible with that other Hypothetical, that we know is *always* true, 'If Allen is out, Brown is *out*'. And it's simply the assumption 'Carr is out' that has caused this absurdity. So there's only *one* possible conclusion. *Carr is in!*"

How long this argument *might* have lasted, I haven't the least idea. I believe *either* of them could argue for six hours at a stretch. But, just at this moment, we arrived at the barber's shop and, on going inside, we found—

WHAT THE TORTOISE SAID TO ACHILLES*

By Lewis Carroll

Achilles had overtaken the Tortoise, and had seated himself comfortably on its back.

"So you've got to the end of our race-course?" said the Tortoise. "Even though it *does* consist of an infinite series of distances? I thought some wiseacre or other had proved that the thing couldn't be done?"

"It *can* be done," said Achilles. "It *has* been done! *Solvitur ambulando.* You see the distances were constantly *diminishing*; and so—"

"But if they had been constantly *increasing*?" the Tortoise interrupted. "How then?"

"Then I shouldn't be *here*," Achilles modestly replied; "and *you* would have got several times round the world, by this time!"

"You flatter me—*flatten*, I mean," said the Tortoise; "for you *are* a heavy weight, and *no* mistake! Well now, would you like to hear of a race-course, that most people fancy they can get to the end of in two or three steps, while it *really* consists of an infinite number of distances, each one longer than the previous one?"

"Very much indeed!" said the Grecian warrior, as he drew from his helmet (few Grecian warriors possessed *pockets* in those days) an enormous note-book and a pencil. "Proceed! And speak *slowly*, please! *Shorthand* isn't invented yet!"

"That beautiful First Proposition by Euclid!" the Tortoise murmured dreamily. "You admire Euclid?"

"Passionately! So far, at least, as one *can* admire a treatise that won't be published for some centuries to come!"

"Well, now, let's take a little bit of the argument in that First Proposition —just *two* steps, and the conclusion drawn from them. Kindly enter them in your note-book. And in order to refer to them conveniently, let's call them *A*, *B*, and *Z*:—

A) Things that are equal to the same are equal to each other.

B) The two sides of this Triangle are things that are equal to the same.

Z) The two sides of this Triangle are equal to each other.

---

* Reprinted from Carroll 1895.

Readers of Euclid will grant, I suppose, that Z follows logically from A and B, so that any one who accepts A and B as true, *must* accept Z as true?"

"Undoubtedly! The youngest child in a High School—as soon as High Schools are invented, which will not be till some two thousand years later—will grant *that*."

"And if some reader had *not* yet accepted A and B as true, he might still accept the *sequence* as a *valid* one, I suppose?"

"No doubt such a reader might exist. He might say 'I accept as true the Hypothetical Proposition that, *if* A and B be true, Z must be true; but, I *don't* accept A and B as true.' Such a reader would do wisely in abandoning Euclid and taking to football."

"And might there not *also* be some reader who would say 'I accept A and B as true, but I *don't* accept the Hypothetical'?"

"Certainly there might. *He*, also, had better take to football."

"And *neither* of these readers," the Tortoise continued, "is *as yet* under any logical necessity to accept Z as true?"

"Quite so," Achilles assented.

"Well, now, I want you to consider *me* as a reader of the *second* kind, and to force me, logically, to accept Z as true."

"A tortoise playing football would be—" Achilles was beginning.

"—an anomaly, of course," the Tortoise hastily interrupted. "Don't wander from the point. Let's have Z first, and football afterwards!"

"I'm to force you to accept Z, am I?" Achilles said musingly. "And your present position is that you accept A and B, but you *don't* accept the Hypothetical—"

"Let's call it C," said the Tortoise.

"—but you *don't* accept

C) If A and B are true, Z must be true."

"That is my present position," said the Tortoise.

"Then I must ask you to accept C."

"I'll do so," said the Tortoise, "as soon as you've entered it in that notebook of yours. What else have you got in it?"

"Only a few memoranda," said Achilles, nervously fluttering the leaves: "a few memoranda of—of the battles in which I have distinguished myself!"

"Plenty of blank leaves, I see!" the Tortoise cheerily remarked. "We shall need them *all!*" (Achilles shuddered.) "Now write as I dictate:—

A) Things that are equal to the same are equal to each other.

B) The two sides of this Triangle are things that are equal to the same.

C) If A and B are true, Z must be true.

Z) The two sides of this Triangle are equal to each other."

"You should call it D, not Z," said Achilles. "It comes *next* to the other three. If you accept A and B and C, you *must* accept Z."

"And why *must* I?"

"Because it follows *logically* from them. If A and B and C are true, Z *must* be true. You don't dispute *that*, I imagine?"

"If A and B and C are true, Z *must* be true," the Tortoise thoughtfully repeated. "That's *another* Hypothetical, isn't it? And, if I failed to see its truth, I might accept A and B and C, and *still* not accept Z, mightn't I?"

"You might," the candid hero admitted; "though such obtuseness would certainly be phenomenal. Still, the event is *possible*. So I must ask you to grant *one* more Hypothetical."

"Very good, I'm quite willing to grant it, as soon as you've written it down. We will call it

D) If A and B and C are true, Z must be true.

Have you entered that in your note-book?"

"I *have!*" Achilles joyfully exclaimed, as he ran the pencil into its sheath. "And at last we've got to the end of this ideal race-course! Now that you accept A and B and C and D, *of course* you accept Z."

"Do I?" said the Tortoise innocently. "Let's make that quite clear. I accept A and B and C and D. Suppose I *still* refused to accept Z?"

"Then Logic would take you by the throat, and *force* you to do it!" Achilles triumphantly replied. "Logic would tell you 'You ca'n't help yourself. Now that you've accepted A and B and C and D, you *must* accept Z!' So you've no choice, you see."

"Whatever *Logic* is good enough to tell me is worth *writing down*," said the Tortoise. "So enter it in your book, please. We will call it

E) If A and B and C and D are true, Z must be true.

Until I've granted *that*, of course I needn't grant Z. So it's quite a *necessary* step, you see?"

"I see," said Achilles; and there was a touch of sadness in his tone.

Here the narrator, having pressing business at the Bank, was obliged to leave the happy pair, and did not again pass the spot until some months afterwards. When he did so, Achilles was still seated on the back of the much-enduring Tortoise, and was writing in his note-book, which appeared to be nearly full. The Tortoise was saying, "Have you got that last step written down? Unless I've lost count, that makes a thousand and one. There are several millions more to come. And *would* you mind, as a personal favour, considering what a lot of instruction this colloquy of ours will provide for the Logicians of the Nineteenth Century—*would* you mind adopting a pun that my cousin the Mock-Turtle will then make, and allowing yourself to be renamed *Taught-Us*?"

"As you please!" replied the weary warrior, in the hollow tones of despair, as he buried his face in his hands. "Provided that *you*, for *your* part, will adopt a pun the Mock-Turtle never made, and allow yourself to be re-named *A Kill-Ease!*"

ANSWERS TO PROBLEMS

Most of the answer sections to logic and mathematics texts contain more than their share of errors, but, unfortunately, my awareness of this does not give me any protection against them. Should any reader find an error here (or for that matter anywhere else in the book), I should be glad to hear about it so that it can be corrected. In the meantime, the reader can compare his "logic" with mine.

1 (§2). The square of the hypotenuse of our beginning 3,4,5 triangle (shaded in one direction) is equal to the total square minus four 3,4,5 triangles (see Fig. 20). The sum of the squares of the legs of the given triangle (areas which have crisscross shading) is equal to the total square minus two congruent rectangles, one of which is obviously equal to two 3,4,5 triangles, so the other must be also. Since things equal to the same thing are equal to each other, the Pythagorean theorem follows for a 3,4,5 triangle.

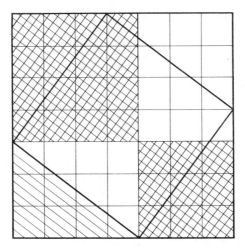

Figure 20

2 (§2). The proof applies to any square, whereas the problem concerned a *unit* square. In that case, $s = 1$ and $d = \sqrt{2}$ and the cancellation in equation (1) is unnecessary. It is probable that the Pythagoreans knew only of the special case—at least until the time of Plato.

3 (§3). *For the defense:* "It is maintained that this man should be sent to the gallows, but this is absurd. For suppose he is sent to the gallows. Then he told the truth and

237

should not be sent by the conditions of the law. On the other hand, if he is not sent to the gallows, then clearly he should not be sent because of the judgment of this court. In either case it follows he should not be sent to the gallows."

For the prosecution: "It is maintained that this man should not be sent to the gallows, but it is this which is absurd. For suppose he is not sent to the gallows. Then he should be sent, since he lied and the law requires this. On the other hand, suppose he is sent to the gallows. Then clearly he should be sent by the judgment of the court. In either case he should be sent to the gallows."

4 (§3). 'Your' can indicate either one of the following:

1) Genitive form of 'you'—that is, it indicates some relation to another person or thing; the 'your' is relative.

2) Possessive form of 'you'—that is, it indicates the property relationship to the person or thing in question. In the final statement of the argument, the occurrence of 'yours' is possessive, that of 'your' is relative.

5 (§4A). An example of a valid argument with a false conclusion is:

> Every man is immortal.
>
> Socrates is a man.
>
> Therefore Socrates is immortal.

Note that although the conclusion follows from the premises, one of the premises is false. It follows from the definition of validity that if the conclusion of a valid argument is false, at least one of the premises must also be false.

An example of an invalid argument with a true conclusion is:

> Every dog is mortal.
>
> Every animal is mortal.
>
> Therefore every dog is an animal.

The invalidity of this argument can be seen in several ways. For example, if a dog were a plant, the premises would be true and the conclusion false. Or you could interchange the premises (in which case they would both remain true) and, keeping an argument of the same structure, derive 'every animal is a dog', which is clearly false. Part of the work of a logician is to determine when two arguments have the same structure.

6 (§4B). *Subcontrary* propositions are such that, while both may be true, they cannot both be false. Both the definition of *contrary* and that of *subcontrary* depend on the assumption that only contingent propositions are involved. If noncontingent propositions are involved, the definitions must be changed to ensure that **A** and **E** (**I** and **O**) are contraries (subcontraries). For example, 'Every man is male' cannot be false. Hence it cannot, by the definition given, be a contrary to 'No man is male'. To take care of this and similar examples, the definitions may be changed as follows: *contrary* (*subcontrary*) propositions are such that, if one may be false (true) and the other may be false (true) then both may be false (true), and they cannot both be true (false). [See Sanford 1968.]

7 (§4B)

	A	E	I	O
A given as T	T	F	T	F
E given as T	F	T	F	T
I given as T	U	F	T	U
O given as T	F	U	U	T
A given as F	F	U	U	T
E given as F	U	F	T	U
I given as F	F	T	F	T
O given as F	T	F	T	F

8 (§4B)

Premiss	Contrapositive
Every S is P	Every non-P is non-S
No S is P	Some non-P is not non-S (by limitation)
Some S is P	No valid contrapositive
Some S is not P	Some non-P is not non-S

9 (§5A). In order, the "justifications" are Postulate 3, Postulate 3, Postulate 1, Definition 15, Definition 15, Common Notion 1.

10 (§5A). There is nothing wrong with the argument except the conclusion! That is, "... thus lines CD and EF cannot meet" does not follow from the above statements. To see this for a special case, suppose that the interior angles are each two-thirds of a right angle (see Fig. 21 on page 240).

It is easy to show that triangles IGG', IHH' and $IG'H'$ are all equilateral and congruent. Thus HH' equals $G'H'$. Now consider repeating the argument, forming successively points H'', H''', H'''', etc. on EF and G'', G''', G'''', ... etc. on CD. Suppose that IH is 1 unit long. HH' will be 1, $H'H''$ will be $\frac{1}{2}$, etc. Thus $G'H'$ will be 1, $G''H''$ will be $\frac{1}{2}$, etc. The distance between the lines will always be 2 minus (the sum of the successive distances HH', $H'H''$, $H''H'''$, etc. to the point in question). Now, letting S stand for the sum, we may suppose that

$$S_n = 1 + x + x^2 + \cdots x^n.$$

Multiplying by x, we get

$$xS_n = x + x^2 + \cdots + x^{n+1}.$$

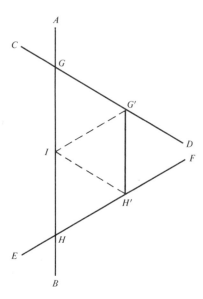

Figure 21

Subtracting the second equation from the first, we get

$$S_n - xS_n = 1 - x^{n+1}$$

$$S_n(1 - x) = 1 - x^{n+1}$$

$$S_n = \frac{1 - x^{n+1}}{1 - x}$$

Now as n approaches infinity, x^{n+1} approaches 0 if $-1 < x < 1$. We express this situation by writing

$$S = \frac{1}{1 - x} = 1 + x + x^2 + x^3 + \cdots.$$

Thus we have

$$1 + \frac{1}{2} + \frac{1}{2^2} + \frac{1}{2^3} + \cdots = \frac{1}{1 - \frac{1}{2}} = \frac{1}{\frac{1}{2}} = 2.$$

From these considerations we can see that CD and EF will meet in such a way as to form an equilateral triangle with one side being GH. Using this information, we can easily see how Achilles can catch the Tortoise (cf. §3). Suppose that Achilles runs just twice as fast as the Tortoise and that he has a lead of just one unit. Under these assumptions and by the above mathematics, Achilles will pass the Tortoise after two units of the race course.

11 (§5B). The additional hypothesis needed for (2) is that lying always involves saying something false. The additional hypothesis for (1) (assuming that the one for (2) is granted) is that the remark is true with respect to all Cretans except the speaker. Were

Table 18

Figure	Conclusion	Premises															
		AA	AE	AI	AO	EA	EE	EI	EO	IA	IE	II	IO	OA	OE	OI	OO
1st	A	V	b	c	bc	b	ab	bc	abc	d	b	cd	bc	bd	ab	bcd	abc
	E	bc	c	bc	c	V	a	c	ac	bcd	c	bcd	c	d	a	cd	ac
	I	V	b	V	b	b	ab	b	ab	d	b	d	b	bd	ab	bd	ab
	O	bc	c	bc	c	V	a	V	a	bcd	c	bcd	c	d	a	d	a
2nd	A	d	b	cd	bc	b	ab	bc	abc	d	b	cd	bc	b	ab	bc	abc
	E	bd	V	bcd	c	V	a	c	ac	bcd	c	bcd	c	c	ac	c	ac
	I	d	b	d	b	b	ab	b	ab	d	b	d	b	b	ab	b	ab
	O	bd	V	bd	V	V	a	V	a	bcd	c	bcd	c	c	ac	c	ac
3rd	A	c	b	c	b	bc	ab	bc	ab	c	b	cd	bd	bc	ab	bcd	abd
	E	bc	c	bc	c	c	a	c	a	bc	c	bcd	cd	c	a	cd	ad
	I	V	b	V	b	b	ab	b	ab	V	b	d	bd	b	ab	bd	abd
	O	bc	c	bc	c	V	a	V	a	bc	c	bcd	cd	V	a	d	ad
4th	A	c	b	cd	bd	bc	ab	bc	ab	c	b	cd	bd	bc	ab	bc	ab
	E	bc	V	bcd	d	c	a	c	a	bc	c	bcd	cd	c	ac	c	ac
	I	V	b	d	bd	b	ab	b	ab	V	b	d	bd	b	ab	b	ab
	O	b	V	bd	d	V	a	V	a	bc	c	bcd	cd	c	ac	c	ac

241

it not for this, the statement could be considered false and no contradiction would follow. However, another unpleasant result would follow; namely, we could conclude that either he or some other Cretan (at some time or other) made a true statement. This conclusion could be drawn without knowing anything at all about other Cretans or about the Cretan in question at other times! Yet does it not seem theoretically possible that there be just one Cretan (= inhabitant of Crete) who makes just one statement (= (1))? Hence, unless we are going to accept the possibility of *a priori* knowledge about Cretans (and, of course, much else) we may conclude—by *reductio ad absurdum* instead of *reductio ad impossibile*—that (1) cannot be false. Of course, by *reductio ad impossibile*, (1) cannot be true either. Paradoxes which involve only the *reductio ad impossibile* are called *antinomies*.

12 (§5B). If (3) is meaningless, then it is neither true nor false. If (3) is neither true nor false, then (3) is not true. But (3) states that it is not true. Therefore (3) is true. Hence (3) cannot be meaningless because it is true. Thus the antimony is not resolved.

13 (§6). Table 18 (see preceding page) indicates the rule or rules broken in each space which represents an invalid syllogistic form. On the limitations of the traditional account of the validity of the syllogism, see Oliver 1967.

14 (§6). The only difference between our new set of conditions and the old one is that the following condition is left out: 'If the conclusion is negative, it contains a negative premiss'. Hence if any invalid syllogism satisfies the new conditions, it must have a negative conclusion. It cannot have an **E** conclusion because the premisses would have to distribute the major and minor terms (c); and the middle term (d). However, an **I** proposition distributes no term and an **A** only one. Therefore two affirmative premisses could not distribute three terms, as required by (c) and (d). If we have an **O** conclusion the major term must be distributed (c) as well as the middle term (d). Hence we need two **A** propositions, with the major term the subject of the major premiss and the middle term the subject of the minor premiss. Hence an **AAO** syllogism in the fourth figure is the only invalid syllogism to satisfy our new set of conditions.

15 (§6). The arguments begs the question, that is, it commits the fallacy of *petitio principii*. If one doubted the truth of the conclusion, an equal doubt would apply to the validity of the argument. This is so because the argument is itself an example of an **AAA** syllogism in the first figure.

16 (§9). Suppose that b evenly divides a_n (but not a_1). We would have $a_n = b \cdot k$, where k is some natural number. But we also know that

$$a_n = a_1 \cdot a_{n-1}$$

so that $b \cdot k = a_1 \cdot a_{n-1}$. That is, $b : a_1 : : a_{n-1} : k$, which by a previous theorem implies that b evenly divides a_{n-1}, since b and a_1 are relatively prime. Applying this argument $n - 1$ times, we conclude that b evenly divides a_1. From our assumption that 'b evenly divides a_n but not a_1' we derive the statement 'if b evenly divides a_n, then it evenly divides a_1'. This latter statement is the contradictory of the former, and thus we conclude that the latter is true.

17 (§9). In the first system the new postulate is not independent, since it can be derived from the other assumptions. The system is consistent if Euclidean geometry without the fifth postulate is consistent. Note that the latter condition is necessary: If Euclidean geometry were inconsistent without the fifth postulate, it would be inconsistent with this new postulate.

The second system is inconsistent if you understand the second Euclidean postulate as asserting the infinitude of the line. One can see this by considering the fact that from it one can derive the fifth postulate, from which follows the hypothesis (now a theorem) of the right angle, which is contrary to that of the obtuse angle. The new postulate—that of the obtuse angle—is thus dependent (its falsity follows from the other axioms) and thus the system is inconsistent.

18 (§9). The most common resolution of the Achilles paradox is the one given in the answer to Problem 10 (§5A). However, there are a number of alternatives. In particular, if "in the indefinitely small the spatial relations of size are not in accord with the postulates of geometry," then it might be that motion is not continuous. What at first appears continuous may on more careful consideration be seen to be discontinuous. Thus the concept of weight as applied to sand seems to be continuous, until one realizes that the weight of a given amount of sand cannot be increased (or decreased) by less than the lightest grain. Similarly, some philosophers—for example, A. N. Whitehead—have argued that it is possible that there is a quantum of space and a quantum of time below which it doesn't make sense to talk about either. Thus motion would be a series of jumps (the analogy of moving pictures suggests itself) and Achilles at one instant would be behind the Tortoise, and at the next, abreast or ahead of it.

19 (§9). The error occurs in the faultily drawn figures. Actually none of the three cases suggested by the figures occur. Part (a) of Fig. 10 (that is, that D lies inside the triangle) is impossible, since by the argument the sum of angles ADG, ADE and EDB equals the sum of the angles CDG, CDF and FDB. But this would make BDG a straight angle (that is, a straight line), contrary to the argument given. Part (b) of Fig. 10 is impossible, since this is contrary to the assumption that ABC is not isosceles. Part (c) of Fig. 10 is faulty in the assumption that E lies *between AB* or F *between BC*.

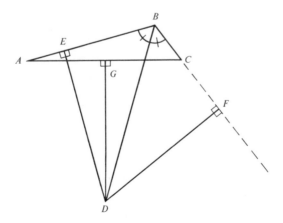

Figure 22

Figure 22 is correctly drawn; in it, right triangles BED and BFD are congruent by the argument given. Thus BE equals BF, but from this we shall not be able to conclude that BA equals BC.

20 (§10A).

$$s < d < 2s$$

$$14s < 10d < 15s$$

$$141s < 100d < 142s$$

$$1414s < 1000d < 1415s$$

$$14142s < 10000d < 14143s$$

$$141421s < 100000d < 141422s$$

$$\vdots$$

$$1 < d/s < 2$$

$$1.4 < d/s < 1.5$$

$$1.41 < d/s < 1.42$$

$$1.414 < d/s < 1.415$$

$$1.4142 < d/s < 1.4143$$

$$1.41421 < d/s < 1.41422$$

$$\vdots$$

By substituting $s = 1$ and $d = \sqrt{2}$, we have the required series.

21 (§10A). Assume that some irrational number has two decimal expansions which are not identical digit by digit. Thus, given that α is that irrational number, we have

$$\alpha = N.a_1a_2a_3\ldots$$

$$\alpha = N.b_1b_2b_3\ldots$$

where N is a natural number and each a_i and b_i is either 0 or a natural number ≤ 9. Now suppose that the first digit at which the two expansions differ is the nth; that is,

$$a_n \neq b_n \quad \text{and} \quad a_n < b_n.$$

Let k be the smallest integer such that $n < k$ and $a_k < 9$. Consider the number β defined as follows:

$$\beta = N.c_1c_2c_3\ldots$$

where $c_i = a_i$ for $i < k$, $c_k = a_{k+1}$, and $c_j = 0$ for $j > k$. It follows that

$$N.a_1a_2a_3\cdots < \beta < N.b_1b_2b_3\cdots.$$

Can you see where in this proof the assumption that we are dealing only with irrational numbers is essential?

22 (§10A). Natural numbers correspond to the positive integers; integers to rational numbers whose denominator is 1 when reduced to the lowest form; rational numbers to real numbers with periodic or terminating decimals; real numbers to complex numbers where the coefficient of i is 0.

23 (§10B). The nonpositive real numbers, since

$$(ai)^2 = (a\sqrt{-1})^2 = a^2(\sqrt{-1})^2 = -a^2.$$

24 (§11). The quotation continues as follows:

For consider: the hundredth day will be described in the hundredth year, the thousandth in the thousandth year, and so on. Whatever day we may choose as being so far on that he cannot hope to reach it, that day will be described in the corresponding year. Thus any day that may be mentioned will be written up sooner or later, and therefore no part of the biography will remain permanently unwritten. This paradoxical but perfectly true proposition depends upon the fact that the number of days in all time is no greater than the number of years.

For another viewpoint, see Diamond 1964.

25 (§11). Require all guests to move simultaneously to the next room, that is, if a guest is in room n he must move into room $n + 1$. Then put the new guest in room 1. Note that even if a (denumerably) infinite number of new guests arrived, there would be room by putting all present guests in even-numbered rooms ($1 \rightarrow 2, 2 \rightarrow 4, 3 \rightarrow 6$, etc.) and the new guests in the odd-numbered rooms. (*Denumerable* is defined in the second paragraph following Problem 25.)

26 (§11). Consider the one-to-one correspondence of the natural numbers and the positive rationals: $1 \leftrightarrow a_1$, $2 \leftrightarrow a_2,\ldots$, where a_1 is a positive rational. From this correspondence form a new one as follows: $1 \leftrightarrow b_1$, $2 \leftrightarrow b_2,\ldots$, where

$$b_1 = a_1, \qquad b_3 = a_2, \qquad b_5 = a_3, \qquad \ldots, \qquad b_{2n-1} = a_n,\ldots;$$

$$b_2 = -a_1, \qquad b_4 = -a_2, \qquad b_6 = -a_3, \qquad \ldots, \qquad b_{2n} = -a_n,\ldots$$

27 (§11). Consider the following number: $N_1.b_1b_2b_3\ldots$, where

$$b_{2i} = 1 \text{ if } a_{i,2i} \neq 1, \qquad b_{2i} = 2 \text{ if } a_{i,2i} = 1 \text{ and } b_{2i-1} = 1.$$

Of course, other diagonals could be chosen.

28 (§11). Suppose that there are k real numbers:

$$r_1 = N_1.c_{1,1}c_{1,2}c_{1,3}c_{1,4}\cdots$$

$$r_2 = N_1.c_{2,1}c_{2,2}c_{2,3}c_{2,4}\cdots$$

$$\vdots$$

$$r_k = N_1.c_{k,1}c_{k,2}c_{k,3}c_{k,4}\cdots$$

Construct the following number: $N_1.d_1d_2d_3\ldots$, where $d_i = 1$ if neither $a_{i,i} = 1$ nor $c_{i,i} = 1$, $d_i = 2$ if either $a_{i,i}$ or $c_{i,i} = 1$ and the other $\neq 2$, and $d_i = 3$ in any other case. Since this new number is not in either list, the set of possible constructed numbers along the original diagonal is infinite; any finite set of possibilities can always be enlarged by one.

29 (§11). The empty set, the set whose only element is 1, the set whose only element is 2, the set whose only element is 3, the set whose only elements are 1 and 2, the set whose only elements are 1 and 3, the set whose only elements are 2 and 3, and the set whose only elements are 1, 2, and 3. If we were going to use such descriptions very often, we would, of course, introduce notation to shorten them.

30 (§12). Let the standard "alphabetical" order be the order suggested in the text. Now one possible enumeration would go by the length of the expression measured in terms of the number of symbols: First, all the expressions of length 1, each standing alone and in alphabetical order; second, all the expressions of length 2, the first symbol followed successively by all expressions of length 1, the second symbol followed successively by all expressions of length 1, etc.; third, all the symbols of length 3, the first symbol followed successively by all expressions of length 2, etc.; fourth, etc. Thus our enumeration would be as follows: $a, b, \ldots$; $aa, ab, \ldots$, $ba, bb \ldots, \ldots$; $aaa, aab, \ldots$, $aba, abb, \ldots, \ldots$; etc.

31 (§12). Consider the problem by putting examples of each in two different columns:

Autological	Heterological
English	green
adjectival	moody
polysyllabic	monosyllabic
⋮	⋮

It seems clear that most adjectives are heterological. But consider that the word 'heterological' is itself an adjective. In which column does it belong? If it were put in the column labeled 'Autological', then it would be heterological, since it would apply to itself. On the other hand, if it were put in the column labeled 'Heterological', then it would be autological. In short, 'heterological' is heterological if and only if it is autological. This paradox is called the *Grelling paradox*, after Kurt Grelling (1886–1941).

32 (§15). (a) is false. Mary begins with a 'J'. To see this, note that Mary is the same as 'Juliet'. Of course, 'Mary' begins with an 'M'. (b) is false. 'Juliet' begins with a 'J' but Juliet presumably only begins on her birthday. (c) and (d) mean exactly the same thing, namely, that Romeo loves Juliet's name. Presumably these are true. Note in this example—as well as elsewhere in this book—the use of letters (or numbers) in parentheses as the names of sentences which follow them.

33 (§15).

	Used but not mentioned	Mentioned but not used	Neither used nor mentioned
a)	(ii)	(i)	(iii) (iv)
b)	(i)	(iii)	(ii) (iv)
c)	(iv)	(i)	(ii) (iii)
d)	(iv)	(i)	(ii) (iii)

34 (§15). In the question the child was *using* the word 'two' to *mention* the numeral '2', and was *using* the word 'and' to *mention* the operation of juxtaposition or concatenation. He was making a metamathematical assertion, whereas the father understood it to be a mathematical one.

35 (§16)

p	q	$p \cdot q$
T	T	T
T	F	F
F	T	F
F	F	F

Hence, $\cdot = f_8^2$. We shall often omit '$\cdot$' and just use juxtaposition for it; that is, 'pq' means the same thing as '$p \cdot q$'.

36 (§16). $\quad \vee = f_{15}^2, \quad | = f_9^2, \quad \supset = f_{12}^2, \quad \equiv = f_{10}^2.$

37 (§16). What we want to capture in our translation is: First, when one component is true the others are false, and second, that at least one is true. The latter we can get by '$(p \vee q) \vee r$'. To get the first condition satisfied, we need only add a component to each disjunct to get

$$((p(\bar{q}\bar{r})) \vee (q(\bar{p}\bar{r}))) \vee (r(\bar{p}\bar{q})).$$

Other correct answers have the same truth table as this one. For example,

$$(((p\bar{q})\bar{r}) \vee ((\bar{p}q)\bar{r})) \vee ((\bar{p}\bar{q})r).$$

Form of propositions

			First premiss $((p(\bar{q}\bar{r})) \vee (q(\bar{p}\bar{r}))) \vee (r(\bar{p}\bar{q}))$	2nd $\bar{p}$	3rd $\bar{q}$	Conclusion r
p	q	r				
T	T	T	F	F	F	T
T	T	F	F	F	F	F
T	F	T	F	F	T	T
T	F	F	T	F	T	F
F	T	T	F	T	F	T
F	T	F	T	T	F	F
F	F	T	T	T	T	T
F	F	F	F	T	T	F

Observe in particular that the following argument is fallacious: Since the form of the first premiss under this interpretation is 'either (p or q, but not both) or r, but not both', this becomes 'either (($p\bar{q}$) $\vee$ ($\bar{p}q$)) or r, but not both' which in turn becomes

$$(((p\bar{q}) \vee (\bar{p}q))\bar{r}) \vee (\overline{((p\bar{q}) \vee (\bar{p}q))r}).$$

The flaw is in the interpretation 'either (p or q, but not both) or r, but not both'. This is true when 'p', 'q', and 'r' are all true.

38 (§16). There are 6 different propositional variables, and therefore the truth table contains $2^6 = 64$ lines. The truth table (Table 19) is on pages 248–249. It illustrates the principal shortcoming of the truth-table method, namely, that tables become unmanageably large with a relatively modest increase in the number of variables. Another even

Table 19

Form of propositions

p	q	r	p_1	q_1	r_1	1st premiss $((p\bar{q}\bar{r}) \vee (\bar{p}q\bar{r})) \vee (\bar{p}\bar{q}r)$	2nd premiss $((p_1(\bar{q}_1\bar{r}_1)) \vee (\bar{p}_1(q_1\bar{r}_1))) \vee (\bar{p}_1(\bar{q}_1 r_1))$	3rd premiss $p \supset p_1$	4th premiss $q \supset q_1$	5th premiss $r \supset r_1$	Conclusion $(p_1 \supset p) \cdot (q_1 \supset q) \cdot (r_1 \supset r)$
T	T	T	T	T	T	F	F	T	T	T	T
T	T	T	T	T	F	F	F	T	T	F	T
T	T	T	T	F	T	F	F	T	F	T	T
T	T	T	T	F	F	F	T	T	F	F	T
T	T	T	F	T	T	F	F	F	T	T	T
T	T	T	F	T	F	F	T	F	T	F	T
T	T	T	F	F	T	F	T	F	F	T	T
T	T	T	F	F	F	F	F	F	F	F	T
T	T	F	T	T	T	F	F	T	T	T	F
T	T	F	T	T	F	F	F	T	T	T	T
T	T	F	T	F	T	F	F	T	F	T	F
T	T	F	T	F	F	F	T	T	F	T	T
T	T	F	F	T	T	F	F	F	T	T	F
T	T	F	F	T	F	F	T	F	T	T	T
T	T	F	F	F	T	F	T	F	F	T	F
T	T	F	F	F	F	F	F	F	F	T	T
T	F	T	T	T	T	F	F	T	T	T	F
T	F	T	T	T	F	F	F	T	T	F	F
T	F	T	T	F	T	F	F	T	T	T	T
T	F	T	T	F	F	F	T	T	T	F	T
T	F	T	F	T	T	F	F	F	T	T	F
T	F	T	F	T	F	F	T	F	T	F	F
T	F	T	F	F	T	F	T	F	T	T	T
T	F	T	F	F	F	F	F	F	T	F	T
T	F	F	T	T	T	T	F	T	T	T	F
T	F	F	T	T	F	T	F	T	T	T	F
T	F	F	T	F	T	T	F	T	T	T	F
T	F	F	T	F	F	T	T	T	T	T	T
T	F	F	F	T	T	T	F	F	T	T	F
T	F	F	F	T	F	T	T	F	T	T	F
T	F	F	F	F	T	T	T	F	T	T	F
T	F	F	F	F	F	T	F	F	T	T	T

```
F F F T    F F F F    T T T    F F F F    F F F T    F F F F    F F T T    F F F F    F F F T

T T T T    T F F F    T F F F    T T T T    T T T T    T F F F    T F F F    T T T T    T T T T
T T T T    T T F F    T T F F    T T F F    T T F F    T T T T    T T T T    T T T T    T T T T
F F F F    T T T T    T T T T    T T T T    T T T T    T T T T    T T T T    T T T T    T T T T

F T T F    F F F T    F T T F    F F F T    F T T F    F F F T    F T T F    F F F T    F T T F

T T T T    F F F F    F F F F    T T T T    T T T T    T T T T    T T T T    F F F F    F F F F

T T F T F  T T T F F  T T F T F  T T F T F  T T F T F  T T F T F  T T F T F  T T F T F  T T F T F
F F F F F  T T T F F  T F F F F  T T F F F  F F F F F  T T T T T  T F F F F  T T T F F  F F F F F
T F F F F  T T T T T  T T T T T  F F F F F  F F F F F  T T T T T  T F F F F  T T T T T  F F F F F
T T F F F  F T T T T  T T T T T  F F F F F  F F F F F  F F F F F  F F F F F  F F F F F  F F F F F
T T F T    F F F F F  F F F F F  F F F F F  F F F F F  F F F F F  F F F F F  F F F F F  F F F F F
```

more extreme example of the same thing may be illustrated by the following argument due to Lewis Carroll:

1) All active old Jews are healthy.
2) All indolent magistrates are unpopular.
3) All rich snuff-takers are unhealthy.
4) All sarcastic magistrates are Jews.
5) All young snuff-takers are pale.
6) All rich old men, who are unhealthy, are sarcastic.
7) All magistrates who are not poets are studious.
8) All rosy magistrates are talented.
9) All talented and popular students are rich.
10) All pale snuff-takers are unpopular.
11) All unpopular magistrates are abstainers from snuff.
12) All talented poets who are active are rich.
13) Therefore no magistrates are snuff-takers.

By techniques which have not been explained, this argument is amenable to the truth-table method. Its form contains 13 variables, and thus the truth table contains $2^{13} = 8192$ lines. The exhibition of the truth table seems to be a job for a computer which, in fact, has been used to solve the problem [cf. Kemeny 1956]. Nevertheless, other techniques make it possible for a human computer to solve the problem in a reasonable time [cf. L. Russell, 1951]. Yet it is perfectly possible for problems to arise with, say, 100 variables. Here even the computer would be inadequate for the job of writing out a full truth table. Thus except for relatively short arguments, the primary function of the truth table is theoretical.

39 (§16). Professor Church, from whose book this problem is taken [cf. Church 1956 105–106], answers that we must suppose that "it is as much a violation of the law to hang an innocent man as it is to let a guilty one go free."

40 (§16). Our problem here is: Can the following three statements all be true?

1) $C \supset (A \supset \bar{B})$
2) $A \supset B$
3) C

The answer would be yes if the conjunction of the statements, that is,

$$((C \supset (A \supset \bar{B})) \cdot (A \supset B)) \cdot C$$

could be true. Constructing the truth table for the form of this statement, we have:

p	q	r	$((p \supset (q \supset \bar{r})) \cdot (q \supset r)) \cdot p$
T	T	T	F
T	T	F	F
T	F	T	T
T	F	F	T
F	T	T	F
F	T	F	F
F	F	T	F
F	F	F	F

The answer to the problem is thus yes, under conditions that 'A' is false and 'C' is true. '$A \supset B$' and '$A \supset \bar{B}$' are not incompatible, provided that 'A' is false. See Burks and Copi 1950 for another analysis.

41 (§16). (1) is incorrect. It should be "$(p \cdot \bar{p})$' is a contradiction'. (2) is correct. (3) and (4) are incorrect, and should respectively be "p' implies '$p \lor q$" and '$p \supset q$'.

42 (§16). The argument has some of the flavor which the ancient controversies had over whether everything can be proved. As we have seen, Aristotle had pointed out that knowledge is impossible if we require a proof for every statement. Similarly, rules of inference are impossible to apply if we always require a new rule of inference to apply a given rule of inference. The first step out of the morass, then, is to distinguish object language from metalanguage. (A), (B), and (Z) belong to the object language, (C) to the metalanguage. Thus (C) is a rule of logic for which no rule of logic is needed for it to apply. Of course, the wily Tortoise might now recast the argument so that we need a meta-metalanguage to express rules for the application of metalanguage rules to the language, and so on. But at this point one must ask whether the Tortoise accepts any rule of logic which can be applied without rules. If not, then he had better give up the game of logic and take up football. Logic, as we have seen, is a social enterprise; it presents a kind of sporting proposition. Like some existentialists, the Tortoise is just a spoilsport. See Rees 1951 and Woods 1965 for further discussion.

43 (§17). **E**: No athlete is brash: $(\forall x)(A(x) \supset \overline{B(x)})$.
I: Some athlete is brash: $(\exists x)(A(x) \cdot B(x))$.

44 (§17).   *One element*

$$A^1(\text{Socrates}) \supset \overline{B^1(\text{Socrates})}$$
$$\underline{A^1(\text{Socrates})}$$
$$\therefore \ \overline{B^1(\text{Socrates})}$$

$$A \supset \bar{B}$$
$$\underline{A}$$
$$\therefore \ \bar{B}$$

Three elements

$$((A^1(\text{Socrates}) \supset \overline{B^1(\text{Socrates})}) \cdot (A^1(\text{Plato}) \supset \overline{B^1(\text{Plato})})) \cdot$$

$$(A^1(\text{Protagoras}) \supset \overline{B^1(\text{Protagoras})})$$

$$(A^1(\text{Socrates}) \lor A^1(\text{Plato})) \lor A^1(\text{Protagoras})$$

$$\therefore \ \overline{(B^1(\text{Socrates}) \lor B^1(\text{Plato})) \lor B^1(\text{Protagoras})}$$

$$((A \supset \bar{B}) \cdot (C \supset \bar{A}_1)) \cdot (B_1 \supset \bar{C}_1)$$

$$(A \lor C) \lor B_1$$

$$\therefore \ \overline{((B \lor A_1) \lor C_1)}$$

One element

p	q	$p \supset \bar{q}$	p	$\bar{q}$
T	T	F	T	F
T	F	T	T	T
F	T	T	F	F
F	F	T	F	T

Two elements

p	q	r	p_1	$(p \supset \bar{q}) \cdot (r \supset \bar{p}_1)$	$p \lor r$	$\overline{q \lor p_1}$
T	T	T	T	F	T	F
T	T	T	F	F	T	F
T	T	F	T	F	T	F
T	T	F	F	F	T	F
T	F	T	T	F	T	F
T	F	T	F	T	T	T
T	F	F	T	T	T	F
T	F	F	F	T	T	T
F	T	T	T	F	T	F
F	T	T	F	T	T	F
F	T	F	T	T	F	F
F	T	F	F	T	F	F
F	F	T	T	F	T	F
F	F	T	F	T	T	T
F	F	F	T	T	F	F
F	F	F	F	T	F	T

The one-element truth table indicates validity, since there is no line where the form of the premiss has a 'T' and the form of the conclusion has an 'F'. The two-element table indicates invalidity, since in line ten the forms of the premisses each contain a 'T' and the form of the conclusion contains an 'F'.

45 (§17). Let the domain be the natural numbers. Let '$A(x, y)$' mean '$x > y$'. Then

$$(\forall y)(\exists x)A(x, y)$$

is true, since it states that for any given number there exists a larger number. On the other hand,

$$(\exists x)(\forall y)A(x, y)$$

is false, since it states that there is a number larger than any number.

46 (§17). For the first, let the domain be Socrates and Plato and the predicate be '$A^1(x)$' meaning 'x is sick'. Assuming that Socrates is not sick and that Plato is,

$$\overline{(((\forall x)(\overline{(A^1(x))}))} \supset (\forall x)(A^1(x)))$$

is false. For the second, let the domain be the natural numbers and if '$A^2(x, y)$' means '$x < y$', then

$$(\forall x)(((\forall y)(\overline{(\overline{A^2(x, y)})})))$$

is true, while

$$((\forall y)(\overline{(\overline{A^2(y, y)})}))$$

is false.

47 (§18). The principle of mathematical induction, that is, the principle which tells us that if a predicate holds for 0, and if, when it holds for any number n, it also holds for $n + 1$, then the predicate holds for all natural numbers.

48 (§18). (1) The 'is' of identity: $a = b$, where 'a' stands for 'Bacon' and 'b' stands for Shakespeare.

(2) The 'is' of predication: $A(a)$, where 'a' stands for 'Bacon' and '$A(x)$' stands for 'x is corrupt'.

(3) The 'is' of membership:

$$(\forall x)(A(x) \supset B(x)),$$

where '$A(x)$' stands for 'x is a man' and '$B(x)$' stands for 'x is corrupt'.

49 (§19). Ticktacktoe.

2	9	4	HOT	FORM	WOES
7	5	3	TANK	HEAR	WASP
6	1	8	TIED	BRIM	SHIP

Since any schoolboy knows the best strategy at ticktacktoe—namely, to start in a corner—the best strategy at ACE (HOT) is to start with either 2 (HOT), 4 (WOES), 6 (TIED), or 8 (SHIP). Then, provided that your opponent doesn't pick 5 (HEAR), you can always win. Otherwise the best you can do is to draw. This problem comes from Martin Gardner's popular column "Mathematical Games" [*Scientific American,* February 1967].

50 (§20). There are a number of possibilities which will work. Here is one: Let '$\supset$' denote f_5^2 as before and let '$^-$' denote f_1^1.

51 (§20). The first factoring out would give us:

$$p \cdot (qr \vee q\bar{r} \vee \bar{q}r \vee \bar{q}\bar{r}) \vee \bar{p} \cdot (qr \vee q\bar{r} \vee \bar{q}r \vee \bar{q}\bar{r}).$$

The second would give us

$$(p \vee \bar{p}) \cdot (qr \vee q\bar{r} \vee \bar{q}r \vee \bar{q}\bar{r}).$$

An exact and detailed example of this transformation would involve many steps, because in the first place, the association laws would have to be applied (by leaving off most parentheses, we have assumed them) and, in the second place, the distribution laws (like the association laws) apply to only two disjuncts, so it would take a number of applications to get our first result. Furthermore, both in the proof and in this problem, it would have to be shown that ' · ' and ' ∨ ' each commute.

52 (§21). At this point we don't. But if the assumption of an infinite domain led to an inconsistency, it would be due to a fault in either the concept of an infinite domain, or in the way in which $\mathscr{L}$ was interpreted. $\mathscr{L}$ itself would still be consistent; and this is all the proof asserts.

53 (§21). The associated formulas for (3(d)) are $((\mathbf{P} \supset \mathbf{Q}) \supset (\mathbf{P} \supset \mathbf{Q}))$. Those for (3(e)) are $(\mathbf{P} \supset \mathbf{P})$. **P** and **Q** are understood to be formulas of $\mathscr{P}$. With regard to (4(b)), note that the associated formulas of $(\forall x)(\mathbf{P})$ will be exactly the same as **P**. Thus the property is preserved.

54 (§21). Consider (1). If there were but one individual, say **a**, (1) is equivalent to '$\mathbf{P(a)} \supset \mathbf{P(a)}$', a substitution instance of a tautology. If there were two individuals, say **a** and **b**, then (1) is equivalent to

$$(\mathbf{P(a)} \vee \mathbf{P(b)}) \supset (\mathbf{P(a)} \cdot \mathbf{P(b)}),$$

which is not valid, since the antecedent can be true and the consequent false. The same argument holds *mutatis mutantis* for (2), (3), etc.

55 (§21). By showing that each of the initial formulas of $\mathscr{L}$ is valid, and that the transformation rules of $\mathscr{L}$ preserve validity, it then follows that all the theorems are valid. (Cf. corresponding proof for $\mathscr{P}$, page 135.)

56 (§21). The error is that while it is true that a formula and its negation cannot both be valid, it is false that one of the formulas has to be. For example, any of the formulas indicated in the list (1), (2), (3), etc., of §21 is invalid. But so are their respective negations. If we happen to be given some arbitrary formula such that both it and its negation are invalid and we applied the procedure suggested in the problem, it would never end. And at no point would we have any way of knowing whether both formulas were invalid or one was valid but we hadn't yet carried out the proofs far enough.

57 (§21). Distinguish between truth and validity. Whereas if a formula is valid interpreted in a set of k elements, it must also be valid in any other set of k elements, but the same cannot be said for truth.

58 (§22). $x^0 = 0'$
$$x^{y'} = x^y \cdot x$$

59 (§22)
$$f_6(y, z, x) = x$$
$$f_7(y, z, x) = f_5(f_3(y, z, x), f_6(y, z, x))$$
$$f_8(x) = 0$$
$$f_9(0, x) = f_8(x)$$
$$f_9(y', x) = f_7(y, f_9(y, x), x)$$
$$f_{10}(y, z, x) = f_9(f_3(y, z, x), f_6(y, z, x))$$
$$f_{11}(x) = 1$$
$$f_{12}(0, x) = f_{11}(x)$$

$f_{12}(y', x) = f_{10}(y, f_{12}(y, x), x)$
f_6 is the identity function with $i = 3$.
f_7 is defined by substitution from $f_5[= g]$, $f_3[= h_1]$, and $f_6[= h_2]$.
f_8 is the constant function with $a = 0$.
f_9 is defined by recursion from $f_8 [= g]$ and $f_7 [= h]$.
f_{10} is defined by substitution from $f_9 [= g]$, $f_3 [= h_1]$, and $f_6 [= h_2]$.
f_{11} is the constant function with $a = 1$.
f_{12} is defined by recursion from $f_{11} [= g]$ and $f_{10} [= h]$.

60 (§22). Let $\overline{Sg}$ be defined as follows:

$$\overline{Sg}(0) = 1$$

$$\overline{Sg}(x') = 0$$

Then we have for all x and y:

$$(x < y) \supset (\overline{Sg}(y \div x) = 0)$$

$$(x \not< y) \supset (\overline{Sg}(y \div x) = 1).$$

61 (§23A). (a) is equivalent to (a') and (b) is equivalent to (b'). Note that the last sentence in (a') is the contrapositive of the last sentence in (a). Similarly for (b') and (b).

62 (§23A). (1) is incorrect because $G_{\mathscr{S}}$ doesn't express the fact that it is unprovable, but only that it is unprovable from the initial formulas of $\mathscr{S}$. In the new system— call it $\mathscr{S}'$—with $G_{\mathscr{S}}$ as an initial formula, $G_{\mathscr{S}}$ expresses the fact that it can't be derived from the other initial formulas. Under the assumption that $\mathscr{S}'$ is consistent, $G_{\mathscr{S}}$ would express its own independence. (2) is correct, as we may make plausible by forming the sentence: 'This sentence is not provable-in-$\mathscr{S}'$'.

63 (§23B). The exponents in the prime factorization of a Gödel number which corresponds to a proof contain composite numbers greater than 9. This is never true for Gödel numbers which correspond to formulas. Hence the Gödel number of an initial formula is different from its one-line proof. Note that not all natural numbers are Gödel numbers.

64 (§23B). (a) '$53P8$' is true if and only if 53 is the Gödel number of a proof-in-$\mathscr{A}$ of the formula with Gödel number 8. Hence '$53P8$' is false.

(b) '$(\forall x)(xPc_4)$' is true if and only if every natural number is a Gödel number of a proof-in-$\mathscr{A}$ of the formula with Gödel number c_4. Hence '$(\forall x)(xPc_4)$' is false.

(c) '$(\exists x)(xPc_4)$' is true if and only if there is a natural number which is the Gödel number of a proof of the formula with Gödel number c_4. Since 'dPc_4' is true, '$(\exists x)(xPc_4)$' is true.

(d) '$Prov(c_4)$' is true if and only if '$(\exists x)(xPc_4)$' is true. Hence '$Prov(c_4)$' is true.

65 (§23B)
a) $(2^{11} \cdot 3^{11} \cdot 5^1 \cdot 7^{13} \cdot 11^3 \cdot 13^{13} \cdot 17^3)$.

b) $(2^5 \cdot 3^{11} \cdot 5^{11} \cdot 7^{11} \cdot 11^1 \cdot 13^{13} \cdot 17^{31} \cdot 19^{11} \cdot 23^1 \cdot 29^{13} \cdot 31^{13} \cdot 37^{29} \cdot 41^{11} \cdot 43^1 \cdot 47^{13} \cdot 53^{13})$.

c) Same as (b).

d) c_4

e) This is true if '$((0) + (0)) = (0)$' has a proof-in-$\mathscr{A}$. Hence it is true.

f) This is true if every natural number is the Gödel number of a proof of '$(0) = ((0)')$'. Hence it is false.

g) The Gödel number of the following formula: '$(((0)') + (0)) = ((0)')$'.

h) c_2

i) c_2

j) This is true if '$\overline{((\forall x)(((x) + (0)) = (x)))}$' is provable. Hence if $\mathscr{A}$ is consistent,

$$Prov\left(Neg\left(Sb\left((17 \; Gen \; c_1) \; \begin{matrix} 19 \\ Nml(1) \end{matrix}\right)\right)\right)$$

is false.

66 (§23B)

$$(1 + 0 = 1) \supset Prov\left(Sb\left(c_1 \; \begin{matrix} 17 \\ Nml(1) \end{matrix}\right)\right).$$

Hence the formula named in the answer to (g) of Problem 65 is provable-in-$\mathscr{A}$.

67 (§23B). The fallacy consists in the confusion of language and metalanguage. The property of being Richardian is a property depending on the language and not a property of the natural numbers (whose properties don't depend on a particular language). To see this, consider the program of the Richard argument actually carried out in two different languages, say, English and French. Evidently the property of being Richardian-in-French will not be equivalent to being Richardian-in-English because the orderings indicated by (2) would not be identical. *Richardian*, then, is a property of the numerals which stand for natural numbers in a particular language. [Cf. §12 and note the same fallacy in our original presentation of the Richardian paradox.]

68 (§23B). Let t be the Gödel number of the formula in $\mathscr{A}$ which corresponds to the following arithmetic sentence:

$$(\forall x)\left(\left(xP\left(Sb\left(y \; \begin{matrix} 19 \\ Nml(y) \end{matrix}\right)\right)\right) \supset (\exists z)\left((z \le x) \cdot zP\left(Neg\left(Sb\left(y \; \begin{matrix} 19 \\ Nml(y) \end{matrix}\right)\right)\right)\right)\right).$$

Then R is the formula in $\mathscr{A}$ which corresponds with R_A, that is, with

$$(\forall x)\left(\left(xP\left(Sb\left(t \; \begin{matrix} 19 \\ Nml(t) \end{matrix}\right)\right)\right) \supset (\exists z)\left((z \le x) \cdot \left(zP\left(Neg\left(Sb\left(t \; \begin{matrix} 19 \\ Nml(t) \end{matrix}\right)\right)\right)\right)\right)\right).$$

If k is the Gödel number of R, then

$$k = Sb\left(t \; \begin{matrix} 19 \\ Nml(t) \end{matrix}\right).$$

69 (§23C). Although it is true that 'if G is provable, then $\overline{G}$ is provable', it does not follow that '$\vdash \overline{G} \supset G$'. The formal expression for 'G is provable' is $\overline{G}$, but the formal expression for '$\overline{G}$ is provable' is not G. It is rather the formula in $\mathscr{A}$ which corresponds to

$$(\exists x)\left(xP\left(Neg\left(Sb\left(p\ {}^{19}_{Nml(p)}\right)\right)\right)\right).$$

70 (§23C). (a) Let '$\mathscr{C}$' denote the new system formed by adding C to the initial formulas of $\mathscr{A}$. $\mathscr{C}$ is consistent, if $\mathscr{A}$ is ω-consistent. To see this, let '**I**' stand for the conjunction of the initial formula schemata of $\mathscr{A}$. Suppose that the conjunction **I** · C is inconsistent. In $\mathscr{A}$ we would then have:

$$\vdash (\mathbf{I} \cdot C) \supset \overline{C}.$$

(Note that the antecedent is always false.) Hence, by the law of exportation, we have:

$$\vdash \mathbf{I} \supset (C \supset \overline{C}).$$

Since (C $\supset \overline{C}$) is equivalent to $\overline{C}$, we have:

$$\vdash \mathbf{I} \supset \overline{C}.$$

So $\overline{C}$ is provable-in-$\mathscr{A}$ because we naturally have $\vdash$ **I**. Now consider that each and every instance of C is provable-in-$\mathscr{A}$. (Try to convince yourself of this before turning to §29(4) for some details.) Hence, if $\mathscr{A}$ is ω-consistent, $\mathscr{C}$ is consistent. Note that the following argument is fallacious: Since $\vdash$ C $\supset$ G holds in $\mathscr{C}$ as well as in $\mathscr{A}$, and since $\vdash$ C because it is an initial formula, we have $\vdash$ G and therefore $\vdash \overline{G}$. The "therefore" inference is incorrect. It does not follow that if for $\mathscr{A}$ '$\vdash$ G' implies '$\vdash \overline{G}$' that the same holds for $\mathscr{C}$.

(b) Let '$\mathscr{C}'$' denote the new system formed by adding $\overline{C}$ to the initial formulas of $\mathscr{A}$. If $\mathscr{A}$ is consistent, $\mathscr{C}'$ is both consistent and ω-inconsistent. For suppose that the conjunction **I** · $\overline{C}$ is inconsistent. In $\mathscr{A}$ we thus have:

$$\vdash (\mathbf{I} \cdot \overline{C}) \supset C.$$

By reasoning similar to (a) above, we conclude that C is provable-in-$\mathscr{A}$. Now the argument of Gödel's second incompleteness theorem gives us the inconsistency of $\mathscr{A}$. Hence if $\mathscr{A}$ is consistent, so is $\mathscr{C}'$. On the other hand, since each and every instance of C is provable-in-$\mathscr{A}$ and therefore in $\mathscr{C}'$, it follows that $\mathscr{C}'$ is ω-inconsistent.

71 (§26). The difficulty is in the fact that the definition of general recursive is not constructive; that is, there is no way to order a set containing all and only (completely defined) general recursive functions. Otherwise, of course, the diagonal method would be sufficient to create an effectively calculable nonrecursive function. [Cf. Kleene 1960 142–143 on this point.]

72 (§26). We cannot (in general) prove M, we can only prove that *if* the machine is consistent M is true. Yet by the argument of Gödel's second incompleteness theorem, the machine can also prove this. That is, if D is the analog of C for the machine, the machine can prove D $\supset$ M. You are hereby warned that this answer is controversial. For discussion see Nagel and Newman 1958 98–102, Putnam 1960 and 1960a, Smart 1961, 1963 116ff, and 1968 317ff, Lucas 1961, Benacerraf 1967, Lucas 1968, and Webb 1968.

73 (§28). Any specific instance of Goldbach's conjecture is recursive. Hence, by the correspondence lemma, the formal equivalent of it is provable-in-$\mathscr{A}$.

74 (§28). If there is a model of $\mathscr{A}$ such that there is an infinite number of natural twin primes but only a finite number of unnatural twin primes, the conjecture would be true but undecidable in $\mathscr{A}$. If there is a model of $\mathscr{A}$ such that there is a finite number of natural twin primes but an infinite number of unnatural twin primes, the conjecture would be false but undecidable in $\mathscr{A}$. I am indebted for this problem and the discussion of this section to Kemeny 1958.

75 (§29). The rules by which we interpret $\mathscr{A}$ may themselves be inconsistent. In a full treatment of mathematical logic, the relation between models and formal systems would be explicated at much greater length than we have in this book. The possibility of such an inconsistency was noted in the proof of the consistency of $\mathscr{L}$. See Problem 52 (§21).

BIBLIOGRAPHY

References in the text to articles and books use the author's (or editor's) name followed by a date. The date usually indicates the date of the quoted edition. Example: "Russell 1929" means "Russell, Bertrand, *Our Knowledge of the External World*, New York: W. W. Norton, 1929." In some cases, if the original article or book is not the actual one from which the quoted material is taken, the bibliography refers first to the original and then to the quoted edition. This procedure was followed wherever it was thought useful for the reader to know the original date of an article, lecture, or book. Example: All references to Gödel 1931 are actually to the translation found in van Heijenoort 1967, as is indicated in the Gödel 1931 listing below. If a bibliographical reference is itself followed by a number, this number indicates the relevant page or pages. Example: "Reichenbach 1962 137" means "Reichenbach, Hans, *The Rise of Scientific Philosophy*, Berkeley, Cal.: University of California Press, 1962, page 137." The comments which have been added to many of the items below are intended to ease the reader's further introduction to mathematical logic. The bibliography is not intended to be complete. Bibliographical references approaching completeness may be found in the *Journal of Symbolic Logic*. This journal has unusually high standards and should be consulted with regard to any serious research in the subject. All acknowledgments of permissions to quote are listed in the front of the book, just following the preface.

Ackermann, Wilhelm
1954   *Solvable Cases of the Decision Problem*, Amsterdam, North-Holland; second printing, 1962. Comprehensive treatment of the known results.

1957   "Philosophical Observations on Mathematical Logic and on Investigations into the Foundations of Mathematics," *Ratio*, **1**, 1–23.

Anderson, Alan Ross
1958   "Mathematics and the 'Language Game'," *Review of Metaphysics*, **11**, 446–458; reprinted in Benacerraf and Putnam 1964 481–490.

1963   "What Do Symbols Symbolize?: Platonism," in *Philosophy of Science: The Delaware Seminar, Volume I, 1961–1962*, edited by Bernard Baumrin, New York: Interscience, 137–158. A droll presentation of some of the contemporary issues of ontology as seen by a platonist. Includes comments (pages 151–158) on platonism by others, and Anderson's replies. *See* Rudner 1963.

1964   (editor) *Minds and Machines*, Englewood Cliffs, N.J.: Prentice-Hall. Contains 8 articles, including Turing 1950, Lucas 1961, Putnam 1960.

Arbib, Michael A.
1964   *Brains, Machines and Mathematics*, New York: McGraw-Hill. This book "is designed for a reader who has heard of such currently fashionable topics as

cybernetics, information theory, and Gödel's theorem and wants to gain from one source more of an understanding of them than is afforded by popularizations." Nevertheless, the book must be read with care, as it is misleading on the subject of Gödel's theorem.

Aristotle

1928 *The Works of Aristotle*, I, translated under the editorship of W. D. Ross, Oxford: Clarendon Press. This work contains the complete *Organon* of Aristotle; that is, *Categoriae* and *De Interpretatione* (translated by E. M. Edghill), *Analytica Priora* (translated by A. J. Jenkinson), *Analytica Posteriora* (translated by G. R. G. Mure), and *Topica* and *De Sophisticis Elenchis* (translated by W. A. Pickard-Cambridge). It is part of a set containing the complete works of Aristotle, but can be purchased separately. It would be well to read a good history of logic (such as Kneale 1962 or Bocheński 1956) in conjunction with the study of *Organon*. There are various textual problems as well as problems of interpretation which make it unlikely that most readers would understand it without such help. Nevertheless, with a good commentary it is for the most part quite intelligible.

1928a *The Works of Aristotle*, VIII, translated under the editorship of W. D. Ross, Oxford: Clarendon Press, second edition. The complete *Metaphysics*, translated by Ross himself.

1941 *The Basic Works of Aristotle*, edited with an introduction by Richard McKeon, New York: Random House. This volume contains all the Aristotle that the nonexpert is likely to need. It runs almost 1500 pages and contains unabridged the following works: *Categories, On Interpretation, Posterior Analytics, Physics, On Generation and Corruption, On the Soul, On Memory and Reminiscence, On Dreams, On Prophesying by Dreams, Metaphysics, Nicomachean Ethics, Politics, Poetics*. There are also excerpts from other works. To fully understand Aristotelian logic, one must understand something of the rest of his thought. Besides the *Organon*, the most important work for this purpose is the *Metaphysics*.

Axinn, Sidney

1968 "Mathematics as an Experimental Science," *Philosophia Mathematica*, **5**, 1–10.

Barbò, Francesca Rivetti

1968 "A Philosophical Remark on Gödel's Unprovability of Consistency Proof," *Notre Dame Journal of Formal Logic*, **9**, 67–74. An argument against the reasoning behind Gödel's second incompleteness theorem. The argument is somewhat obscure, but appears to derive its plausibility from confusing the distinctions made clear in Lacey and Joseph 1968.

Bar-Hillel, Yehoshua: *See* Fraenkel and Bar-Hillel.

Barker, Stephen

1969 "Realism as a Philosophy of Mathematics," in Bulloff, Holyoke, and Hahn 1969 1–9. An examination of objections to realism; a position in the philosophy of mathematics which Gödel has espoused. *See also* Silvers 1966.

Basson, A. H.

1957 "Unsolvable Problems," *Proceedings of the Aristotelian Society*, **57**, 269–280. A nontechnical explanation of the spirit of Church's theorem.

Bauer-Mengelberg, Stefan
1965 Review of translation of Gödel 1931 in Gödel 1962; *Journal of Symbolic Logic*, **30**, 359–362.

1966 Review of Davis 1965, *Journal of Symbolic Logic*, **31**, 484–494.

Beiler, Albert H.
1964 *Recreations in the Theory of Numbers: The Queen of Mathematics Entertains*, New York: Dover. An excellent and entertaining introduction to the subject. Furthermore, it is useful for reference, since it contains much material that is not otherwise easily available.

Benacerraf, Paul
1967 "God, the Devil, and Gödel," *Monist*, **51**, 9–32. An essay which aims at refuting most of the arguments in Lucas 1961. Benacerraf ends, however, at a weakened thesis which brings him—with certain qualifications—to the following restatement of the import of Gödel's theorems: "If I am a Turing machine, then I am barred by my very nature from obeying Socrates' profound philosophic injunction: KNOW THYSELF." *See* Boolos 1968 and Lucas 1968.

Benacerraf, Paul, and Putnam, Hilary
1964 (editors) *Philosophy of Mathematics: Selected Readings*, Englewood Cliffs, N.J.: Prentice-Hall. A very useful anthology which contains 30 selections, including Hilbert 1925 (in part), Bernays 1934, Gödel 1944, Hempel 1945, Anderson 1958, Bernays 1959.

Benardete, José A.
1964 *Infinity: An Essay in Metaphysics*, Oxford: Clarendon Press. One of the very few books in metaphysics which takes into account the results of mathematical logic. At times wild and extravagant, it is nevertheless always interesting. It shows, I believe, that mathematical logic has not only enlarged our knowledge but has also enlarged our imagination.

Beneš, Vaclav Edvard
1953 "On Some Alleged Implications of Mathematical Logic," *Philosophical Studies*, **4**, 56–58. *See* Myhill 1952 and Leblanc 1957.

Bernays, Paul
1934 "Sur le platonisme dans les mathématiques," *L'enseignement mathématique*, **34** (1935), 52–69; translated in Benacerraf and Putnam 1964 274–286.

1959 "Comments on Ludwig Wittgenstein's *Remarks on the Foundations of Mathematics*," *Ratio*, **2**, 1–22; reprinted in Benacerraf and Putnam 1964.

1965 "Some Empirical Aspects of Mathematics," in *Information and Prediction in Science*, edited by S. Dockx and P. Bernays, New York, Academic Press, 1965, 123–128. Includes a discussion of the analytic-synthetic question.

Berry, George D. W.
1953 "Symposium: On the Ontological Significance of the Löwenheim–Skolem Theorem, I," *Academic Freedom, Logic, and Religion*, Philadelphia: American Philosophical Association (Eastern division), 39–55. The article concludes that the Löwenheim–Skolem theorem "is irrelevant to nominalism and whatever

262

support it offers conceptualism against platonism is slight in comparison with the forces, on both sides, already in the field." *See* Myhill 1953.

Beth, Evert Willem

1959   *The Foundations of Mathematics*, Amsterdam: North-Holland. An encyclopedic work, very useful for reference.

Blanché, Robert

1962   *Axiomatics*, translated by G. B. Keene, New York: Free Press.

Bocheński, Inocenty M.

1956   *Formale Logik*, Freiburg, Verlag Karl Alher; translation by Ivo Thomas, *A History of Formal Logic*, Notre Dame, Indiana: University of Notre Dame Press, 1961.

Bonola, Roberto

1955   *Non-Euclidean Geometry: A Critical and Historical Study of Its Developments*, translated by H. S. Carslaw, introduction by Federigo Enriques; New York: Dover. This text—first published in 1906—is out of date in a number of respects. Nevertheless, it is a useful reference work to supplement Wolfe's *Non-Euclidean Geometry*. It becomes even more useful in the Dover edition because bound with it are supplements containing John Bolyai's "The Science of Absolute Space" and Nicholas Lobachevski's "The Theory of Parallels." Both have been translated by G. B. Halsted.

Boolos, George S.

1968   Review of Lucas 1961 and Benacerraf 1967, *Journal of Symbolic Logic*, **33**, 613–615.

Braffert, P., and Hirschberg, D.

1963   (editors) *Computer Programming and Formal Systems*, Amsterdam: North-Holland.

Bronowski, J.

1966   "The Logic of the Mind," *American Scholar*, **35**, 233–242. "The logic of the mind differs from formal logic in its ability to overcome and indeed to exploit the ambivalences of self-reference, so that they become the instruments of imagination." *See* Lucas 1961 for more on this subject.

Bulloff, Jack, Holyoke, Thomas C., and Hahn, S. W.

1969   (editors) *Foundations of Mathematics: Symposium Papers Commemorating the Sixtieth Birthday of Kurt Gödel*, New York: Springer-Verlag. Includes a bibliography of the works of Gödel as well as, among other things, Barker 1969.

Burks, A. W., and Copi, I. M.

1950   "Lewis Carroll's Barber Shop Paradox," *Mind*, **59**, 219–222. This paper offers an analysis different from the one given in the answer to Problem 40 (§16). It also gives further bibliographic references for Carroll 1894.

Butrick, Richard

1965   "The Gödel Formula: Some Reservations," *Mind*, **74**, 411–414. I have more than a few reservations about this article, which claims that "the Gödel formula is not decidable or demonstrable in any system, and that in that sense it is a pseudo-sentence." The article is apparently based only on Nagel and Newman 1958, and it fails to make the crucial distinctions which are necessary in order to understand Gödel's proof. *See* Lacey and Joseph 1968.

Cantor, Georg

1895– Beiträge zur Begründung der transfiniten Mengenlehre," *Mathematische Annalen*,
1897 **46,** 481–512; **49,** 207–246; translation and introduction by Philip E. B. Jourdain, *Contributions to the Founding of the Theory of Transfinite Numbers*, La Salle, Illinois: Open Court, 1941. Unfortunately this is the only work of Cantor's translated into English; there should be many more. All references in the present text are made to the translated edition, which is very good. The book is quite difficult; reading Fraenkel 1953 will reduce that difficulty.

Carroll, Lewis

1894 "A Logical Paradox," *Mind*, **3,** 436–438. See Appendix A; *also see* Burks and Copi 1950.

1895 "What the Tortoise said to Achilles," *Mind*, **4,** 278–280. *See* Appendix B; also see Rees 1951 and Woods 1965.

Cervantes Saavedra, Miguel de

1949 *The Ingenious Gentleman Don Quixote de la Mancha*, II, translated by Samuel Putnam, New York: Viking Press.

Chari, C. T. K.

1963 "Further Comments on Minds, Machines and Gödel," *Philosophy*, **38,** 175–178. *See* Lucas 1961.

Chihara, Charles S.

1963 "Mathematical Discovery and Concept Formation," *Philosophical Review*, **72,** 17–34. "I shall consider the question: Is the mathematician a discoverer or a creator?"

Church, Alonzo

1933 "The Richard Paradox," *American Mathematical Monthly*, **41** (1934), 356–361.

1935 "An Unsolvable Problem of Elementary Number Theory," *American Journal of Mathematics*, **58** (1936), 345–363; reprinted in Davis 1965 89–107.

1936 "A Note on the Entscheidungsproblem," *Journal of Symbolic Logic*, **1,** 40–41. "Correction," **1,** 101–102; reprinted in Davis 1965 110–115. Entscheidungsproblem = Decision problem.

1942 Review of Findlay 1942, *Journal of Symbolic Logic*, **7,** 129.

1956 *Introduction to Mathematical Logic*, I, Princeton, N.J.: Princeton University Press. A difficult but very careful presentation of the subject. The formal systems in Chapter 3 of the present book are very much indebted to this book.

1962 "Logic," *Encyclopedia Britannica*, **14,** 295–305; also see "Logic, History of (IV, Modern Logic)," 317–323. A bird's-eye view of the whole subject. At times somewhat unintelligible because of its condensation, it is nevertheless a superb summary. Only the experts would fail to learn something from these articles.

1962a Review of Hanson 1961, *Journal of Symbolic Logic*, **27,** 471–472.

1965 Review of introduction to Gödel 1962, *Journal of Symbolic Logic*, **30,** 357–359.

1966   "Paul J. Cohen and the Continuum Problem," in *Proceedings of International Congress of Mathematics*, Moscow, 1966, 15–20. A brief expository account of Cohen's work on the occasion of an award to him for outstanding achievements in mathematics. Church concludes by stating that "... if a choice must in some sense be made among the rival set theories, rather than merely and neutrally to develop the mathematical consequences of the alternative theories, it seems that the only basis for it can be the same informal criterion of simplicity that governs the choice among rival physical theories when both or all of them equally explain the experimental facts."

Coder, David
1969   "Goedel's Theorem and Mechanism," *Philosophy*, **44**, 234–237. Still another critical article on Lucas 1961.

Cohen, Paul J., and Hersh, Reuben
1967   "Non-Cantorian Set Theory," *Scientific American*, **217**, No. 6, 104–116. A popular account of one of the most important results of set theory. For some reactions to these results, see Church 1966 and the article "Set Theory" by A. A. Fraenkel in Edwards 1967.

Copi, Irving M.
1949   "Modern Logic and the Synthetic *a Priori*," *Journal of Philosophy*, **46**, 243–245. *See* Turquette 1950 and Copi 1950. Copi argues that Gödel's incompleteness theorems imply that the theory that "all *a priori* truths are analytic" is wrong or at least must be very carefully reformulated. Turquette thinks the theorems irrelevant to the issue. For my part, I agree with Turquette on the empirical nature of certain mathematical statements, but with Copi that Turquette's positive suggestions are wrong and that Gödel's results (and the generalizations thereof) imply that the widely accepted theory that all mathematical statements are analytic must be abandoned. However, the reader should check for himself.

1950   "Gödel and the Synthetic *a Priori*: A Rejoinder," *Journal of Philosophy*, **47**, 633–636; *see* Copi 1949 and Turquette 1950. *See also* Burks and Copi.

Courant, Richard, and Robbins, Herbert
1953   *What is Mathematics?* New York: Oxford University Press.

Crosson, Frederick J.: *See* Sayre and Crosson.

Davis, Martin
1958   *Computability and Unsolvability*, New York: McGraw-Hill.

1965   (editor) *The Undecidable: Basic Papers on Undecidable Propositions, Unsolvable Problems, and Computable Functions*, Hewlett, N.Y.: Raven Press. Includes many of the important papers of metalogic, beginning with a translation of Gödel 1931. Also included are Church 1935, Church 1936, Turing 1936–7, Rosser 1936, Post 1936, Rosser 1939, Post 1944. One should read Bauer-Mengelberg 1966 before any serious use of this volume.

DeLong, Howard
1968   *Notes Toward a Profile of Mathematical Logic*, Hartford, Conn.: Trinity College. An earlier version of the present book.

DeSua, Frank
1956 "Consistency and Completeness—a Résumé," *American Mathematical Monthly*, **63**, 295–305. An elementary survey which is now somewhat out of date, especially with respect to comments on Gödel's second incompleteness theorem.

Diamond, R. J.
1964 "Resolution of the Paradox of Tristram Shandy." *Philosophy of Science*, **31**, 55–58. An analysis which I believe is somewhat faulty. See Problem 24 (§11).

Dummett, Michael
1963 "The Philosophical Significance of Gödel's Theorem," *Ratio*, **5**, 140–155. A nontechnical article which considers the relation of Gödel's theorem to the philosophical doctrine that the meaning of an expression is to be explained in terms of its use.

Edwards, Paul
1967 (editor) *The Encyclopedia of Philosophy*, 8 volumes, New York: Macmillan and Free Press. The only first-rate philosophical encyclopedia in English. The articles on logic and related topics are especially good. They are also very substantial, involving hundreds of pages. Were it not for their large number, they would be listed separately in this bibliography. Hence, this blanket suggestion: On research concerning any logical or philosophical topic, always (at least) consult this work.

Emch, Arnold F.
1935 "The *Logica Demonstrativa* of Girolamo Saccheri," *Scripta Mathematica*, **3**, 51–60, 143–152, 221–233. The most complete description of *Demonstrative Logic* in English. This series of articles should be the starting point for anyone wishing to investigate Saccheri's contribution to logic. The articles contain a good number of references as to where further information about Saccheri may be obtained. Since *Demonstrative Logic* itself was not available to me, my own comments depended on those of Emch, Kneale 1962, Enriques 1929, and Halsted's introduction to Saccheri 1733. The first two were the most helpful, although I think Emch's treatment of the *consequentia mirabilis* is incorrect.

Enriques, Federigo
1929 *The Historic Development of Logic*, translated by Jerome Rosenthal, New York: Henry Holt.

Euclid
1926 *The Thirteen Books of Euclid's Elements*, 3 volumes, second edition, translated with introduction and commentary by T. L. Heath; Cambridge: Cambridge University Press; reprinted by Dover Publications, New York, 1956. A first-rate translation and commentary—by far the best way to read the *Elements*. The *Elements* contains much besides geometry, and there are few who won't learn a great deal by working their way through these volumes.

Fang, J.
1964 Review of Henkin 1962, *Philosophia Mathematica*, **1**, 45–50. *Also see* Henkin 1964.

Feferman, Solomon

1960 "Arithmetization of Metamathematics in a General Setting," *Fundamenta Mathematicae*, **49**, 35–92. This paper should be studied by anyone who wishes to draw philosophical consequences from Gödel's second incompleteness theorem. It is, however, very difficult. *See* Mendelson 1964, Mostowski 1966, and Webb 1968 for an introduction to some of the main conceptions.

Feigenbaum, Edward A., and Feldman, Julian

1963 (editors) *Computers and Thought*, New York: McGraw-Hill. A score of articles which concentrate mostly on the practical problems of increasing the capacities of present-day machines. Marvin Minsky's bibliography on artificial intelligence is extremely useful. It includes a "descriptor index" with such headings as "What Can a Machine Know?" "Free Will in Man and Machines," and "The Mind-Brain Problem." The book includes Gelernter 1959, Gelernter, Hansen, and Loveland 1960, Newell, Shaw, and Simon 1957, Newell and Simon 1961, Minsky 1961, Turing 1950.

Feldman, Julian: *see* Feigenbaum and Feldman.

Findlay, J.

1942 "Goedelian Sentences: a Non-Numerical Approach," *Mind*, **51**, 259–265. An informal explanation; should be read in conjunction with Church 1942.

Fine, Arthur

1969 Review of Schlegel 1967, *Philosophical Review*, **78**, 528–531.

Finkelstein, David

1969 "Matter, Space and Logic," in Cohen, Robert S., and Wartofsky, Marx W. (editors), *Boston Studies in the Philosophy of Science*, **5**, Dordrecht, Holland: D. Reidel. "It is now commonplace that Mechanics and Geometry are empirical sciences insofar as they deal with reality. I shall emphasize the empirical aspects of Logic in this talk." *See* Putnam 1969.

Fitch, Frederic B.

1946 "Self-Reference in Philosophy," *Mind*, **55**, 64–73. This paper discusses some of the relationships between self-reference in comprehensive philosophical theories (such as Whitehead's or Descartes') and self-reference in mathematics and logic. The "logic" of such theories is found to be different from that of most ordinary theories. For example, the *ad hominem* type of argument is normally invalid, but in certain types of philosophical speculation it is "perfectly valid." This is an interesting paper for anyone who wishes to connect arguments in traditional philosophy with those in mathematical logic.

Fitzpatrick, P. J.

1966 "To Gödel via Babel," *Mind*, **75**, 332–350. A clever exposition of Gödel's proof through the use of English, French, and Latin to make distinctions between recursive arithmetic, metalanguage, and object language. *Also see* Lacey and Joseph 1968.

Fraenkel, Abraham A.

1953 *Abstract Set Theory*, Amsterdam: North Holland; second revised edition 1961; third revised edition 1966. A very clear and comprehensive presentation of the subject. The second and third editions each include new material, but, unfortunately, at the price of leaving out useful material in the first edition. In particular, the 130-page bibliography of the first edition is drastically reduced in the second and third.

1955   *Integers and Theory of Numbers* (Scripta Mathematica Studies, Number Five), New York: Scripta Mathematica. A short, elementary introduction.

Fraenkel, Abraham A., and Bar-Hillel, Yehoshua
1958   *Foundations of Set Theory*, Amsterdam: North-Holland. An excellent summary, very useful for finding further references on a wide range of topics in the foundation of mathematics.

Frege, Gottlob
1879   *Begriffsschrift, eine der arithmetischen nachgebildete Formelsprache des reinen Denkens*, Halle, Germany: Nebert; translated in van Heijenoort 1967 5–82. The work which established Frege as the greatest logician of all times. *See*, for example, Kneale 1962 510ff, van Heijenoort 1967 1.

Freud, Sigmund
1953   *The Complete Psychological Works of Sigmund Freud*, XIV, standard edition, edited by L. Strachey, London: Hogarth.

Gamow, George
1947   *One two three ... infinity: Facts and Speculations of Science*, New York: Viking Press.

Gardner, Martin
1958   *Logic Machines and Diagrams*, New York: McGraw-Hill. An excellent popular account. *Also see* Gardner's articles on "Logic Diagrams" and "Logic Machines" in Edwards 1967.

Gelernter, H.
1959   "Realization of a Geometry-Theorem Proving Machine," *Proceedings of an International Conference on Information Processing*, Paris: UNESCO House, 273–282; reprinted in Feigenbaum and Feldman 1963 134–152. "... our ultimate goal stands clearly before us; it is the design of an efficient theorem-prover in some undecidable system." *See* Wang 1960.

Gelernter, H., Hansen, J. R., and Loveland, D. W.
1960   "Empirical Exploration of the Geometry-Theorem Proving Machine," *Proceedings of the Western Joint Computer Conference*, **17,** 143–147; reprinted in Feigenbaum and Feldman 1963 153–163. "... the geometry-theorem proving machine ... has found solutions to a large number of problems taken from high-school textbooks and final examinations in plane geometry. Some of these problems would be considered quite difficult by the average high-school student."

Gellius, Aulus
1927   *The Attic Nights of Aulus Gellius*, I, translated by J. C. Rolfe, New York: Putnam.

George, F. N.
1962   "Minds, Machines, and Gödel," *Philosophy*, **37,** 62–63. *See* Lucas 1961.

Goddard, L.

1958 "'True' and 'Provable'," *Mind*, **67**, 13–31. A rather involved argument on the way in which Gödel's theorems relate to the notions of true and provable. Parsons and Kohl 1960 criticize this treatment; for a reply see Goddard 1962.

1962 "Proof-Making," *Mind*, **71**, 74–80. A reply to Parsons and Kohl 1960.

Gödel, Kurt

1930 "Die Vollständigkeit der Axiome des logischen Funktionenkalküls," *Monatshefte für Mathematik und Physik*, **37**, 349–360; translated in van Heijenoort 1967 583–591.

1931 Über formal unentscheidbare Sätze der Principia Mathematica und verwandter Systeme I," *Monatshefte für Mathematik und Physik*, **38**, 173–198. (English translations in Gödel 1962 35–72 and Davis 1965 5–38. The best translation, however, is in van Heijenoort 1967 596–616, to which all references in the present text are made.) This is the most important single paper in metatheory; it contains the first use of what is now known as Gödel numbering.

1944 "Russell's Mathematical Logic," in Schilpp, Paul A., *The Philosophy of Bertrand Russell*, New York: Tudor, 125–153; reprinted in Benacerraf and Putnam 1964 211–232.

1962 *On Formally Undecidable Propositions of Principia Mathematica and Related Systems* I. Translation of Gödel 1931 by B. Meltzer, with an introduction by R. B. Braithwaite; Edinburgh: Oliver and Boyd. Before using, check Church 1965 and Bauer-Mengelberg 1965.

Good, I. J.

1967 "Human and Machine Logic," *British Journal for the Philosophy of Science*, **18**, 144–147. An answer to Lucas 1961; *see* Good 1969.

1969 "Gödel's Theorem Is a Red Herring," *British Journal for the Philosophy of Science*, **19**, 357–358. An answer to Lucas 1967.

Goodstein, Reuben Louis

1957 Recursive Number Theory: *A Development of Recursive Arithmetic in a Logic-Free Equation Calculus*, Amsterdam: North-Holland. For the most part, a quite readable introduction.

1965 *Essays in the Philosophy of Mathematics*, Leicester, England: Leicester University Press. A very readable series of eleven essays, on such topics as "Proof by *Reductio ad Absurdum*," "Logical Paradoxes," "Mathematical Systems," and "The Significance of the Incompleteness Theorems." Goodstein is often unorthodox, yet (perhaps for this very reason) well worth reading.

Grelling, Kurt

1936 "The Logical Paradoxes," *Mind*, **45**, 481–486.

Grünbaum, Adolf

1963 *Philosophical Problems of Space and Time*, New York: Alfred A. Knopf. Chapter 5 discusses the question of the geometry of visual space, in particular the claim in Luneberg 1947 and elsewhere that the geometry of visual space is non-Euclidean. Many further references are given.

Hahn, S. W.: *See* Bulloff, Holyoke, and Hahn.

Hansen, J. R.: *See* Gelernter, Hansen, and Loveland.

Hanson, Norwood Russell
1961   "The Gödel Theorem," *Notre Dame Journal of Formal Logic*, **2**, 94–100, 228. An expository article which should only be read in conjunction with Church 1962a.

Hatcher, William S.
1968   *Foundations of Mathematics*, Philadelphia: W. B. Saunders. A survey which includes chapters on Frege's system, Hilbert's program and Gödel's incompleteness theorems, Quine's systems, and categorical algebra.

Heath, Sir Thomas
1921   *A History of Greek Mathematics*, 2 volumes, Oxford: Clarendon Press. This is still the standard work on Greek mathematics, and is an excellent reference work. Although somewhat out of date here and there, this should not prevent the reader from considering it as the first place to consult on Greek mathematics, just so long as it is not the last. Heath has a very good sense of what is historically probable and his explanations are always clear.

Helmer, Olaf
1937   "The Significance of Undecidable Sentences," *Journal of Philosophy*, **34**, 490–494. "In this paper I propose to deal with the following problem. Positivism rejects statements as metaphysical and void of sense, if they do not admit of a truth decision. On the other hand, Gödel has shown that the majority of calculi— among them classical mathematics—contain sentences whose undecidability is demonstrable. How can it possibly be compatible with the positivistic point of view, to admit the significance of sentences of this kind?" A somewhat dated but worthwhile discussion.

Hempel, Carl G.
1945   "On the Nature of Mathematical Truth," *American Mathematical Monthly*, **52**, 543–556; reprinted in Benacerraf and Putnam 1964 366–381. A very lucid statement of a positivist position. Nevertheless, it is unclear how the author would deal with the complications revealed by the limitative theorems.

Henkin, Leon
1950   "Completeness in the Theory of Types," *Journal of Symbolic Logic*, **15**, 81–91.

1962   "Are Logic and Mathematics Identical?" *Science*, **138**, 788–794. A survey of logicism and its vicissitudes. It is authoritative and includes a popular description of current (that is, 1936–1962) research in mathematical logic. *See* Fang 1964. Putnam 1967a also explores this topic.

1964   "A Letter to Reviewer," *Philosophia Mathematica*, **1**, 118–119. A reply to Fang 1964.

Hersh, Reuben: *See* Cohen and Hersh.

Heyting, Arend
1956   *Intuitionism: an Introduction*, Amsterdam: North-Holland.

Hilbert, David

1899   *Grundlagen der Geometrie*, Leipzig: Teubner; translated as *The Foundations of Geometry* by E. J. Townsend, LaSalle, Illinois: Open Court, 1938. A book based on lectures which Hilbert delivered in 1898–1899. These lectures are universally recognized as one of the first complete and successful attempts to put Euclid on a sound logical basis. Most of it is quite readable, but be prepared for a high degree of abstraction.

1925   "Über das Unendliche," *Mathematische Annalen*, **95** (1926), 161–190. All references to translation in van Heijenoort 1967 369–392. There is also a partial translation in Benacerraf and Putnam 1964 134–151.

Hintikka, Jaakko

1969   (editor) *The Philosophy of Mathematics*, London: Oxford University Press. A collection of 11 essays, most of which are quite technical. It includes Kreisel 1967 (in part), Robinson 1967 (in part), and Tarski 1959.

Hirschberg, D.: *See* Braffert and Hirschberg.

Holyoke, Thomas C.: *See* Bulloff, Holyoke, and Hahn.

Joseph, Geoffrey: *See* Lacey and Joseph.

Kac, Mark, and Ulam, Stanislaw

1969   *Mathematics and Logic: Retrospect and Prospects*, New York: New American Library. "... computing machines already have made significant contributions both to the *problematics* and to the *methodology* of mathematics. It is inconceivable that they will not continue to do so to an ever-increasing extent in spite of invective against them by some contemporary followers of Plato."

Kalmár, László

1959   "An Argument Against the Plausibility of Church's Thesis," in *Constructivity in Mathematics; Proceedings of the Colloquium Held at Amsterdam, 1957*, edited by A. Heyting, Amsterdam: North-Holland, 1959, 72–80. *See* Kalmár 1967.

1967   "Foundations of Mathematics—Whither Now?" in Lakatos 1967 187–194. Here Kalmár qualifies his 1959 article by saying, "I have no objection against Church's thesis if it is taken as an *empirical* one, confirmed several times in practice, but, like any other empirical thesis, to be abandoned if a counterexample is found in the future" (193). This admission leaves unclear (at least to me) how he differs from Kleene, except in their expectations about future mathematical discoveries. See Kleene 1967 240–241.

Kant, Immanuel

1885   *Introduction to Logic*, London: Longmans, Green.

1961   *Critique of Pure Reason*, translated by F. M. Müller, New York: Doubleday, Dolphin Books.

Kemeny, John G.

1956   "An Experiment in Symbolic Logic on the IBM 704," Rand Corporation Report P-966, September 7.

1958   "Undecidable Problems of Elementary Number Theory," *Mathematische Annalen*, **135**, 160–169. "Recent developments in mathematical logic offer a new method for proving that certain problems are undecidable in elementary number theory. It is possible that some of the famous unsolved problems will fall into this category. The purpose of this paper is to summarize the known results, describe the new method, and use it to establish some new results."

1962 "Semantics in Logic," *Encyclopedia Britannica,* **20,** 313C–313G. A very readable discussion of such topics as the semantic paradoxes, the means that have been used to avoid them, Tarski's theory of truth, the concept of analyticity, and Frege's distinction between sense and denotation.

Kempner, A. J.

1936 "Remarks on 'Unsolvable' Problems," *American Mathematical Monthly,* **43,** 467–473. A discussion of the meaning of 'unsolvable' when applied to problems such as Goldbach's conjecture. Since the article makes no reference to the then-known incompleteness results, its value is somewhat limited. Nevertheless, it can be recommended for its attempt at isolating some of the possibilities which face us with regard to unsolved problems. *See also* Turing 1954.

Kleene, Stephen Cole

1952 *Introduction to Metamathematics,* New York: Van Nostrand. This is the classic textbook on the subject (including the theory of recursive functions). However, it is not easy. Reading Kleene 1967 and Goodstein 1957 first reduces the difficulty considerably.

1960 "Mathematical Logic: Constructive and Non-Constructive Operations," in *Proceedings of the International Congress of Mathematicians, 14–21 August 1958,* edited by J. A. Todd, Cambridge: Cambridge University Press, 137–153.

1962 "Mathematics, Foundations of," *Encyclopedia Britannica,* **15,** 82B–83 (6 pages). An excellent summary which comments on the nature of the axiomatic method; the crisis caused by the paradoxes; the philosophical positions of logicism, intuitionism, and formalism; computability and decidability; Gödel's and Church's theorems. The treatment is of necessity condensed, but is as clear as can be expected in such short compass.

1967 *Mathematical Logic,* New York: John Wiley. In my opinion, this is the best single introduction to the techniques of mathematical logic.

Kneale, William and Martha

1962 *The Development of Logic,* Oxford: Clarendon Press. A magnificent book. In a clear and readable style, it traces the history of logic from the beginning to the present day. The Kneales are not afraid to take positions on controversial subjects, and although they are not always convincing they are quite fair to their opponents. First-rate scholarship in the history of logic is really just beginning, so it is likely that a fair number of their statements will have to be revised in light of later evidence. Nevertheless, no subsequent history of logic is likely to be worth reading unless it takes this one into account. In any case, the historical comments in the present book have relied heavily on their judgment.

Kneebone, G. J.

1963 *Mathematical Logic and the Foundation of Mathematics: An Introductory Survey,* London: D. Van Nostrand. A very competent survey which contains much historical and bibliographic material. It is especially useful as a reference text, since numerous points of view, proofs, and philosophical issues are summarized.

Kohl, Herbert R.: *See* Parsons and Kohl.

Körner, Stephen

1960 *The Philosophy of Mathematics,* London: Hutchinson University Library. A good introduction. May be read with Benacerraf and Putnam 1964.

1965 "An Empiricist Justification of Mathematics," in *Logic, Methodology and Philosophy of Science,* edited by Y. Bar-Hillel, Amsterdam: North-Holland, 222–227. "... it will be argued that scientific theories embedded in mathematics function, and are justified, together with their mathematical framework as syncategorematic constituents of empirical propositions."

1967 "On the Relevance of Post-Gödelian Mathematics to Philosophy," in Lakatos 1967 118–137. "... one must conclude that the metamathematical discoveries of the present century imply the falsehood of the common doctrines shared by the classical philosophies of non-competitive mathematical theories..." Includes a discussion by Gert H. Müller, Y. Bar-Hillel, and a reply by the author.

Kreisel, Georg

1958 Review of Ludwig Wittgenstein's *Remarks on the Foundations of Mathematics, British Journal for the Philosophy of Science,* **9,** 135–158.

1967 "Mathematical Logic: What has it done for the Philosophy of Mathematics?" in *Bertrand Russell: Philosophy of the Century,* edited by Ralph Schoenman, Boston: Little, Brown, 201–272, 315–316; reprinted (in part) in Hintikka 1969. A long, rambling, difficult essay which covers an extremely wide range of topics. (Mastering Myhill 1952 and Webb 1968 first would probably greatly reduce that difficulty.) Perhaps the following is a representative sample of Kreisel's conclusions: "I do not think that [Gödel's incompleteness theorem] establishes the non-mechanistic character of mathematical activity... what it establishes is the non-mechanistic character of the laws satisfied by, for instance, the natural numbers..."

Lacey, Hugh, and Joseph, Geoffrey

1968 "What the Gödel Formula Says," *Mind,* **77,** 77–83. An article which refutes Butrick 1965 by distinguishing arithmetic, metalanguage, and object language. Much the same ground is covered in Fitzpatrick 1966. The explication of Gödel's theorem in the present book has been influenced by these two articles.

Lakatos, Imre

1967 (editor) *Problems in the Philosophy of Mathematics: Proceedings of the International Colloquium in the Philosophy of Science, London, 1965, Volume 1,* Amsterdam: North-Holland. See Kalmár 1967, Körner 1967, Robinson 1967, Szabó 1967.

Langer, Susanne K.

1925 "A Set of Postulates for the Logical Structure of Music," *Monist,* **39,** 561–570. A dated paper, but in a realm which is still largely unexplored. It closes with the following question: "Is it possible that music is not the only interpretation for this algebra, but that some logician versed in the arts, especially in arts other than music, might trace similar structures in some other form of aesthetic expression?" The reader interested in the relations of logic and music might also find Myhill 1952 helpful.

Leblanc, Hugues

1957 Review of Myhill 1952, Beneš 1953, Myhill 1954, *Journal of Symbolic Logic*, **22**, 314–316.

1965 Review of Wilson 1964, *Journal of Symbolic Logic*, **30**, 366.

Lewis, David

1969 "Lucas Against Mechanism," *Philosophy*, **44**, 231–233. A partial defense of Lucas 1961, but watch out for *petitio principii*.

Loveland, D. W.: *See* Gelernter, Hansen, and Loveland.

Löwenheim, Leopold

1915 "Über Möglichkeiten im Relativkalkül," *Mathematische Annalen*, **76**, 447–470; translated in van Heijenoort 1967 232–251. The first proof of a limitative theorem.

Lucas, J. R.

1961 "Minds, Machines and Gödel," *Philosophy*, **36**, 112–127; reprinted in Anderson 1964 43–59. "Gödel's theorem seems to me to prove that Mechanism is false, that is, that minds cannot be explained as machines." A careful presentation of this thesis, but one which depends, I believe, on not making the kind of distinctions emphasized by Fitzpatrick 1966 and Lacey and Joseph 1968. Nevertheless, Lucas' article is well worth studying; it has generated more discussion than any other article on the philosophical import of Gödel's theorem. *See* Whiteley 1962, George 1962, Chari 1963, Smart 1963, Benacerraf 1967, Good 1967, Lucas 1967, Sayre 1967, Boolos 1968, Lucas 1968, Smart 1968, Webb 1968, Coder 1969, Good 1969, Lewis 1969.

1967 "Human and Machine Logic: A Rejoinder," *British Journal for the Philosophy of Science*, **19**, 155–156. An answer to Good 1967.

1968 "Satan Stultified: A Rejoinder to Paul Benacerraf," *Monist*, **52**, 145–158. *See* Benacerraf 1967.

Luneburg, Rudolph K.

1947 *Mathematical Analysis of Binocular Vision*, Princeton, N.J.: Princeton University Press. *See* Grünbaum 1963.

Mates, Benson

1953 *The Logic of the Stoa*, Berkeley, Cal.: University of California Press. Second edition entitled *Stoic Logic*, 1961. This book shows—in a definitive way, I believe—that Stoic logic differed from Aristotelian logic and was in a number of ways a precursor of parts of mathematical logic. Brings up to date the pioneering work of Łukasiewicz.

1962 "Logic, History of (I. Ancient Logic)," *Encyclopedia Britannica*, **14**, 306–311. A good summary if you don't have time for the longer way of Kneale 1962, Bocheński 1956, and Mates 1953.

Mehlberg, Henryk

1962 "The Present Situation in the Philosophy of Mathematics," in *Logic and Language: Studies Dedicated to Professor Rudolf Carnap on the Occasion of his*

274

Seventieth Birthday, Dordrecht, Holland: Reidel, 69–103. Useful for its comments on the relation between the empirical sciences and mathematics.

Mendelson, Elliott
1964   *Introduction to Mathematical Logic*, Princeton, N.J.: Van Nostrand.

Menger, Karl
1933   "Die neue Logik," *Krise und Neuaufbau in den exakten Wissenschaften*, Leipzig, 93–122; translated in *Philosophy of Science*, **4**, 299–366, 1937. One of the very few early accounts which is readable, up to date (at the time it was written), and as little misleading as is possible for a popular account.

Meschkowski, Herbert
1965   *Evolution of Mathematical Thought*, San Francisco, Cal.: Holden-Day. A readable, relatively short (153 pages), and sophisticated survey.

Minsky, Marvin L.
1961   "Steps Toward Artificial Intelligence," *Proceedings of the Institute of Radio Engineers*, **49**, 8–30; reprinted in Feigenbaum and Feldman 1963 406–450. A difficult, technical paper, but one which has some philosophical interest, especially on the mind-body problem. Cf. Putnam 1960.

1968   "Matter, Mind, and Models," in Marvin Minsky (editor), *Semantic Information Processing*, Cambridge, Mass.: M.I.T. Press, 425–432. An essay which "attempts to explain why people become confused by questions about the relation between mental and physical events." It concludes with the observation that "when intelligent machines are constructed, we should not be surprised to find them as confused and as stubborn as men in their convictions about mind-matter, consciousness, free will, and the like." Cf. Putnam 1960.

Morgenbesser, Sidney
1967   (editor) *Philosophy of Science Today*, New York: Basic Books. A quite elementary but very authoritative collection of 17 essays. On deductive logic there are pieces on "Truth and Provability," "Completeness" (both by Leon Henkin), "Computability" (Stephen Kleene), "Necessary Truth" (W. V. O. Quine).

Mostowski, Andrzej
1955   "The Present State of Investigations on the Foundations of Mathematics," *Rozprawy Matematyczne*, **9**, Warszawa. ". . . the search for a definition of arithmetic only by means of mathematical methods is not possible without having recourse to the origin of the notion of a natural number based on experience."

1957   *Sentences Undecidable in Formalized Arithmetic: An Exposition of the Theory of Kurt Gödel*, Amsterdam: North-Holland.

1966   *Thirty Years of Foundational Studies: Lectures on the Development of Mathematical Logic and the Study of the Foundations of Mathematics in 1930–1964*, New York: Barnes and Noble. An excellent, but very sophisticated, survey.

Myhill, John
1952   "Some Philosophical Implications of Mathematical Logic. I. Three Classes of Ideas," *Review of Metaphysics*, **6**, 165–198. A profound and exciting exposition,

much influenced by Post. The bibliography, which is annotated, is very helpful. Beneš 1953, Myhill 1954, Leblanc 1957, Wilson 1964, Leblanc 1965 should be read for further discussion.

1953 "Symposium: On the Ontological Significance of the Löwenheim–Skolem Theorem, II," *Academic Freedom, Logic, and Religion,* Philadelphia: American Philosophical Association (Eastern division), 57–70. Myhill concludes that a "philosophical lesson of the Löwenheim–Skolem theorem is that the formal communication of mathematics presupposes an informal community of understanding." This is a well-written article and deserves careful study. *See* Berry 1953.

1954 "Retort to Mr. Beneš," *Philosophical Studies,* **5,** 47–8. *See* Myhill 1952, Beneš 1953, and Leblanc 1957.

1960 "Some Remarks on the Notion of Proof," *Journal of Philosophy,* **57,** 461–471. "... I hope to have at least laid the groundwork for a method which can make some progress in solving certain problems lying on the boundary of philosophy and technical logic."

1961 Review of Nagel and Newman 1958, *Journal of Philosophy,* **58,** 209–218.

Nagel, Ernest, and Newman, James R.

1958 *Gödel's Proof,* New York: New York University Press. (An enlarged version of an article of the same title which appeared in *Scientific American,* **194,** No. 6, 1956, 71–86.) In spite of the fact that it contains a number of minor errors and misleading statements, this book remains one of the clearest popular explanations of Gödel's first and second incompleteness theorems. A good article to read in conjunction with this book is Lacey and Joseph 1968. For overly critical reviews, see Myhill 1961 and Putnam 1960a. See below for Nagel and Newman's answers.

1961 "Discussion: Putnam's Review of Gödel's Proof," *Philosophy of Science,* **28,** 209–211.

1961a "Communication," *Journal of Philosophy,* **58,** 218–220.

Newell, Allen, Shaw, J. C., and Simon, H. A.

1957 "Empirical Explorations with the Logic Theory Machine: A Case Study in Heuristics," *Proceedings of the Western Joint Computer Conference,* **15,** 218–239; reprinted in Feigenbaum and Feldman 1963 109–133. "We have specified a system for finding proofs of theorems in elementary symbolic logic, and by programming a computer to these specifications, have obtained empirical data on the problem-solving process in elementary logic. The program is called the Logic Theory Machine ..." *See* Wang 1960.

Newell, Allen, and Simon, H. A.

1961 "GPS, a Program that Simulates Human Thought," in *Lernende Automaten,* Munich: R. Oldenbourg KG; reprinted in Feigenbaum and Feldman 1963 279–293. "... the free behavior of a reasonably intelligent human can be understood as the product of a complex but finite and determinate set of laws. Although we know this only for small fragments of behavior, the depth of the explanation is striking."

276

Newman, James R.: *See* Nagel and Newman.

Nietzsche, Friedrich
1954   *The Portable Nietzsche*, translated by Walter Kaufmann, New York: Viking Press.

1956   *The Birth of Tragedy and The Genealogy of Morals*, translated by Francis Golffing, New York: Doubleday.

Ogilvy, C. Stanley
1962   *Tomorrow's Math: Unsolved Problems for the Amateur*, New York: Oxford University Press. An excellent survey of mathematical problems whose statements are elementary, but whose solutions are unknown.

Oliver, J. Willard
1967   "Formal Fallacies and Other Invalid Arguments," *Mind*, **76**, 463–478. A well-organized article exposing some simple errors committed by most introductory logic texts. For more on this topic, see Sanford 1968.

Ortega Y. Gasset, José
1961   *History As a System*, New York: W. W. Norton.

Parsons, Charles, and Kohl, Herbert R.
1960   "Self-Reference, Truth and Provability," *Mind*, **69**, 69–73. *See* Goddard 1958 and 1962.

Patton, Thomas E.
1965   "Church's Theorem on the Decision Problem," *Notre Dame Journal of Formal Logic*, **6**, 147–153. An expository paper.

Plato
1937   *The Dialogues of Plato*, translated by B. Jowett, 2 volumes. New York: Random House.

1955   *The Republic of Plato*, translated by F. M. Cornford. New York: Oxford University Press.

Popper, Karl R.
1952   "The Nature of Philosophical Problems and Their Root in Science," *British Journal for the Philosophy of Science*, **3**, 124–156; reprinted in Popper 1963. Quite useful for its interpretation of Plato's cosmology and of Euclid as a cosmologist. Cf. Szabó 1967.

1963   *Conjectures and Refutations: The Growth of Scientific Knowledge*, London: Routledge and Kegan Paul.

Post, Emil Leon
1921   "Introduction to a General Theory of Elementary Propositions," *American Journal of Mathematics*, **43**, 163–185; reprinted in van Heijenoort 1967 265–283.

1936   "Finite Combinatory Processes, Formulation I," *Journal of Symbolic Logic*, **1**, 103–105; all references made to reprint in Davis 1965 288–291. "…a fundamental discovery in the limitations of the mathematicizing power of Homo Sapiens has been made…"

1944   "Recursively Enumerable Sets of Positive Integers and Their Decision Problems," *Bulletin of the American Mathematical Society*, **50**, 284–316; all references to reprint in Davis 1965 305–337. A very important paper. The informal expository character of much of it adds to its usefulness as an introduction. Post's approach to the limitative theorems is very simple and elegant.

Putnam, Hilary

1960   "Minds and Machines," in Hook, Sidney (editor), *Dimensions of Mind: A Symposium*, New York: New York University Press, 148–179; reprinted in Anderson 1964 183–196. Cf. Minsky 1961 and 1968.

1960a Review of Nagel and Newman 1958, *Philosophy of Science*, **27**, 205–207.

1967   "Mathematics Without Foundations," *Journal of Philosophy*, **64**, 5–22. Among other things, this article considers the possibility that Fermat's last theorem might be (what we have called) *practically undecidable*.

1967a "The Thesis that Mathematics Is Logic," in *Bertrand Russell: Philosophy of the Century*, edited by Ralph Schoenman, Boston: Little, Brown, 273–303. A study of what Russell meant by 'logic' and the claim that mathematics can be reduced to logic (in that sense). May be read in conjunction with Henkin 1962.

1969   "Is Logic Empirical?" in Cohen, Robert S., and Wartofsky, Marx W. (editors), *Boston Studies in the Philosophy of Science*, **5**, Dordrecht, Holland: D. Reidel. "...could some of the 'necessary truths' of logic ever turn out to be false *for empirical reasons*? I shall argue that the answer to this question is in the affirmative, and that logic is, in a certain sense, a natural science." *See* Finkelstein 1969. *See also* Benacerraf and Putnam.

Quine, Willard Van Orman

1949   *Methods of Logic*, revised edition, New York: Henry Holt. One of the relatively few elementary textbooks which give students "glimpses beyond" elementary logic.

1961   *From a Logical Point of View*, second revised edition, New York: Harper and Row. A collection of 9 essays which includes "Two Dogmas of Empiricism." This famous essay argues against a sharp distinction between the analytic and the synthetic.

1966   *Selected Logic Papers*, New York: Random House. A collection of 23 technical papers on logic. Some become quite difficult at points, but most are readable. More important, many are devoted to clearing up confusions and therefore are especially useful, if not to the beginner, at least to the nonexpert.

1966a *The Ways of Paradox and Other Essays*, New York: Random House. A very readable collection of 21 essays, mostly of a nontechnical sort. Read together, they impress one not only with Quine's famous elegant style, but also with his uncommonly good judgment. Many philosophical issues are discussed with essays such as "Foundations of Mathematics," "On the Application of Modern Logic," "Truth by Convention," and "The Scope and Language of Science."

Rees, W. J.

1951   "What Achilles Said to the Tortoise," *Mind*, **60**, 241–246. An article about Carroll 1895.

278

Reichenbach, Hans
1962 *The Rise of Scientific Philosophy*, Berkeley, Cal.: Univ. of California Press.

Rescher, Nicholas
1969 *Many-Valued Logic*, New York: McGraw-Hill. A survey which contains a comprehensive bibliography.

Resnik, Michael David
1966 "On Skolem's Paradox," *Journal of Philosophy*, **63**, 425–438. An extended discussion. For an answer to Resnik, see Thomas 1968. Further comments are in Resnik 1969.

1969 "More on Skolem's Paradox," *Nous*, **3**, 185–196.

Richard, Jules
1905 "Les principes des mathématiques et le problème des ensembles," *Revue générale des sciences pures et appliquées*, **16**, 541–543; translated in van Heijenoort 1967 143–144.

Robinson, Abraham
1965 "Formalism 64," in *Logic, Methodology and Philosophy of Science*, edited by Y. Bar-Hillel, Amsterdam: North-Holland, 228–246. "... Infinite totalities do not exist in any sense of the word (i.e., either really or ideally).... Nevertheless, ... we should act *as if* infinite totalities really existed."

1967 "The Metaphysics of the Calculus," in Lakatos 1967 28–46; reprinted (in part) in Hintikka 1969. Includes a discussion by Peter Geach, Hans Freudenthal, A. Heyting, Y. Bar-Hillel, M. Bunge, and a reply by the author.

Rogers, Hartley, Jr.
1967 *Theory of Recursive Functions and Effective Computability*, New York: McGraw-Hill. A text of wide scope; however, it is not an easy one.

Rosser, John Barkley
1936 "Extensions of Some Theorems of Gödel and Church," *Journal of Symbolic Logic*, **1**, 87–91; reprinted in Davis 1965 231–235.

1939 "An Informal Exposition of Proofs of Gödel's Theorems and Church's Theorem," *Journal of Symbolic Logic*, **4**, 53–60; reprinted in Davis 1965 223–230.

Rudner, Richard S.
1963 "What Do Symbols Symbolize?: Nominalism," in Bernard Baumrin (editor), *Philosophy of Science: The Delaware Seminar, Volume 1, 1961–1962*, New York: Interscience, 159–186. An answer to Anderson 1963. Included is a discussion (pages 176–186) of nominalism, and Rudner's replies.

Russell, Bertrand
1929 *Our Knowledge of the External World*, New York: W. W. Norton.

1957 *Mysticism and Logic*, New York: Doubleday. *See also* Whitehead and Russell.

Russell, L. J.
1951 "A Problem of Lewis Carroll," *Mind*, **60**, 394–396.

Saccheri, Giralamo
1733 *Euclides ab Omni Naevo Vindicatus*, Milan; introduction and partial translation by G. B. Halsted, Chicago, Open Court, 1920. The translation—to which all references in the present text are made—is only of the first part of Saccheri's

work (about two-thirds of the whole). The introduction by Halsted makes some extravagant claims and is careless. The original Latin text is given with the English translation on facing pages. Emch 1935 should be read before this. It is too bad that we do not have a critical edition and translation of all of Saccheri's works. In the meantime one must put up with this rather second-rate scholarship.

Sanford, David H.
1968   "Contraries and Subcontraries," *Nous*, **2**, 95–96. *See* Oliver 1967.

Sayre, Kenneth M.
1967   "Philosophy and Cybernetics," in *Philosophy and Cybernetics*, edited by Frederick J. Crosson and Kenneth M. Sayre, Notre Dame, Indiana: University of Notre Dame Press, 3–33. An essay which concludes that "cybernetics renders determinism in human behavior a highly unlikely thesis." Along the way there is an attempted refutation of Lucas 1961; an attempt which is, I believe, less successful than Webb 1968.

Sayre, Kenneth M., and Crosson, Frederick J.
1963   *The Modelling of Mind: Computers and Intelligence*, Notre Dame, Indiana: University of Notre Dame Press. A series of fourteen essays, including reprints of Lucas 1961 and Wang 1960.

Schlegel, Richard
1967   *Completeness in Science*, New York: Appleton-Century-Crofts. A treatment of different kinds of completeness in science. Chapter 5 (pages 61–83) discusses the Gödel incompleteness theorem, but is limited by its failure to explain the concept of recursiveness, as well as its failure to note that the Gödelian sentence is false under some interpretation of the system. Nevertheless it can be used with caution, say, in conjunction with Fine 1969.

Shaw, J. C.: *See* Newell, Shaw, and Simon.

Silvers, Stuart
1966   "On Gödel's Philosophy of Mathematics," *Philosophia Mathematica*, **3**, 1–8. *See also* Barker 1969.

Simon, H. A: *See* Newell, Shaw, and Simon; and Newell and Simon.

Skolem, Thoralf
1920   "Logisch-kombinatorische Untersuchungen über die Erfüllbarkeit oder Beweisbarkeit mathematischer Sätze nebst einem Theoreme über dichte Mengen," *Skrifter utgit av Videnskapsselskapet i Kristiania, I. Mathematisk-naturvidenskabelig klasse*, No. 4, 1–36; translated in van Heijenoort 1967 254–263.

1923   "Begründung der elementaren Arithmetik durch die rekurrierende Denkweise ohne Anwendung scheinbarer Veränderlichen mit unendlichem Ausdehnungsbereich," *Skrifter utgit av Videnskapsselskapet i Kristiania, I. Mathematisk-naturvidenskabelig klasse*, No. 6, 1–38; all references to translation in van Heijenoort 1967 303–333.

1933   "Über die Unmöglichkeit einer vollständigen Charakterisierung der Zahlenreihe mittels eines endlichen Axiomensystems, *Norsk matematisk forenings skrifter*, series 2, no. 10, 73–82. *See* Skolem 1955.

1934 Über die Nicht-charakterisierbarkeit der Zahlenreihe mittels endlich oder abzählbar unendlich vieler Aussagen mit ausschliesslich Zahlenvariablen, *Fundamenta mathematicae*, **23**, 150–161. *See* Skolem 1955.

1946 "The Development of Recursive Arithmetic," *Dixième Congrès des Mathématiciens Scandinaves*, Copenhagen, 1–16.

1955 "Peano's Axioms and Models of Arithmetic," in *Mathematical Interpretations of Formal Systems*, Amsterdam, North-Holland, 1–14 "... I will ... give an account as short as possible of my old proof of [the non-categoricity of Peano's or similar axiom systems], my exposition now being a little different in some respects."

Smart, J. J. C.
1961 "Gödel's Theorem, Church's Theorem and Mechanism," *Synthèse*, **13**, 105–110. *See* Lucas 1961 and Smart 1963.

1963 *Philosophy and Scientific Realism*, New York: Humanities Press. *See* Lucas 1961.

1968 *Between Science and Philosophy: An Introduction to the Philosophy of Science*, New York: Random House. *See* Lucas 1961.

Smith, David Eugene
1929 (editor) *A Source Book in Mathematics*, New York: McGraw-Hill.

Smullyan, Raymond M.
1969 "The Continuum Hypothesis," in *The Mathematical Sciences: A Collection of Essays*, edited by the National Research Council's Committee on Support of Research in the Mathematical Sciences (COSRIMS), with the collaboration of George A. W. Boehm, Cambridge, Mass.: M.I.T. Press, 252–260. A popular account. *See also* Cohen and Hersh 1967.

Szabó, Árpód
1967 "Greek Dialectic and Euclid's Axiomatics," in Lakatos 1967 1–27. "... why did the Greeks not rest satisfied with practical or empirical mathematical knowledge? ... My problem is to explain the change in the criterion of truth in mathematics from justification by practice or experience to justification by theoretical reasons. My solution is that this change was due to the impact of philosophy, and more precisely of Eleatic dialectic upon mathematics." Includes a very interesting discussion by W. C. Kneale, L. Kalmár, A. Robinson, J. R. Lucas, P. Bernays, G. J. Whitrow, K. R. Popper, and a reply by the author. Cf. Popper 1952 and 1963.

Tarski, Alfred
1948 *A Decision Method for Elementary Algebra and Geometry*, Santa Monica, Cal.: RAND Corporation; second revised edition, Berkeley, Cal.: University of California Press, 1951.

1953 *Undecidable Theories*, in collaboration with Andrzej Mostowski and Raphael M. Robinson, Amsterdam: North-Holland.

1959 "What Is Elementary Geometry?" in *The Axiomatic Method, with Special Reference to Geometry and Physics*, edited by Leon Henkin, Patrick Suppes, and Alfred Tarski, Amsterdam: North-Holland, 1959, 16–29; reprinted in Hintikka

1969. This paper demonstrates that the traditional notion of elementary geometry is vague but that, with the help of modern logic and metamathematics, it can be exactly defined. Reading this paper first will make easier the reading of Tarski 1948.

1969   "Truth and Proof," *Scientific American*, **220**, No. 6, 63–77. An excellent expository discussion of the semantic conception of truth, the notion of a formal proof, and the relation between the two. Tarski emphasizes the continuity between Aristotle and the semantic conception, and, after introducing a Gödel-type argument, he concludes that "the notion of a true sentence functions thus as an ideal limit which can never be reached but which we try to approximate by gradually widening the set of provable sentences."

Thomas, William J.
1968   "Platonism and the Skolem Paradox," *Analysis*, **28**, 193–196. A reply to Resnik 1966.

Tóth, Imre
1969   "Non-Euclidean Geometry Before Euclid," *Scientific American*, **221**, No. 5, 87–98. On the meager evidence we have of the discussion concerning the possibility of a non-Euclidean geometry before Euclid's definitive formulation of the fifth postulate.

Turing, Alan Mathison
1936–7   "On Computable Numbers, with an Application to the Entscheidungsproblem," *Proceedings of the London Mathematical Society*, 2nd series, **42**, 230–265; "Correction," *ibid.*, **43**, 544–546; reprinted in Davis 1965 116–154.

1950   "Computing Machinery and Intelligence," *Mind*, **59**, 433–460; reprinted in Anderson 1964 4–30 and Feigenbaum and Feldman 1963 11–35.

1954   "Solvable and Unsolvable Problems," *Science News* (Penguin Books), **31**, 7–23. An excellent elementary article. *See also* Kempner 1936.

Turquette, Atwell R.
1950   "Gödel and the Synthetic *a Priori*," *Journal of Philosophy*, **47**, 125–129. *See* Copi 1949 and 1950.

Ulam, Stanislaw: *See* Kac and Ulam.

Ullian, Joseph
1963   "Mathematical Objects" in *Philosophy of Science: The Delaware Seminar, Volume I, 1961–1962*, edited by Bernard Baumrin, New York: Interscience, 187–205. An illuminating discussion of Gödel's theorems and their relevance to the problem of the nature of mathematical objects. Included are some comments on the twin-prime conjecture and the possibility of its being undecidable in some system.

Unamuno, Miguel de
1954   *The Tragic Sense of Life*, New York: Dover.

van Dantzig, D.
1955   "Is $10^{10^{10}}$ a Finite Number?" *Dialectica*, **9**, 273–277. "... the question put in the title of this paper does not admit a unique and unambiguous answer."

Vandiver, H. S.

1946 "Fermat's Last Theorem: Its History and the Nature of the Known Results Concerning It," *American Mathematical Monthly*, **53,** 555–578. Illustrates the extreme difficulty of solving the problem.

van Heijenoort, Jean

1967 (editor) *From Frege to Gödel: A Source Book in Mathematical Logic,* 1879–1931, Cambridge, Mass.: Harvard University Press. First-rate scholarship in the introductions combined with a large number of important papers makes this book—in the words of Quine—"far and away the most important publishing event in mathematical logic in 31 years" (that is, since the founding of the *Journal of Symbolic Logic*). It includes, among other things, reprints or translations of Frege 1879, Richard 1905, Löwenheim 1915, Skolem 1920, Post 1921, Skolem 1923, Hilbert 1925, Gödel 1930, and Gödel 1931.

Wang, Hao

1960 "Toward Mechanical Mathematics," *IBM Journal of Research and Development,* **4,** 2–22; reprinted (in part) in Sayre and Crosson 1963. Using an IBM 704, the author was able to prove over 200 theorems of *Principia Mathematica* in less than three minutes of computer time. He concludes: "While a universal decision procedure for all mathematical problems is not possible, formalization does seem to promise that machines will do a major portion of the work that takes up the time of research mathematicians today." *See* Newell, Shaw and Simon 1957 and Gelernter 1959.

1962 *A Survey of Mathematical Logic,* Peking, China: Science Press.

1965 "Logic and Computers," *American Mathematical Monthly,* **72,** Part II, No. 2, 135–140. A nontechnical survey of the successes and current problems in the area of computer logic.

Watson, A.

1938 "Mathematics and its Foundations," *Mind,* **7,** 440–451. An interesting and substantial article which emphasizes the importance of the diagonal argument to the Gödel theorem.

Webb, Judson

1968 "Metamathematics and the Philosophy of Mind," *Philosophy of Science,* **35,** 156–178. A difficult and penetrating inquiry which aims at refuting Lucas 1961. However, it contains much else and should be studied by anyone interested in relating philosophical questions to results of metamathematics. Besides Lucas 1961 the following should be read before this article: Smart 1961, Benacerraf 1968, Myhill 1952, Beneš 1953, Myhill 1954, Copi 1949, Turquette 1950, Copi 1950.

Weyl, Hermann

1944 "David Hilbert and His Mathematical Work," *Bulletin of the American Mathematical Society,* **50,** 612–654. An appreciation of David Hilbert written shortly after his death on February 14, 1943. It describes very well Hilbert's part in the rise of mathematical logic.

Whitehead, Alfred North, and Russell, Bertrand
1910– *Principia Mathematica*, I (1910, second edition 1925), II (1912, second edition
1913   1927), III (1913, second edition 1927), Cambridge, England: Cambridge University Press.

Whiteley, C. H.
1962   "Minds, Machines and Gödel," *Philosophy*, **37**, 61–62. See Lucas 1961.

Wilder, R. L.
1959   "Axiomatics and the Development of Creative Talent," in *The Axiomatic Method,
with Special Reference to Geometry and Physics*, edited by Leon Henkin, Patrick
Suppes, and Alfred Tarski, Amsterdam, North-Holland, 474–488. ". . . I believe
that the great advances that the [axiomatic] method has made in mathematical
research during the past 50 years can, to a considerable extent, find a parallel in
the teaching of mathematics, and that its wise and strategic use, at special times
along the line from elementary teaching to the first contacts with the frontiers
of mathematics, will result in the discovery and development of much creative
talent that is now lost to mathematics."

Wilson, N. L.
1964   "Psychologism, Logic, and Mr. Myhill," *Philosophia Mathematica*, **1**, 1–4. See
Myhill 1952 and Leblanc 1965.

Wittgenstein, Ludwig
1921   "Logisch-philosophische Abhandlung," *Annalen der Naturphilosophie*, **14,**
185–262; all references to translation in *Tractatus Logico-Philosophicus*, London:
Routledge and Kegan Paul, 1922.

Wolfe, Harold E.
1945   *Introduction to Non-Euclidean Geometry*, New York: Dryden Press. A clearly
written and well-organized introduction to non-Euclidean geometry. Although
intended as a textbook, it is clearly within the competence of anyone who can
understand Euclid. Therefore it can be read without the help of a teacher. It
won't be easy going all the way, but Wolfe's explanations are clear enough to
prevent a reader from getting permanently stalled.

Woods, John
1965   "Was Achilles' 'Achilles' Heel' Achilles' Heel?" *Analysis*, **25**, 142–146. A discussion of Carroll 1895.

Yourgrau, Wolfgang
1969   "Gödel and Physical Theory," *Mind*, **78**, 77–90. A rather rambling discussion
of the relation of mathematics and physical theory in light of recent work in
logic.

INDEXES

SYMBOL INDEX

INDEX

abbreviations, 154

abnormal sets, 82, 123–124

absolute standard of measure, 47, 49, 52

absolute value, 158

accidentally true statement, 204
(*See also* Truth)

ACE, 130–131, 164, 204, 253

Achilles' argument, 9, 52, 108, 233–236, 240, 243, 251

Adam of Balsham, 72

addition, 59–60, 153–154, 158, 212
logical, 105

aggregate (*See* Set)

aleph-null, 78

algebra, 54, 69, 93–94, 131, 152, 163–164, 176, 212

algebra of logic, 70

algebraic numbers, 75

algebraic techniques, 70, 86

algorithm, 132, 149, 156–157
(*See also* Church's thesis, Decision procedure, Effective finite method)

alphabetic order, 83, 108, 120, 246

alternation, 96

analytic *a priori*, 39, 205, 222

analytic geometry, 65–70, 93
(*See also* Cartesian coordinate system)

analytic judgments, 39, 40

analytic–synthetic distinction, 39–40, 50, 205, 222, 228–229, 261, 271, 277
(*See also a priori* intuitions)

and, 97, 247

angle, bisection of, 186
trisection of, 29–30, 69, 186, 193

angle of parallelism, 48–49

antecedent, 98–99

antinomies, 242 (*See also* Paradoxes)

a posteriori judgments, 39, 50

apparent variable, 112, 165

a priori intuitions, 49, 58, 88
(*See also* Analytic–synthetic distinction)

a priori judgments, 39, 50

A proposition, 16–18, 20–22, 112–113

Aquinas, Thomas, 228

Archimedes, 46, 227

argument forms, 100, 104–105

Aristotelian logic, 14–24, 89, 110, 273
(*See also* Syllogisms, Traditional logic)

Aristotle, 1–2, 6, 8–9, 13–24, 27–28, 30–31, 34–36, 37, 41, 55, 93, 94, 124, 136, 164–165, 223, 229, 251, 281

Aristotle's theory of definition, 23, 27

arithmetic, 25, 28, 59, 127, 152, 159–160, 163–164, 175, 178–179, 182, 184–187, 200, 203–205, 223
fundamental theorem of, 167
unsolved problems of, 192–193, 205–216, 271
(*See also* Recursive (primitive) arithmetic)

artificial languages, 92, 99

associated formula of the propositional calculus, 141–142, 143

association, laws of, 105, 253–254

autological adjectives, 84, 246